*Progress in
Cancer Research and Therapy
Volume 4*

PROGESTERONE RECEPTORS
IN
NORMAL AND NEOPLASTIC TISSUES

Progress in
Cancer Research and Therapy

Vol. 1: Control Mechanisms in Cancer
Wayne E. Criss, Tetsuo Ono, and John R. Sabine, editors, 1976

Vol. 2: Control of Neoplasia by Modulation of the Immune System
Michael A. Chirigos, editor, 1977

Vol. 3: Genetics of Human Cancer
John J. Mulvihill, Robert W. Miller, and Joseph F. Fraumeni, Jr., editors, 1977

Vol. 4: Progesterone Receptors in Normal and Neoplastic Tissues
William L. McGuire, Jean-Pierre Raynaud, and Etienne-Emile Baulieu, editors, 1977

Vol. 5: Immunotherapy of Cancer: Present Status of Trials in Man
William D. Terry and Dorothy Windhorst, editors, due 1977

Progress in
Cancer Research and Therapy
Volume 4

Progesterone Receptors in Normal and Neoplastic Tissues

Edited by

William L. McGuire, M.D.
*Professor of Medicine
University of Texas Health
  Science Center at San Antonio
San Antonio, Texas*

Jean-Pierre Raynaud, Ph.D.
*Director of Biochemical Pharmacology
Centre de Recherches Roussel-Uclaf
Romainville, France*

Etienne-Emile Baulieu, M.D., Ph.D.
*Professor of Biochemistry
Faculté de Médecine
Université Paris-Sud
and
Scientific Director
U33 INSERM
Bicêtre, France*

Raven Press ■ New York

Raven Press, 1140 Avenue of the Americas, New York, New York 10036

© 1977 by Raven Press Books, Ltd. All rights reserved. This book is protected by copyright. No part of it may be reproduced, stored in a retrieval system, or transmitted, in any form or by any means, electronic, mechanical, photocopying, recording, or otherwise, without the prior written permission of the publisher.

Made in the United States of America

Library of Congress Cataloging in Publication Data

Main entry under title:

Progesterone receptors in normal and neoplastic tissues.
  (Progress in cancer research and therapy; v. 4)
  Includes bibliographical references and index.
  1. Breast--Cancer--Congresses. 2. Progesterone--Congresses. 3. Hormone receptors--Congresses. 4. Uterus--Cancer--Congresses. I. McGuire, William L. II. Raynaud, Jean Pierre. III. Baulieu, E. E. IV. Roussel (Canada) Ltd. V. Series.
  [DNLM: 1. Receptors, Progesterone--Analysis--Congresses. 2. Receptors, Estrogen--Analysis--Congresses. 3. Breast neoplasms--Congresses. 4. Uterine neoplasms--Congresses. 5. Progestational hormones, Synthetic--Diagnostic use--Congresses. W1 PR667M v. 4 / WP870 P964 1976]
  RC280.B8P76   616.9'94'49   77-72065
  ISBN 0-89004-163-6

Preface

Reproductive biologists have known for a long time that a progestational agent is maximally effective if preceded by a series of estrogen injections. However, the biochemical basis for the estrogen "priming" effect has been elucidated only within the past decade. It was discovered that for each class of steroid hormones there exists in the cytoplasm of the target cell a unique protein referred to as the "receptor." The steroid enters the cell, binds to the receptor, the receptor-steroid complex translocates to the nucleus, where it initiates the events characteristic for that hormone. In certain animal species such as guinea pig, the receptor for progesterone was easily identified using radioactive progesterone as the ligand. It was then determined that the synthesis of progesterone receptor was dependent on prior estrogen stimulation. Thus, the well-known estrogen priming effect required for a maximal progestational response could be explained on biochemical basis.

A major obstacle to our complete understanding of progesterone receptors was that in many species, including man, radioactive progesterone was unsuitable as a ligand for measuring the receptor. Only within the past few years has this problem been overcome. Scientists at Roussel-Uclaf published their findings on the use of R5020 (17,21-dimethyl-10-nor-4,9-pregnadiene-3,20-dione), a synthetic progesterone that appears to be an ideal radioactive ligand for progesterone receptor in all species tested. The subsequent use of R5020 in studying progesterone receptors during the estrous cycle, pregnancy, and lactation has been a major contribution to our understanding of reproductive biology. Other studies have indicated additional uses for the compound in selecting proper endocrine therapy for patients with breast cancer and these principles may also apply to endometrial carcinoma.

In the fall of 1976, two years after R5020 had become available, a meeting of those investigators who had accumulated significant data using this compound was organized. This meeting was sponsored by Roussel (Canada) Ltd. and was held in Montebello, Quebec. The organizing committee (E. E. Baulieu, J. P. Raynaud, and W. L. McGuire) selected 20 investigators to participate in the program. Each was asked to write a paper, the sum of which make up the present volume. We believe that this volume, which emphasizes sound biological principles concerning progesterone receptors, will be particularly useful to both students and investigators in the fields of reproductive biology and breast cancer.

Finally, on behalf of the contributors in this volume, the scientific organizing committee wishes to thank D. Buxton and J. Gareau from Roussel

(Canada) Ltd. and J. Mathieu from the Roussel-Uclaf Research Centre (Romainville, France) for their excellent administrative support and organization.

W. L. McGuire
J. P. Raynaud
E. E. Baulieu
(*January 1977*)

Contents

1   Progesterone Receptors: Introduction and Overview
    W. L. McGuire, J. P. Raynaud, and E. E. Baulieu

9   R5020, a Tag for the Progestin Receptor
    J. P. Raynaud

23 Characterization of the Progestin Receptor from Lactating Mammary Glands of the Goat
    F. S. Markland, Jr., and T. W. Hutchens

39 Interaction of R5020 with Binding Sites in Normal and Neoplastic Mammary Tissues
    J. L. Wittliff, R. G. Mehta, and T. E. Kute

59 Determinations of High-Affinity Gestagen Receptors in Hormone-Responsive and Hormone-Independent GR Mouse Mammary Tumors by an Exchange Assay
    J. L. Daehnfeldt and P. Briand

71 Progesterone and Estradiol Receptors in DMBA-Induced Mammary Tumors Before and After Ovariectomy and After Subsequent Estradiol Administration
    A. J. M. Koenders, A. Geurts-Moespot, S. J. Zolingen, and Th. J. Benraad

85 Regulation of Hormone Receptor Levels and Growth of DMBA-Induced Mammary Tumors by RU16117 and Other Steroids in the Rat
    P. A. Kelly, J. Asselin, F. Labrie, and J. P. Raynaud

103 Estrogen and Progesterone: Their Relationship in Hormone-Dependent Breast Cancer
    K. B. Horwitz and W. L. McGuire

125 Clinical Correlations of Endocrine Ablation with Estrogen and Progesterone Receptors in Advanced Breast Cancer
    N. Bloom, E. Tobin, and G. A. Degenshein

141 Estrogen and Progesterone Receptors in Human Breast Cancer
    G. Leclercq, J. C. Heuson, M. C. Deboel, N. Legros, E. Longeval, and W. H. Mattheiem

155 Estrogen and Progestin Receptors in Normal and Cancer Tissue
    B. Ramanath Rao and J. S. Meyer

CONTENTS

171 Estrogen and Progestin Receptors in Human Breast Cancer
J. P. Raynaud, T. Ojasoo, J. C. Delarue, H. Magdelenat, P. Martin, and D. Philibert

193 Interactions of R5020 with Progesterone and Glucocorticoid Receptors in Human Breast Cancer and Peripheral Blood Lymphocytes *In Vitro*
M. Lippman, K. Huff, G. Bolan, and J. P. Neifeld

211 Estrogen and Progesterone Receptors and Glucose Oxidation in Mammary Tissue
J. Levy and S. M. Glick

227 Cytoplasmic Progestin Receptors in Mouse Uterus
D. Philibert and J. P. Raynaud

245 Changes in Progesterone and Estrogen Receptors in the Rat Uterus During the Estrous Cycle and the Puerperium
F. Gómez, H. G. Bohnet, and H. G. Friesen

261 Use of [^{3}H]R5020 for the Assay of Cytosol and Nuclear Progesterone Receptor in the Rat Uterus
E. Milgrom, M. T. Vu Hai, and F. Logeat

271 Nuclear and Cytoplasmic Progesterone Receptors in the Rat Uterus: Effects of R5020 and Progesterone on Receptor Binding and Subcellular Compartmentalization
M. R. Walters and J. H. Clark

287 Measurement of the Progesterone Receptor in Human Endometrium Using Progesterone and R5020
F. Bayard, B. Kreitmann, and B. Derache

299 Progestin Receptor in Endometrial Carcinoma
J. Å. Gustafsson, N. Einhorn, G. Elfström, B. Nordenskjöld, and Ö. Wrange

313 Progesterone Receptors in Normal Human Endometrium and Endometrial Carcinoma
K. Pollow, M. Schmidt-Gollwitzer, and J. Nevinny-Stickel

339 Subject Index

Contributors

J. Asselin
Medical Research Council Group in
 Molecular Endocrinology
Le Centre Hospitalier de l'Université
 Laval
Quebec G1V 4G2, Canada

Etienne-Emile Baulieu
Département de Chimie Biologique
Faculté de Médecine Paris-Sud
94270 Bicêtre, France

F. Bayard
Laboratoire d'Endocrinologie
 Expérimentale
C.H.U. de Rangueil
Toulouse 31052 Cedex, France

Th. J. Benraad
Department of Medical Biology
University of Nijmegen
Nijmegen, The Netherlands

Norman Bloom
Department of Surgery
Maimonides Medical Center
Brooklyn, New York 11219

Heinz G. Bohnet
Department of Physiology
University of Manitoba
Winnipeg, Manitoba R3E OW3,
 Canada

Gail Bolan
Medical Breast Cancer Section
Medicine Branch
National Cancer Institute
National Institutes of Health
Bethesda, Maryland 20014

P. Briand
The Fibiger Laboratory
DK 2100 Copenhagen, Denmark

James H. Clark
Department of Cell Biology
Baylor College of Medicine at
 Houston
Houston, Texas 77030

J. L. Daehnfeldt
The Fibiger Laboratory
DK 2100 Copenhagen, Denmark

M. C. Deboel
Service de Médecine et
 Laboratoire d'Investigation Clinique
Institut Jules Bordet
1000 Brussels, Belgium

George A. Degenshein
Department of Surgery
Maimonides Medical Center
Brooklyn, New York 11219

J. C. Delarue
Institut Gustave Roussy
94330 Villejuif, France

B. Derache
Laboratoire d'Endocrinologie
 Expérimentale
C.H.U. de Rangueil
Toulouse 31052 Cedex, France

Nina Einhorn
Radiumhemmet
Karolinska Sjukhuset
S-104 01 Stockholm 60, Sweden

CONTRIBUTORS

Gunilla Elfström
Department of Chemistry
Karolinska Institutet
S-104 01 Stockholm 60, Sweden

Henry G. Friesen
Department of Physiology
University of Manitoba
Winnipeg, Manitoba R3E OW3,
  Canada

A. Geurts-Moespot
Department of Medical Biology
University of Nijmegen
Nijmegen, The Netherlands

Seymour M. Glick
Soroka Medical Center and University
Center for the Health Sciences
Ben Gurion University of the Negev
Beersheva, Israel

Fulgencio Gomez
Department of Physiology
University of Manitoba
Winnipeg, Manitoba R3E OW3,
  Canada

Jan-Åke Gustafsson
Department of Chemistry
Karolinska Institutet
S-104 01 Stockholm 60, Sweden

J. C. Heuson
Service de Médecine et
  Laboratoire d'Investigation Clinique
Institut Jules Bordet
1000 Brussels, Belgium

K. B. Horwitz
University of Texas Health Science
  Center at San Antonio
San Antonio, Texas 78284

Karen Huff
Medical Breast Cancer Section
Medicine Branch
National Cancer Institute
National Institutes of Health
Bethesda, Maryland 20014

T. William Hutchens
Department of Biochemistry
Cancer Research Institute
University of Southern California
  School of Medicine
Los Angeles, California 90033

P. A. Kelly
Medical Research Council Group in
  Molecular Endocrinology
Le Centre Hospitalier de l'Université
  Laval
Quebec G1V 4G2, Canada

A. J. M. Koenders
Department of Medical Biology
University of Nijmegen
Nijmegen, The Netherlands

B. Kreitmann
Laboratoire d'Endocrinologie
  Expérimentale
C.H.U. de Rangueil
Toulouse 31052 Cedex, France

Timothy E. Kute
Department of Biochemistry
University of Louisville
  School of Medicine
Health Sciences Center
Louisville, Kentucky 40201

F. Labrie
Medical Research Council Group in
  Molecular Endocrinology
Le Centre Hospitalier de l'Université
  Laval
Quebec G1V 4G2, Canada

G. Leclercq
Service de Médecine et Laboratoire
  d'Investigation Clinique
Institut Jules Bordet
1000 Brussels, Belgium

N. Legros
Service de Médecine et Laboratoire
  d'Investigation Clinique
Institut Jules Bordet
1000 Brussels, Belgium

CONTRIBUTORS

Joseph Levy
Soroka Medical Center and University
Center for the Health Sciences
Ben Gurion University of the Negev
Beersheva, Israel

Marc Lippman
Medical Breast Cancer Section
Medicine Branch
National Cancer Institute
National Institutes of Health
Bethesda, Maryland 20014

F. Logeat
Groupe de Recherches sur la
 Biochimie Endocrinienne et la
 Reproduction (U 135 INSERM)
Faculté de Médecine Paris-Sud
94 Bicêtre, France

E. Longeval
Service de Médecine et Laboratoire
 d'Investigation Clinique
Institut Jules Bordet
1000 Brussels, Belgium

H. Magdelenat
Fondation Pierre-Marie Curie
75005 Paris, France

Francis S. Markland
Department of Biochemistry and
 Cancer Research Institute
University of Southern California
School of Medicine
Los Angeles, California 90033

P. Martin
Centre Hospitalier Régional
13005 Marseille, France

W. H. Mattheiem
Service de Chirurgie
Institut Jules Bordet
1000 Brussels, Belgium

W. L. McGuire
University of Texas Health Science
 Center at San Antonio
San Antonio, Texas 78284

Rajendra G. Mehta
Department of Biochemistry
University of Louisville
 School of Medicine
Health Sciences Center
Louisville, Kentucky 40201

John S. Meyer
Department of Pathology
Jewish Hospital
St. Louis, Missouri 63110

E. Milgrom
Groupe de Recherches sur la
 Biochimie Endocrinienne et la
 Reproduction (U 135 INSERM)
Faculté de Médecine Paris-Sud
94 Bicêtre, France

James P. Neifeld
Department of Surgery
Medical College of Virginia
Richmond, Virginia 23225

Josef Nevinny-Stickel
Universitäts-Frauenklinik und
 -Poliklinik Charlottenburg der
 Freien Universität Berlin
1000 Berlin 19, Germany

Bo Nordenskjöld
Radiumhemmet
Karolinska Sjukhuset
S-104 01 Stockholm 60, Sweden

T. Ojasoo
Centre de Recherches Roussel-Uclaf
93230 Romainville, France

D. Philibert
Centre de Recherches Roussel-Uclaf
93230 Romainville, France

Kunhard Pollow
Institut für Molekularbiologie und
 Biochemie der Freien Universität
 Berlin
1000 Berlin 33, Germany

B. Ramanath Rao
Section of Cancer Biology
Division of Radiation Oncology
Mallinckrodt Institute of Radiology
Washington University School of
   Medicine
St. Louis, Missouri 63110

Jean-Pierre Raynaud
Centre de Recherches Roussel-Uclaf
93230 Romainville, France

Mannfred Schmidt-Gollwitzer
Universitäts-Frauenklinik und
   Poliklinik Charlottenburg der
   Freien Universität Berlin
1000 Berlin 19, Germany

Ellis Tobin
Department of Surgery
Maimonides Medical Center
Brooklyn, New York 11219

M. T. Vu Hai
Groupe de Recherches sur la
   Biochimie Endocrinienne et la
   Reproduction (U 135 INSERM)
Faculté de Médecine Paris-Sud
94 Bicêtre, France

Marian R. Walters
Department of Cell Biology
Baylor College of Medicine
Houston, Texas 77030

James L. Wittliff
Department of Biochemistry
University of Louisville
   School of Medicine
Health Sciences Center
Louisville, Kentucky 40201

Örjan Wrange
Department of Chemistry
Karolinska Institutet
S-104 01 Stockholm 60, Sweden

S. J. Zolingen
Department of Medical Biology
University of Nijmegen
Nijmegen, The Netherlands

Progesterone Receptors: Introduction and Overview

W. L. McGuire, J. P. Raynaud, and E. E. Baulieu

University of Texas Health Science Center at San Antonio, San Antonio, Texas 78284; Centre de Recherches Roussel Uclaf, 93230 Romainville, France; and INSERM, Département de Chimie Biologique, Faculté de Médecine de Bicêtre, 94270 Bicêtre, France

Recently, the subject of progesterone and progesterone receptors has received a great deal of attention from both reproductive biologists and tumor biologists. The measurement of progesterone receptor (PgR), however, has been hampered by the lack of a suitable radioactive ligand. In 1974, scientists at Roussel Uclaf published their findings on the use of a new synthetic progestin, R5020 (17,21-dimethyl-19-nor-4,9-pregnadiene-3,20-dione), to measure PgR. The high binding affinity, the steroid specificity, and the demonstration of a 7 to 8S peak on sucrose density gradient centrifugation in all species studied seemed to indicate that R5020 would be an ideal radioactive ligand for this measurement. These initial reports led more than 150 investigators throughout the world to request samples of radioactive R5020 for testing in their own laboratories. Subsequently, several publications have confirmed the early results obtained by the Roussel scientists and indicated possible additional uses for the compound in selecting proper endocrine therapy for patients with breast cancer.

The opening chapter of this volume by Raynaud provides the rationale behind the development of synthetic steroids such as R5020. The author details the biochemical properties of the compound (lack of tight plasma binding, high affinity for the progestin receptor, slow dissociation rate of the complex, and negligible interference with other steroid receptors) that permit its successful use as a radioactive ligand for the detection of PgR in target tissues (uterus, pituitary, and hypothalamus). The biological potency of R5020 in peripheral and central target tissues is discussed. It is 10- to 100-fold more potent than progesterone depending on the bioassay used. It has high antiestrogenic activity and is devoid of androgen and glucocorticoid activities.

The next chapter by Markland and Hutchens from Los Angeles characterizes PgR from the lactating mammary gland of the goat. The authors were able to demonstrate a specific PgR that bound [^{3}H]R5020 with a sedimentation coefficient of 7.9S and a dissociation constant of 4×10^{-10} M. In addition, they showed that nonradioactive dexamethasone, cortisol, or estradiol could not compete for the binding of [^{3}H]R5020 to PgR. On the other hand, both progesterone and triamcinolone acetonide were effective competitors. These studies demonstrated for the first time that a PgR distinct from glucocorticoid receptor was present in normal mammary tissue. This demonstration was particularly important since others had previously reported that the same receptor molecule in lactating mammary glands could bind both progesterone and glucocorticoids and therefore that both of these steroids shared a common receptor and possibly a common mechanism of action.

This point is further developed in the chapter by Wittliff and colleagues from Rochester, who studied the interaction of [^{3}H]R5020 with binding sites in normal and neoplastic mammary tissues from the rat. They demonstrated that [^{3}H]R5020 binding could be detected in lactating rat mammary gland, in R3230AC rat mammary tumor, and in DMBA-induced rat mammary tumor, as well as in human breast cancer tissue. They noted that virtually every tumor from rodents and 81% of human breast tumors contained R5020 binding sites. Under their experimental conditions, the majority of [^{3}H]R5020 binding sites sedimented at approximately 4S on sucrose gradients regardless of the buffer, the type of homogenization, or other experimental manipulations. They also found that [^{3}H]R5020 binding was decreased by a single high concentration (10^{-5} M) of all classes of hormone (progestins, estrogens, glucocorticoids, and androgens). These results were discussed extensively by the various investigators and were considered inconclusive because the true competitive effect could not be deduced without using a wide range of radioinert steroid concentrations. The inability to demonstrate 8S binding peaks for R5020 in human breast cancer tissue could not be explained.

Daehnfeldt and Briand from Copenhagen have established that in the GR mouse mammary tumor model system, PgR can be readily demonstrated in those tumors that are hormone dependent. In contrast, PgR cannot be demonstrated in hormone-independent tumors of the same strain.

Several chapters consider estrogen and progesterone receptors in DMBA-induced rat mammary tumors. Koenders and colleagues from The Netherlands demonstrate the wide range of PgR and estrogen receptor (ER) in growing DMBA tumors. After ovariectomy, ER declined to approximately 50% of its pretreatment value, whereas PgR fell to very low or undetectable levels. Estrogen administration to rats bearing regressed tumors further decreased the ER levels, but PgR increased dramatically in all of the tumors. Tumor regrowth occurred with estradiol administration in those tumors that had relatively high preovariectomy ER and PgR levels. The authors concluded that PgR synthesis in experimental mammary carcinomas, as in normal target tissues, is estrogen dependent.

Kelly and colleagues from Laval, Quebec reported on studies of ER, PgR, and prolactin receptors in DMBA-induced rat mammary tumors. They were able to demonstrate that daily injection of RU16117 (11α-methoxy-17α-ethynyl-1,3,5(10-estratriene-3,17-diol), an antiestrogen, led to a reduction in size and number of established DMBA tumors equivalent to that observed after ovariectomy. At the highest RU16117 dose used, cytoplasmic ER and PgR were decreased approximately 50% in the tumors. Ovariectomy alone reduced ER, PgR, and prolactin receptors, whereas estrogen treatment increased tumor size, ER, and PgR. Prolactin treatment also increased tumor size and ER but had no effect on PgR. Progesterone administration had no effect on tumor size or receptor levels. The authors therefore confirmed the stimulatory effect of prolactin on ER levels and the stimulatory effect of estrogen on PgR levels in DMBA tumors.

The chapter by Horwitz and McGuire from San Antonio reviews in some detail the clinical effects of progesterone in breast cancer as well as the demonstration of PgR in mammary tissues. They present data derived from the MCF-7 human breast tumor cell line that has receptors for estrogen, progesterone, androgens, and glucocorticoids. R5020 in high molar ratios can inhibit both [^{3}H]dexamethasone and [^{3}H]triamcinolone acetonide binding. In contrast, triamcinolone acetonide, but *not* dexamethasone, inhibits [^{3}H]R5020 binding to PgR. The authors conclude that PgR and glucocorticoid receptor are distinct molecules and, furthermore, that triamcinolone acetonide is a progestin in addition to a glucocorticoid.

Horwitz and McGuire hypothesize that PgR measurements in human breast cancer tissue might be useful in selecting patients for endocrine therapy. They provide additional data drawn from experimental DMBA tumors to show that PgR is indeed estrogen dependent and therefore may be considered an end-product of estrogen action in mammary tumor cells. They have measured ER and PgR in more than 500 human breast cancer specimens, finding that ER is present in 77% of primary tumors and in 66% of metastatic tumors. PgR is found in only a small proportion of either primary or metastatic tumors that lack ER, but is present in the majority of those tumors containing ER. Clinical correlation data in 44 patients show that therapy was uniformly unsuccessful in those patients lacking both receptors. In patients whose metastatic tumor tissue contained ER but not PgR, there was a 40% response rate; however, when both ER and PgR were present, the response rate was 92%. These data support the authors' original hypothesis that PgR may be a sensitive marker for predicting response to endocrine therapy.

The use of PgR and ER assays to predict the response to endocrine therapy is further supported by the data of Bloom et al. from New York. In a small series of 13 patients, they found that six out of seven patients whose tumors contained both ER and PgR responded to endocrine ablative therapy. Two patients having ER but not PgR in the tumor failed to respond, as did four patients whose tumors contained neither receptor. Clearly, the data in the last two studies are sufficiently encouraging to warrant larger clinical trials.

ER and PgR were also studied in human breast cancer tissue by Leclercq and colleagues in Brussels, who agreed that [^{3}H]R5020 is a convenient ligand for assaying PgR in human breast tumor specimens. ER and PgR were found to be present in a large proportion of both malignant and benign breast mammary tumors, the presence of PgR being essentially restricted to ER-positive specimens. PgR was also more frequent in ER-positive primary tumors than in metastatic tumors, suggesting a loss of PgR during the metastatic process. There were too few responsive breast tumor patients in this series to evaluate.

Additional studies of steroid receptors in normal uterine tissues from a variety of species as well as in human breast cancers were carried out by Rao and Meyer from St. Louis. They also showed the advantages of [^{3}H]R5020 in measuring PgR in these various tissues. Interestingly, ER and PgR were present in human breast cancer specimens irrespective of the thymidine labeling index, apparently indicating that the rate of tumor growth did not correlate with the presence or absence of these receptors.

Receptor studies of human breast cancer specimens were also performed by Raynaud and colleagues in collaboration with teams from the Institut Gustave Roussy, Fondation Pierre-Marie Curie, and Centre Hospitalier Régional de Marseille (France). They found that tumors from male patients tended to have relatively high levels of both ER and PgR, whereas levels of both receptors were low in specimens of gynecomastia from men and in benign breast tumors from women. They could not find any correlation between the plasma progesterone level and the PgR site concentration. Specificity studies on mammary tumor cytosol confirmed that progestin receptor was being measured with [^{3}H]R5020 since unlabeled nonprogestins, including dexamethasone, did not compete for this binding. These authors also presented data on the control experiments required to develop an exchange assay for the measurement of both total estrogen and progestin binding sites (determination of saturating ligand concentrations and of complex dissociation rates). The advantages of a potent, new, highly specific estrogen, R2858 (11β-methoxy-17α-ethynyl-1,3,5(10)-estratriene-3,17-diol), which does not bind to human plasma proteins, are described. Their data suggest that [^{3}H]-labeled natural hormones in ER and PgR assays might give artificially high values due to serum protein binding sites, which could be avoided with the use of [^{3}H]R2858 and [^{3}H]R5020, respectively.

Lippman and colleagues from Bethesda have used a different approach to study progesterone and glucocorticoid receptors in breast cancer tissue. They have studied the binding and biological activity of R5020 in MCF-7 human breast cancer cell culture and human peripheral blood lymphocytes. They demonstrate that both R5020 and progesterone, but not dexamethasone, compete with [^{3}H]progesterone for binding sites. They obtained a two-component Scatchard plot which suggests that R5020 binds to two classes of progesterone binding sites. In addition, they show that R5020, like the antiglucocorticoid cortexolone, inhibits DNA synthesis (a characteristic glucocorticoid action in the cell line), whereas progesterone does not have the same effect. They use human blood lymphocytes to show that R5020 and progesterone inhibit PHA stimulation in these lymphocytes as do glucocorticoids. They have previously shown that these lymphocytes do not contain PgR.

In keeping with the previous suggestion by Horwitz and colleagues that it would be preferable to measure an end-product of estrogen action rather than the initial binding step in order to predict tumor responsiveness to endocrine therapy, Levy and Glick from Beer Sheva propose that glucose oxidation might be such an end-point. They have measured ER and PgR in human breast cancer tissues and have correlated these data with the ability of estrogen, progesterone, or R5020 to influence glucose oxidation in these tumors. They interpret their results as showing that tumors in which glucose oxidase responds to hormones usually contain PgR in addition to ER, whereas nonresponsive tumors do not contain PgR, irrespective of the ER content. The authors clearly point out the preliminary nature of these studies and note the conflicting results in the literature concerning the use of products of glucose metabolism as a correlate of endocrine responsiveness in breast cancer.

Turning to the equally important subject of progesterone receptors in the uterus, Philibert and Raynaud from the Roussel Research Centre at Romainville presented their data on PgR in mouse uterus. They first demonstrated that [^{3}H]R5020 can exchange for occupied PgR filled with progesterone. Receptor sites filled with hormone were stable for at least a 24-hr period at 0°C but rapidly degraded in the absence of hormone at any temperature. It is of interest that both progesterone and R5020 associated with the receptor molecule at approximately the same rate but that progesterone dissociated 8 to 10 times faster than R5020. Complete replacement of bound progesterone by excess labeled R5020 was achieved in 4 hr at 0°C. This exchange assay enabled them to study the receptor sites in mice under different hormonal environments. For example, they found that the highest levels of PgR in adult animals were at proestrus. PgR values during pregnancy were usually two to three times lower than during the estrous cycle and reached a minimum at midpregnancy. Furthermore, the authors confirmed the estrogen dependency of PgR and showed that androgens could not substitute for estrogen in this priming effect and that progestins like R5020 could inhibit the priming effect.

Gomez and colleagues from Winnipeg carried out similar studies in rat uterus following changes in ER and PgR during the estrous cycle, pregnancy, and lactation. They found that PgR levels showed a nadir at metestrus and a peak at proestrus. On the other hand, cytoplasmic ER was lowest at proestrus. During pregnancy, and immediately after parturition, the PgR level was in the metestrous range. PgR then rose progressively to very high values by day 20 of normal lactation. In contrast, in nonlactating animals from whom the litter had been removed within 12 hr of delivery, PgR increased much more rapidly, reaching the same high plateau by day 5. It is of interest that uterine involution was significantly retarded in the nonlactating animals and that the ER was similar in both the lactating and the nonlactating groups. The authors suggest that at least part of the difference observed between these two groups may be due to endogenous levels of progesterone occupying the receptor sites. The authors also point out, however, that there are undoubtedly other factors that may have influenced the absolute PgR levels in these studies.

Milgrom and colleagues from Bicêtre also studied cytoplasmic PgR in rat uterus during the estrous cycle and pregnancy, but, in addition, investigated the concentration of PgR in the nuclei. They found that during the estrous cycle both cytosol and nuclear PgR were at the maximal concentration at proestrus, but that the ratio of nuclear receptor to cytoplasmic receptor was the highest at metestrus. During pregnancy the concentration of the cytoplasmic receptor dropped slightly on day 5 but then increased progressively during the entire pregnancy to attain a six-fold higher value on day 22. Meanwhile, the nuclear receptor increased on day 5, dropped slightly on day 6, and then increased again to attain the highest values between days 9 and 15. Afterward, the concentration of nuclear PgR began to decrease rapidly and by day 22 the concentration was very low. In fact, in some animals on the verge of parturition no nuclear receptor could be detected.

Additional data on nuclear and cytoplasmic PgR in rat uterus were presented by Walters and Clark from Houston. They were able to validate their exchange methodology and show similar hormonal specificity for the cytoplasmic and nuclear assays. Injections of either progesterone or R5020 resulted in the depletion of cytosol PgR and the simultaneous accumulation of nuclear PgR. Following nuclear translocation, they observed partial replenishment of cytosol receptor within 3 to 9 hr. Estradiol injection increased the number of cytosol PgR per gram of uterus by 24 hr. Estrogen withdrawal resulted in a relatively rapid decrease in the number of progesterone binding sites.

Bayard and colleagues from Toulouse have studied the kinetics of R5020 binding in human endometrium. They calculated the equilibrium constants and relative rates of dissociation of progesterone and R5020 binding in the absence and presence of glycerol. The authors then used saturating amounts of labeled hormones and determined the PgR content of 17 samples of human endometrium. An excellent correlation was found when [^{3}H]progesterone was used as a ligand and the isotope dilution by endogenous progesterone was considered. With [^{3}H]R5020 it was not necessary to correct for the endogenous progesterone content; therefore the authors suggest that [^{3}H]R5020 might be the ideal ligand for measuring PgR in human endometrium in both pre- and postmenopausal patients; they caution, however, that the high endogenous levels of progesterone found during pregnancy might still interfere with the absolute determination.

Gustafsson and colleagues from Stockholm used a new approach to measure PgR in endometrial carcinoma. They pointed out that isoelectric focusing had been used successfully for measuring estradiol receptor in human mammary carcinoma and that since larger sample volumes can be applied to isoelectric focusing columns, this might have some advantage over other methodologies having limited sample capacities. They found that PgR-positive endometrial carcinoma specimens have a sharp [^{3}H]R5020 peak at pH 5.5 that is completely displaceable by an excess of unlabeled R5020 or progesterone, but not by an excess of cortisol. They studied specimens from 11 patients and found that the number of specific cytoplasmic R5020 binding sites ranged from 0 to 62 fmoles/mg of protein. They also have a study in progress which is attempting to correlate the quantity of PgR in tumor specimens with the responsiveness of these tumors to progesterone therapy. One limitation of the isoelectric focusing method for measuring PgR is proposed by the authors: certain R5020 binding components of serum also focus in the pH 5 region of the gradient, and a few of the endometrial cancer specimens did show nonspecific binding in this region. Apparently, then, the isoelectric focusing technique does not offer the same advantages as that with the estradiol receptor.

In the final chapter by Pollow and colleagues from Berlin, ER and PgR were studied in both normal human endometrium and endometrial carcinoma. They used a variety of techniques including sucrose gradient centrifugation, ion-exchange chromatography, gel filtration, gradient polyacrylamide gel electrophoresis, and isoelectric focusing. They found that PgR levels in normal human endometrial tissue were directly dependent on the stage of the menstrual cycle. The level of cytoplasmic PgR was low during the early proliferative stage but increased toward the 14th day of the menstrual cycle, reaching its highest value around the time of ovulation and then declining sharply during the secretory phase. Cytosols of 30 endometrial carcinomas were examined for both ER and PgR. The levels correlated with the histological type of the tumor: ER was highest in nondifferentiated carcinomas, whereas PgR was lowest in this type. In fact, PgR could not be detected in 6 of the 11 undifferentiated specimens. The authors also studied the specific activity of 17β-hydroxysteroid dehydrogenase in various subcellular fractions of normal human endometrium and found that enzyme activity was 10-fold higher during the early secretory phase than during the proliferative phase. They also made the interesting observation that the cytosol PgR level highly correlated with the 17β-hydroxysteroid dehydrogenase activity of microsomes, and that there was a significant inverse correlation between the cytosol ER level and the enzyme activity. In the subcellular fractions of endometrial carcinoma, the enzyme activity decreased with decreasing differentiation of the tumor, as did the PgR level. After treatment of patients with progestins, enzyme activity increased only in the well-differentiated carcinomas, suggesting that the hormonal mechanism responsible for the stimulation of enzyme activity had been retained along with PgR in these well-differentiated tumors.

In summary, we can conclude that [^{3}H]R5020 is an excellent ligand for measuring PgR. Total receptor sites can be conveniently measured by exchange assays at low temperature to avoid receptor degradation. The stoichiometry of cytoplasmic depletion, translocation, and nuclear alteration of PgR needs further study.

The early suggestion that PgR measurements might be useful in a clinical setting are amply supported by studies from many independent laboratories of both experimental and human breast cancer. The same considerations may even now apply to endometrial carcinoma. The studies of PgR and ER during the estrous cycle, pregnancy, and lactation are a major contribution to our understanding of normal reproductive biology.

Progesterone Receptors in Normal and Neoplastic Tissues, edited by W. L. McGuire et al. Raven Press, New York © 1977.

R5020, a Tag for the Progestin Receptor

J. P. Raynaud

Centre de Recherches Roussel-Uclaf, 93230 Romainville, France

AN INTRODUCTION TO "TAGS"

It is now a well-accepted notion that steroid hormones act by binding to a protein "receptor" in the cytoplasm of the target cell; the complex formed is translocated into the nucleus where it triggers the sequence of events constituting the biological response. The detection and assay of these cytoplasmic receptors have not been without difficulty, since natural hormones are often also tightly bound by other proteins located within the cell or present by plasma contamination (progesterone is bound not only by the progestin receptor but also by a cortisol binder) and/or are metabolized under *in vitro* experimental conditions (dihydrotestosterone is metabolized to androstanediol even at 0°C).

To avoid the problems arising from the use of natural hormones, we have focused our attention over the last few years on highly potent synthetic

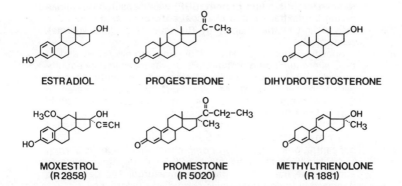

FIG. 1. Formulas of synthetic steroids ("tags") used to detect and assay hormone receptors. R2858 = 11β-methoxy-17-ethynyl-1,3,5(10)-estratriene-3,17β-diol; R5020 = 17,21-dimethyl-19-nor-pregna-4,9-diene-3,20-dione; R1881 = 17β-hydroxy-17α-methyl-estra-4,9,11-trien-3-one. For biochemical studies, these compounds were labeled with tritium in positions 6 and 7 with a specific activity of approximately 50 Ci/mmole according to batch (A. Jouquey, Roussel-Uclaf). Their purity is checked by thin-layer chromatography (TLC) using the following solvent systems: benzene:ethyl acetate (6:4, vol/vol) and methylene chloride:methanol (9:1, vol/vol) for [³H]R2858; benzene:ethyl acetate (7:3, vol/vol) for [³H]R5020 and benzene:ethyl acetate (1:1, vol/vol) for [³H]R1881.

hormones without these disadvantages and have used them as tools to track down and assay elusive receptors. These "tags" have been chosen on the basis of their lack of tight binding in plasma, their resistance to degradation under *in vitro* incubation conditions, their hormone specificity for the receptor under study, the slow dissociation rate of the complex they form with this receptor, and the possibility of labeling them with high specific activity. At present, among available molecules, the ones which seem to meet these requirements best are the estrogen R2858 (Moxestrol) (1,2), the progestin R5020 (Promestone) (3–6), and the androgen R1881 (methyltrienolone) (7–9) (Fig. 1). All these compounds are highly potent molecules, at least 10 times as active as the natural hormones in routine biological activity tests; none bind specifically to plasma proteins (Table 1); none are metabo-

TABLE 1. *Competition for specific plasma binding*

	CBG	SBP	EBP	PBG
Cortisol	100	<0.1	–	<0.1
Progesterone	30	<0.1	<0.1	100
R5020	<0.1	<0.1	<0.1	10
R1881	<0.1	1	<0.1	5
DHT	0.3	100	<0.1	20
Testosterone	10	–	<0.1	5
Estradiol	<0.1	10	100	<0.1
Moxestrol	<0.1	<0.1	<0.1	<0.1

Specific binding to corticosteroid binding globulin (CBG) and to sex steroid binding protein (SBP) was measured on human serum by dextran-coated charcoal adsorption as described in ref. 10 and in the chapter by Raynaud et al. (*this volume*), respectively.

Specific binding to estradiol binding protein (EBP) and to progesterone binding globulin (PBG) was measured by equilibrium dialysis on immature rat plasma and pregnant guinea pig serum, respectively. Dilute plasma (1:200) was prepared in 0.15 M phosphate buffer (pH 7.4) and dilute serum (1:2,500) in 0.01 M Tris-HCl (pH 7.4), 2.5 M sucrose buffer. Dilute plasma or serum (1 ml) was dialyzed against 15 ml of labeled estradiol (0.5 nM) or progesterone (2.5 nM) in the presence of various concentrations of radioinert competitor. After magnetic stirring at 4°C for 48 hr, the radioactivity of 0.2-ml samples from inside and outside the dialysis bag was measured. The percentage decrease in bound radioactive steroid was determined.

The above results have been confirmed by D. W. Chan and W. R. Slaunwhite (*personal communication*), who recorded the following association constants (K_a) for binding to purified human CBG at 4°C, pH 7.4: 4×10^8 M^{-1} for progesterone, 2×10^8 M^{-1} for cortisol, and 1×10^6 M^{-1} for R5020. Similar data have also been recorded by U. Westphal (*personal communication*). R5020 binds to human CBG with a K_a of 8×10^5 M^{-1}, to human serum albumin with a K_a of 6×10^5 M^{-1}, and to guinea pig PBG with a K_a of 1×10^8 M^{-1}.

TABLE 2. "In vitro" relative binding affinity for specific tissue receptors

	Est	Pro	And	Glu	Min
Estradiol	100	1.2	4.2	0.4	0.2
Progesterone	<0.1	100	4.8	0.4	7.5
R5020	<0.1	230	2.8	17	0.3
Testosterone	<0.1	0.8	100	0.2	1.0
R1881	<0.1	190	220	40	12
Cortisol	<0.1	<0.1	<0.1	38	21
Dexamethasone	<0.1	0.5	<0.1	100	17
Aldosterone	<0.1	0.7	<0.1	1.5	100

Biological origin of receptors: immature mouse uterus (Est), uterus from estrogen-primed rabbits (Pro), prostate of castrated rats (And), liver and kidney from adrenalectomized rats (Glu and Min, respectively). *Homogenization* of 1 g tissue in 24 ml (Est), 49 ml (Pro), 3 ml (And), 9 ml (Glu), and 2 ml (Min) of 0.01 M Tris-HCl (pH 7.4), 0.25 M sucrose buffer (Est, Pro, And, Glu) or of Krebs-Ringer phosphate glucose buffer (pH 7.4) (Min). *Centrifugation* of homogenate at 105,000 × g for 1 hr (Est, Pro, And, Glu). *Incubation* of cytosol for 2 hr at 0°C with 1 nM [^{3}H] estradiol (Est), 2 nM [^{3}H]R5020 (Pro), 1 nM [^{3}H]R1881 (And), or for 4 hr at 0°C with 2.5 nM [^{3}H]dexamethasone (Glu) in the absence or presence of 1 to 2,500 nM radioinert competitor. Incubation of kidney homogenate for 30 min at 25°C with 2.5 nM [^{3}H]aldosterone (Min) followed by centrifugation at 800 g for 10 min at 4°C. *Measurement of bound radioactivity* by dextran-coated charcoal adsorption as described in the legend of Fig. 3 and determination of the competitor concentration required for 50% displacement of radioligand from its specific binding sites. Results are expressed as the ratio of reference compound concentration to test compound concentration.

lized *in vitro*. Except for R2858, however, none are totally specific to the receptor corresponding to a particular class of hormone. R5020 competes to 10% for labeled dexamethasone binding in rat liver cytosol, and R1881 binds as markedly as R5020 to the progestin receptor in rabbit uterus (Table 2) (11,12). This lack in R1881 specificity has been a temporary advantage in studies on human hyperplastic prostate. R1881 has not only enabled the assay of androgen receptor in hyperplastic prostate, but it has also been the basis of the observation that a receptor more progestin- than androgen-like is sometimes present in variable amounts in this tissue (13).

A large screening program, involving competition for the binding of available tags (R2858 or estradiol, R1881, R5020, dexamethasone, and aldosterone) to five different receptors (estrogen, androgen, progestin, gluco-, and mineralocorticoid) has been set up in order to try and obtain yet more specific molecules. Several compounds in a study of over 400 molecules (11,12) seem to be highly suitable candidates to tag the progestin receptor and may yet supersede R5020. A slight increase in specificity does not, however, necessarily justify the experimental load involved in the develop-

BIOCHEMICAL STUDIES ON R5020

R5020 has enabled the characterization and assay of progestin receptor binding sites in the 105,000 × g supernatant (cytosol) of uterine homogenate from several species since, unlike progesterone, it does not bind specifically to serum corticosteroid binding globulin (Table 1) and since it forms a more stable, slow-dissociating complex with the progestin receptor than progesterone (14; Raynaud et al. and Philibert and Raynaud, *this volume*). Thus, for instance, it has been possible to establish by sucrose density gradient analysis of cytosol incubated with [^{3}H]R5020 that a similar 7 to 8S progestin-specific component is present in the uterine cytoplasm of several

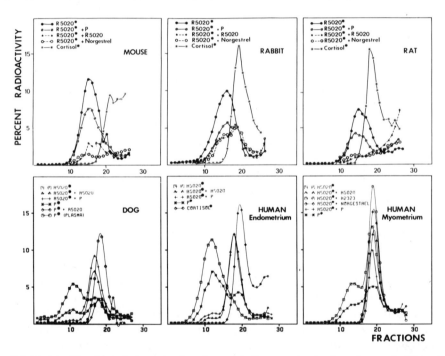

FIG. 2. Identification of the cytoplasmic progestin receptor in the uterus of different species by sucrose density gradient analysis. Uteri were excised, weighed, crushed, pooled, then homogenized (1/3, wt/vol) with a Teflon-glass homogenizer in 0.01 M Tris-HCl (pH 7.4), 1 mM EDTA, 12 mM thioglycerol buffer containing 10% glycerol. The homogenate was centrifuged at 105,000 × g for 90 min at 0 to 4°C. The supernatant (cytosol) was incubated for 1 hr at 0°C with 1 to 6 nM [^{3}H]R5020 in the presence or absence of 100 nM radioinert competitor, layered (0.3 ml) on a linear 5 to 20% sucrose gradient prepared in homogenization buffer and centrifuged at 45,000 rpm for 16 hr at 4°C (SW 50.1 rotor, Beckmann L$_3$50 centrifuge).

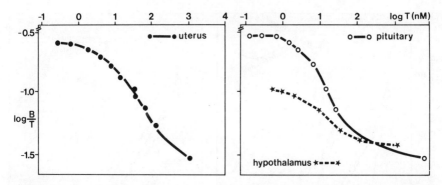

FIG. 3. Proportion graphs of the binding of [^{3}H]R5020 to the cytoplasmic progestin receptor in different organs of the rat. The uterus, pituitary, and hypothalamus were excised, weighed, and homogenized with a Teflon-glass homogenizer in 0.01 M Tris-HCl (pH 7.4), 0.25 M sucrose buffer. The homogenate was centrifuged at 105,000 × g for 60 min at 0 to 4°C and incubated for 4 hr at 0°C with 0.1 to 10 nM [^{3}H]R5020 in the presence or absence of increasing concentrations of radioinert compound. Bound radioactivity was measured by a dextran-coated charcoal adsorption technique: 0.1 ml of incubated cytosol was stirred in the presence of 0.1 ml Norit A–dextran (1.25 to 0.625%) for 10 min at 0°C and centrifuged for 10 min at 1,500 g. The radioactivity of a 0.15-ml supernatant sample was counted. Results are represented in a proportion graph: log fraction bound steroid (log B/T) versus log total steroid concentration (log T). The intrinsic dissociation constant (K_d) and number of binding sites (N_s) were evaluated as described previously (15). (K_d = 12, 6, 6 nM, and N_s = 32, 4, 0.3 pmoles/g tissue for the uterus, pituitary, and hypothalamus, respectively).

species including the rat (5) and DMBA tumors in the rat (16), mouse (5), rabbit (4), guinea pig (4), human (6), and dog (Fig. 2). This receptor is latent in the unprimed animal (4,5) and has been shown to have the same specificity in the rabbit and mouse (17).

Measurement of specific [^{3}H]R5020 binding by a dextran-coated charcoal adsorption technique has furthermore revealed that the above-identified progestin receptor is present in different organs of the same animal. [^{3}H]R5020 binds to a progestin receptor in the uterus, pituitary, and hypothalamus of the rat (Fig. 3) with comparable intrinsic dissociation constants; the number of specific binding sites varies considerably, however, according to the target tissue (M. Moguilewsky, *personal communication*). The hormone specificity of this binding is similar in peripheral (uterus) and central (pituitary) organs (Fig. 4). Potent progestins such as norgestrel and norprogesterone competed markedly for [^{3}H]R5020 binding; other classes of hormone (estradiol, testosterone, aldosterone, and dexamethasone) had little or no effect. The above experiments thus establish that the progestin receptor is similar in all species studied and in different target organs of the same species.

Like most pregnane derivatives, R5020 is highly specific, competing only for the progestin receptor and very little for binding to other receptors. In

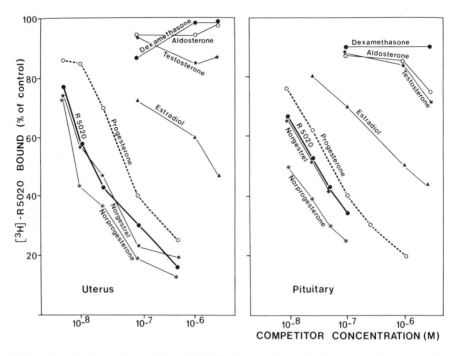

FIG. 4. Specificity of the binding of [³H]R5020 to the cytoplasmic progestin receptor in rat uterus and pituitary. After cytosol was incubated with 2.5 nM (pituitary) or 5 nM (uterus) [³H]R5020 in the presence of various concentrations of radioinert competitor, bound radioactivity was measured as described in the legend of Fig. 3.

Fig. 5 the relative binding affinities of a series of progestins for different receptors have been recorded as well as their progestational activity in rabbit uterus. This figure shows that if factors such as pharmacokinetics and metabolism are taken into account, the relative binding affinity of a molecule for a given receptor can be related to its biological potency (11,12,17) and that, on the basis of structure-activity relationships, R5020 should be a highly potent progestin.

BIOLOGICAL ACTIVITY OF R5020

The biological activity profile of R5020 is quite consistent with the above determinations on relative binding affinity. In animal studies (G. Azadian-Boulanger, *personal communication*), R5020 is fundamentally a highly potent progesterone-like agent. Its ability to induce endometrial proliferation on local and subcutaneous administration to the estrogen-primed rabbit, to form deciduomata, and to maintain pregnancy in the castrated rat is 10 to 100 times that of progesterone (Fig. 6). Its antiovulatory activity, when measured by the decrease in the number of corpora lutea following

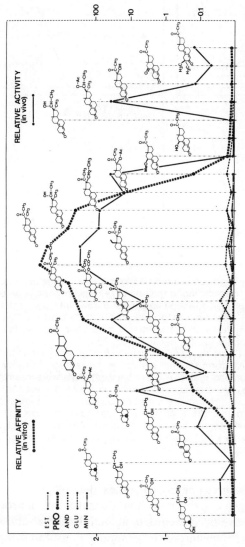

FIG. 5. Relationship between the relative binding affinity of progestins to the cytoplasmic progestin receptor in rabbit uterus and their progestational activity. Progestational activity was evaluated according to Clauberg's test (18) as described in ref. 17. The endometrial proliferation on subcutaneous administration of test compound to estrogen-primed immature rabbits was graded histologically according to McPhail's scale. The dose giving rise to a response of 2 McPhail units was taken as the active dose and used to calculate the relative potency of each steroid in comparison to progesterone (=1). Relative binding affinity was measured as described in the legend of Table 2.

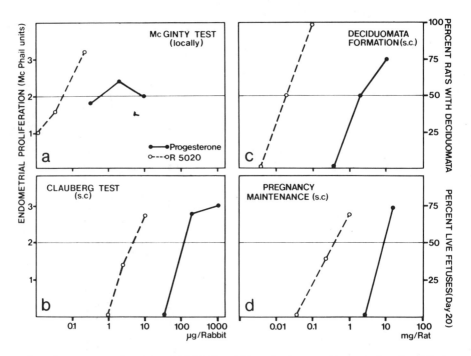

FIG. 6. Progestational activity of R5020 in rabbit and rat uterus. Progestational activity in the rabbit was measured either according to the test of **(a)** McGinty et al. (19) or **(b)** Clauberg (18). The animals were primed subcutaneously for 5 days with 5 µg estradiol **(a)** or 10 µg estradiol benzoate **(b)**. The test compound was then administered either locally (i.e., deposited within two ligatures of the uterus) 3 days after priming **(a)** or subcutaneously for 5 days after priming **(b)**. On decapitation, 72 hr **(a)** or 24 hr **(b)** after treatment, the uterus was fixed in Bouin fluid and its transformation was graded histologically according to McPhail's scale. Progestational activity in the rat was given by the ability to induce deciduomata **(c)** and to maintain pregnancy **(d)**. Rats were castrated on the day of estrus (day 1) and treated subcutaneously for 10 days with the test compound. On day 5, a thread was passed through one uterine horn. On day 11, the percentage of rats with deciduomata was determined **(c)**. Pregnant rats were castrated on day 8 of pregnancy, day 1 being determined by the presence of a vaginal plug. The test compound was administered subcutaneously from days 8 to 19. The animals were killed 24 hr after the last treatment and the percentage of live fetuses was determined **(d)**.

treatment of the rat, is approximately 20 times that of progesterone (*not shown*), and its antiestrogenic activity is about 50 times that of progesterone (Fig. 7). Similar results have been recorded, at slightly higher doses, following oral administration. Progesterone is generally considered inactive by the oral route.

Unlike progesterone, which is approximately 50 times less androgenic than testosterone in the castrated rat, R5020 is totally devoid of androgenic activity up to a dose of 25 mg. It is also totally devoid of uterotrophic activity (Fig. 7).

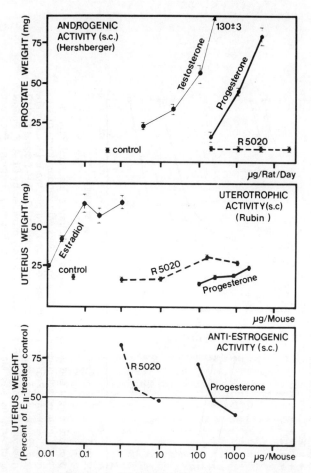

FIG. 7. Androgenic, uterotrophic, and antiestrogenic activities of R5020. Castrated immature male rats received subcutaneous injections of test compound for 7 days. The prostate gland was excised and weighed 24 hr after the last treatment and the increase in weight with respect to controls was determined (20). Immature female mice received subcutaneous injections of test compound alone or simultaneously with 0.09 μg estradiol for 3 days. The uterus was excised and weighed 24 hr after the last injection. The increase in weight with respect to controls gave an indication of uterotrophic activity, the inhibition of the estradiol-induced increase gave an indication of antiestrogenic activity (21).

Unlike progesterone, androgens are potent inhibitors of LH secretion in the castrated animal (22). Since R5020 is a highly potent progestin totally devoid of androgenic activity, its action on gonadotropin secretion was investigated in comparison with an androgen (dihydrotestosterone, DHT) and two classes of progestins, namely, pregnane and estrane derivatives (F. Labrie, *personal communication*). As illustrated in Fig. 8, all estrane derivatives exerted pronounced androgenic activity and inhibited LH secre-

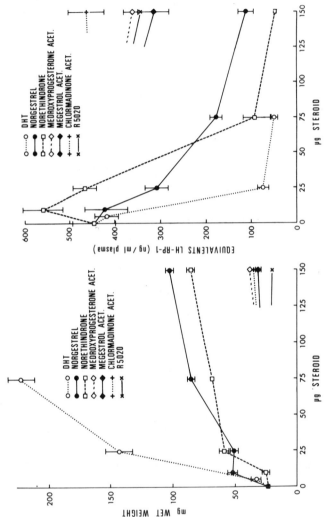

FIG. 8. Effect of R5020 on prostate weight and on plasma LH levels in castrated rats. Adult rats, castrated 2 weeks previously, were injected subcutaneously, daily for 7 days with various doses of synthetic progestins or dihydrotestosterone. Prostate weight was recorded and plasma LH levels were measured by double-antibody radioimmunoassay using rat hormones (NIAMD Rat LH-1-3 and NIAMD Rat LH-RP-1) and rabbit antiserum (NIAMD Anti-Rat LH Serum I) kindly provided by Dr. A. F. Parlow from the National Institute of Arthritis and Metabolic Diseases, Rat Pituitary Hormone Program. Pregnane derivatives: Medroxyprogesterone acetate (17α-acetoxy-6α-methyl-pregn-4-ene-3,20-dione); megestrol acetate (17α-acetoxy-6-methyl-pregna-4,6-diene-3,20-dione); chlormadinone acetate (6-chloro-17α-hydroxy-pregna-4,6-diene-3,20-dione acetate). Estrane derivatives: Norgestrel (13-ethyl-17-hydroxy-18,19-dinor-17α-pregn-4-en-20-yn-3-one); norethindrone (17-hydroxy-19-nor-17α-pregn-4-en-20-yn-3-one).

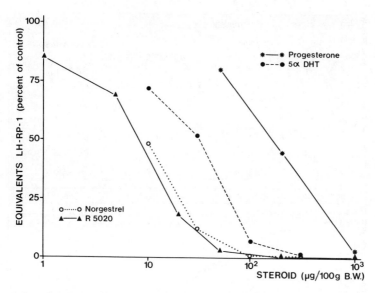

FIG. 9. Effect of R5020 injected at 9 A.M. and 4 P.M. on estrus on the plasma LH concentration measured in the afternoon (4 P.M.) of expected proestrus in 4-day cycling rats.

tion; pregnane derivatives, on the other hand, had virtually no androgenic potency and no significant effect on LH secretion up to a daily dose of 150 µg.

The action of R5020 on the proestrus surge of gonadotropin secretion was investigated in comparison with progesterone, norgestrel, and DHT (Fig. 9). A dose of 50 µg of R5020 totally inhibited LH secretion as did norgestrel, a compound having both progestin and androgenic activities. A similar dose of DHT and progesterone exerted only partial inhibition (50% and 25%, respectively).

Finally, although R5020 competes to approximately 10% for labeled dexamethasone binding to rat liver cytosol, its ability to induce tyrosine aminotransferase (TAT) activity in hepatoma tissue culture cells (G. Beck, *personal communication*) is extremely low and no different from that of progesterone (Fig. 10). It is, however, a somewhat more potent inhibitor of the dexamethasone- or cortivazol-induced TAT activity than progesterone, a result which might be in relation to its competition for the glucocorticoid receptor.

The clinical activity of R5020 is at present under investigation. Preliminary results (H. Rozenbaum, *personal communication*) indicate that this compound is a pure progestin, often effective at a dose as low as 125 µg/day, in the treatment of women with menstrual irregularities, fibroadenomas, and other indications of progesterone deficiency.

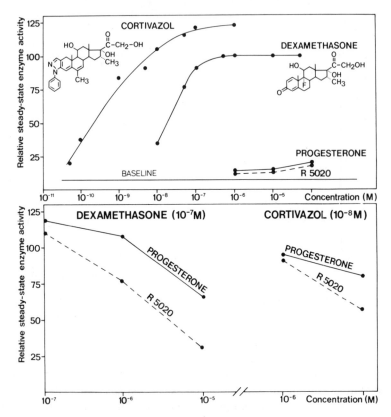

FIG. 10. Glucocorticoid and antiglucocorticoid activity of R5020 as given by the induction and inhibition of induction of tyrosine aminotransferase activity in hepatoma tissue culture (HTC) cells. HTC cells were grown at 37°C in suspension cultures in SWIM's medium supplemented with 10% calf serum, and TAT activity was assayed according to the method of Diamondstone (23). The plotted values are the means of measurements at 9 and 24 hr. The induction obtained with 10^{-6} M dexamethasone was taken as 100% and corresponded to 142 mU/mg protein.

REFERENCES

1. Azadian-Boulanger, G., and Bertin, D. (1973): Synthèse et activité utérotrophique du 11β-méthoxy estradiol, 11β-méthoxy estriol et 11β-méthoxy 17α-éthynyl estradiol. *Chim. Therap.*, 8:451–454.
2. Raynaud, J. P., Bouton, M. M., Gallet-Bourquin, D., Philibert, D., Tournemine, C., and Azadian-Boulanger, G. (1973): Comparative study of estrogen action. *Mol. Pharmacol.*, 9:520–533.
3. Joly, R., Warnant, J., Jolly, J., and Farcilli, A. (1973): Préparation du 17-méthyl-19-norpregna-4,9-diène-3, 20-dione. *Bull. Soc. Chim.*, 9,10:2694–2697.
4. Philibert, D., and Raynaud, J. P. (1974): Progesterone binding in the immature rabbit and guinea pig uterus. *Endocrinology*, 94:627–632.
5. Philibert, D., and Raynaud, J. P. (1973): Progesterone binding in the immature mouse and rat uterus. *Steroids*, 22:89–98.

6. Philibert, D., and Raynaud, J. P. (1974): Binding of progesterone and R5020, a highly potent progestin, to human endometrium and myometrium. *Contraception,* 10:457-466.
7. Velluz, L., Nominé, G., Bucourt, R., and Mathieu, J. (1963): Un analogue triénique de la testostérone. *C. R. Acad. Sci. [D] (Paris),* 257:569-570.
8. Bonne, C., and Raynaud, J. P. (1975): Methyltrienolone, a specific ligand for cellular androgen receptors. *Steroids,* 26:227-232.
9. Bonne, C., and Raynaud, J. P. (1976): Assay of androgen binding sites by exchange with methyltrienolone (R1881). *Steroids,* 27:497-507.
10. Raynaud, J. P., Bouton, M. M., Philibert, D., Delarue, J. C., Guérinot, F., and Bohuon, C. (1977): Progesterone and estradiol binding sites in human breast carcinoma. *Res. Steroids (in press).*
11. Raynaud, J. P. (1977): Une stratégie de recherche pour les hormones de synthèse: leurs interactions avec les récepteurs hormonaux. *Actualités Pharmacologiques,* Vol. 29, edited by J. Cheymol, J. R. Boissier, and P. Lechat, pp. 49-64. Masson et Cie, Paris.
12. Raynaud, J. P. (1977): A strategy for the design of potent hormones. Proceedings of the Vth International Symposium on Medicinal Chemistry, Paris, July 1976 *(in press).*
13. Asselin, J., Labrie, F., Gourdeau, Y., Bonne, C., and Raynaud, J. P. (1976): Binding of [^{3}H]-methyltrienolone (R1881) in rat prostate and human benign prostatic hypertrophy (BPH). *Steroids,* 28:449-459.
14. Jänne, O., Kontula, K., and Vihko, R. (1976): Kinetic aspects in the binding of various progestins to the human uterine progesterone receptor. Abstracts, Vth International Congress of Endocrinology, Hamburg, July 1976, p. 151, No. 367.
15. Baulieu, E. E., and Raynaud, J. P. (1970): A "proportion graph" method for measuring binding systems. *Eur. J. Biochem.,* 13:293-304.
16. Asselin, J., Labrie, F., Kelly, P. A., Philibert, D., and Raynaud, J. P. (1976): Specific progesterone receptors in dimethylbenzanthracene (DMBA)-induced mammary tumors. *Steroids,* 27:395-404.
17. Raynaud, J. P., Philibert, D., and Azadian-Boulanger, G. (1974): Progesterone-progestin receptors. In: *The Physiology and Genetics of Reproduction, Part A,* edited by E. M. Coutinho and F. Fuchs, pp. 143-160. Plenum Press, New York.
18. Clauberg, C. (1930): Zur Physiologie und Pathologie der Sexualhormone, im besonderen des Hormons des Corpus Luteum. 1-Mitt: der biologische Test für das Luteohormon (das spezifische Hormon des Corpus Luteum) am infantilen Kaninchen. *Zbl. Gyn.,* 54:2757-2770.
19. McGinty, D. A., Anderson, L. P., and McCullough, N. B. (1939): Effect of local application of progesterone on the rabbit uterus. *Endocrinology,* 24:829-832.
20. Hershberger, L. G., Shipley, E. G., and Meyer, R. K. (1953): Myotrophic activity of 19-nortestosterone and other steroids determined by modified levator ani muscle method. *Proc. Soc. Exp. Biol. Med.,* 83:175-180.
21. Rubin, B. L., Dorfman, A. S., Black, L., and Dorfman, R. I. (1951): Bioassay of estrogens using the mouse uterine response. *Endocrinology,* 49:429-439.
22. Ferland, L., Drouin, J., and Labrie, F. (1976): Role of sex steroids on LH and FSH secretion in the rat. In: *Hypothalamus and Endocrine Functions,* edited by F. Labrie, J. Meites, and G. Pelletier, pp. 191-209. Plenum Press, New York.
23. Diamondstone, T. I. (1966): Assay of tyrosine transaminase activity by conversion of p-hydroxyphenylpyruvate to p-hydroxybenzaldehyde. *Anal. Biochem.,* 16:395-401.

*Progesterone Receptors in Normal and Neoplastic Tissues,* edited by W. L. McGuire et al. Raven Press, New York © 1977.

Characterization of the Progestin Receptor from Lactating Mammary Glands of the Goat

Francis S. Markland, Jr., and T. William Hutchens

Department of Biochemistry and Cancer Research Institute, University of Southern California School of Medicine, Los Angeles, California 90033

Previous studies by various workers (12,21) have provided extensive evidence for the existence of specific cytoplasmic receptor proteins that interact with steroid hormones as a prerequisite for eliciting hormonal response. Specific progesterone receptors have been demonstrated in the reproductive tracts of many different species (2,6,13,14,16,23–26,31) as well as in human and rat mammary tumors (1,7,9,10,19,32,34). O'Malley and his collaborators (4,29,30) have extensively purified and characterized progesterone receptors in human uterus and chick oviduct. However, there had been little work on the characterization and/or purification of specific progesterone receptor proteins from normal mammary tissue until the recent reports from Wittliff's laboratory on the characterization of the progesterone receptor in rat lactating mammary gland (7,19,34).

Identification of the specific progesterone receptor has been difficult since it is known that progestins bind to the corticosteroid binding globulins (CBG) (9,10) and to other nonspecific components which sediment in the 4S region during sucrose density gradient centrifugation. Progestins have also been shown in mammary tissue to compete with glucocorticoids for binding to the glucocorticoid receptor (34). In fact, it has been suggested by Wittliff (34) that these two classes of steroids act through a common mechanism.

The use of the highly potent synthetic progestin R5020 (22), which does not bind to plasma CBG (9,10,23,24), has helped resolve the problem of identification of the specific receptor. R5020 specifically bound to the cytoplasmic receptor has been shown to give rise to a 7 to 8S peak on density gradient centrifugation that is distinct from the CBG peak (22). In uterine systems (22,23) specific R5020 binding is found only in the 7 to 8S region, and noninhibitable binding accounts for a separate peak seen in the 4 to 5S region. However, in human mammary tumors (10) and the MCF-7 human breast cancer cell line (9) there is R5020 binding of specific nature in both the 7 to 8S and 4 to 5S regions. Interestingly, competitive inhibition studies with the MCF-7 cell line (9) and in human breast cancer tissue (10) suggest separate and distinct receptors for both progestins and glucocorticoids, which is in contrast to the studies described above for normal mammary tissue.

This laboratory, aside from clinical aspects of receptor analysis, is interested in the isolation and characterization of steroid hormone receptors. There have been no published reports of attempts to purify the progestin receptor from mammary tissue; generally the oviduct or uterus has been used as starting material. However, because of the possible relevance to breast carcinoma, we are using mammary tissue for isolation and characterization of the receptor. Due to the obvious difficulty in obtaining sufficient human tissue, and since large-scale purification and characterization of the progestin receptor are our ultimate goal, we have chosen to work with goats since their mammary glands during lactation are large enough to provide sufficient tissue for our studies.

We have used R5020 in this study to aid in the identification and characterization of the progestin receptor found in cytosol from lactating goat mammary tissue.

MATERIALS AND METHODS

Reagents

Reagent grade Trisma base and dithiothreitol (DTT) were obtained from Sigma Chemical. Ethylenediaminetetraacetic acid (EDTA) was purchased from J. T. Baker. Ultrapure, RNase-free sucrose used for density gradient centrifugation, radioimmunoassay-grade charcoal, and ultrapure ammonium sulfate were purchased from Schwarz/Mann. Dextran T-70 and Sephadex G-25 were obtained from Pharmacia Fine Chemicals. ScintillAR-grade toluene was obtained from Mallinkrodt. Triton X-100 and Omnifluor were from New England Nuclear. All other reagents were of analytical reagent grade.

Steroids

Unlabeled dexamethasone (9α-fluoro-16α-methyl-$11\beta,17\alpha,21$-trihydroxy-1,4-pregnadiene-3,20-dione), cortisol ($11\beta,17\alpha,21$-trihydroxy-4-pregnene-3,20-dione), and dihydrotestosterone (17β-hydroxy-5α-androstan-3-one) were purchased from Steraloids. Unlabeled progesterone (4-pregnene-3,20-dione) was from Calbiochem. 17β-Estradiol and triamcinolone acetonide (9α-fluoro-11β, 16α, 17α, 21-tetrahydroxy-pregna-1,4-diene-3,20-dione-16,17-acetonide) were from Sigma Chemical. Tritium-labeled triamcinolone acetonide (TA) was purchased from Amersham/Searle. Tritium-labeled estradiol (E_2) and dihydrotestosterone (DHT) were from New England Nuclear. Tritium-labeled synthetic R5020 (17,21-dimethyl-19-nor-pregna-4,9-diene-3,20-dione-6,7-^{3}H) and radioinert R5020 were supplied by Roussel-

Uclaf, Paris, France. Stock solutions of all steroids were made up in absolute ethanol and stored at 0 to 4°C.

All steroids are checked for purity periodically by thin-layer chromatography.

Animals and Tissue Storage

Lactating goats were purchased from local farmers. The goats were anesthetized with ether and sacrificed. Mammary glands were quickly excised and placed into a beaker of ice cold TED buffer (0.01 M Tris-HCl, 1.5 mM EDTA, 0.5 mM DTT, pH 7.4). After rinsing, the mammary glands were trimmed of fat and excess connective tissue, cut into small pieces (~ 500 mg), and frozen in liquid nitrogen. The frozen pieces can be stored for extended periods of time at liquid nitrogen temperatures without losing specific binding activity.

Preparation of Cytosol

Frozen pieces of lactating mammary gland were pulverized in a Microdismembrator at liquid nitrogen temperatures. The frozen, pulverized tissue was diluted with 2 to 4 volumes of ice cold TED buffer and homogenized using a Polytron PT-10 ST tissue homogenizer at a setting of 5. Several 10-sec bursts with interim 50-sec cooling periods were usually needed for the homogenization. The homogenate was centrifuged in a Beckman L5-65 ultracentrifuge using an SW 56Ti rotor at 40,000 rpm ($\sim 208,000 \times g$) for 1 hr. The supernatant was drawn from beneath the lipid layer with a syringe and filtered through a 45-μm Millex disposable filter (Millipore). The resultant cytosol was stored in an ice bath until use, no more than 1 hr later. These cytosols yield an average protein concentration of 6 to 7 mg/ml as determined by the method of Lowry et al. (15).

Dextran-Coated Charcoal Assay

Cytosol (0.2 ml) was added to 1.5-ml microfuge tubes in which varying quantities of ^{3}H-R5020 (total binding) or ^{3}H-R5020 in the presence of 100- or 1,000-fold molar excess unlabeled R5020 (nonspecific binding) had been previously dried down under nitrogen. Duplicate reactions were incubated for either 3 or 18 hr at 0 to 4°C. To terminate the reaction, we added 1.0 ml of a dextran-coated charcoal (DCC) suspension (0.5% radioimmunoassay-grade charcoal and 0.05% dextran T-70 in TED buffer) to each of the microfuge tubes and allowed it to incubate with periodic agitation for 10 min at 0 to 4°C. The tubes were then centrifuged twice in a Beckman Microfuge B at full speed ($\sim 10,000 \times g$) for 5 min each at 0 to 4°C. A 0.5-ml aliquot was taken from each tube and placed in a counting vial with 5.0 ml of a toluene-based scintillation cocktail (700 ml toluene; 300 ml Triton

X-100; 4.0 g Omnifluor). The samples were counted at 46% efficiency in a Nuclear Chicago Isocap 300 liquid scintillation counter or at 35% efficiency in a Beckman LS 3150 T liquid scintillation counter. Charcoal efficiency was determined for each concentration of ^{3}H-R5020 used in the assay and was always maintained above 99.5 ($\pm$0.5)%. Efficiency was calculated from the difference between the counts from ^{3}H-R5020 in 0.5 ml TED buffer (total counts) and the counts from ^{3}H-R5020 in TED buffer after treatment with DCC (background counts). Specific binding was calculated from the total binding less nonspecific binding.

Temperature Stability of Specifically Bound ^{3}H-R5020 with Time

Cytosol (10.0 ml) was preincubated with 6.0 nM ^{3}H-R5020 in the presence or absence of 100-fold molar excess radioinert R5020 for 18 hr at 0 to 4°C and treated with a DCC pellet to remove free steroid. The supernatants were then incubated at 0°, 15°, or 37°C and subsequently assayed at various times (1 min to 26 hr) thereafter by removing duplicate 0.2-ml samples from both the ^{3}H-R5020 and ^{3}H-R5020 plus radioinert R5020 tubes. Aliquots were counted and the amount of specifically bound ^{3}H-R5020 remaining was determined.

Sucrose Density Gradient Centrifugation (SG)

Cytosol (0.4 ml) was added to each of several 1.5-ml microfuge tubes containing either 6.0 nM ^{3}H-R5020 or 6.0 nM ^{3}H-R5020 in the presence of 100- to 1,000-fold molar excess radioinert R5020 (or other competitor as noted), which had been dried down under nitrogen just prior to incubation. The reactions were incubated for either 3 or 18 hr at 0 to 4°C. Labeled cytosols were then mixed with the pellet from 1.5 ml of DCC, vortexed, and incubated with periodic agitation for 10 min at 0 to 4°C to remove free steroids. The charcoal was removed by centrifugation (twice, 5 min each) in the Beckman Microfuge B. Aliquots (0.2 ml) of clear supernatant were removed and layered onto 5 to 40% linear sucrose gradients. The gradients were prepared using TED buffer and formed in 4.0-ml polyallomer tubes using a Buchler gradient former. Occasionally, TDK buffer (0.01 M Tris-HCl, 0.5 mM DTT, and 0.4 M KCl, pH 7.4) was used for centrifugation in 10 to 30% sucrose density gradients. The gradients were centrifuged in a Beckman L5-65 ultracentrifuge using an SW 56Ti rotor at 56,000 rpm (407,000 $\times$ g) for 15 to 16 hr at 0 to 4°C. Fractions (0.1 ml) were collected from the bottom of each tube into scintillation vials. The same toluene-based scintillation cocktail described for DCC assay was added (10 ml) to each vial and the radioactivity measured in a Nuclear Chicago Isocap 300 liquid scintillation counter at 32% efficiency or in a Beckman LS 3150 T liquid scintillation counter at 41% efficiency.

Human serum albumin (4.6S) and human γ-globulin (7.1S) were used as marker proteins to estimate the sedimentation coefficient of the receptor proteins according to the method of Martin and Ames (18). Fractions (0.1 ml) were collected after SG of a mixture of the two proteins, and protein concentrations were determined by the method of Lowry et al. (15).

Partial Purification of R5020 Receptor by Ammonium Sulfate Precipitation

Frozen pieces of the lactating goat mammary tissue were partially thawed at 0 to 4°C, minced, and placed immediately into 3 volumes of ice cold TED buffer. This was immediately homogenized with a 10-sec on and 50-sec off cycle using a Polytron PT-35 tissue homogenizer at a setting of 6. The resultant homogenate was treated as described above to obtain the cytosol. Solid ultrapure ammonium sulfate was slowly added to the cytosol while stirring at 0 to 4°C to give an ammonium sulfate saturation of 26%. After 60 min the precipitate was collected by centrifugation at 30,000 × g for 20 min. The supernatant was decanted and the precipitate was stored in the cold overnight. The pellet was completely dissolved in 0.2 volumes of TED buffer and applied to a Sephadex G-25 column (1.6 × 40 cm) equilibrated with TED buffer. Elution was with the TED buffer at a flow rate of 72 ml/hr and fractions of 1.2 ml were collected. Aliquots from each fraction were assayed for sulfate by precipitation with 1.0 M barium chloride. Protein content was monitored in each fraction by measuring the absorbence at 280 nM. Fractions containing protein were pooled and assayed by DCC for R5020 binding activity.

RESULTS

Protein Concentration and Specific ^{3}H-R5020 Binding

Specific ^{3}H-R5020 binding increases linearly with cytosol protein concentrations up to at least 8 mg/ml (Fig. 1). This ensures that our assay is

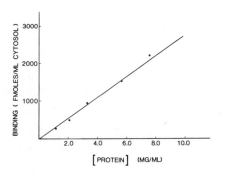

FIG. 1. Protein concentration and specific ^{3}H-R5020 binding. Several different dilutions of cytosol (0.2 ml) were added in duplicate to vials containing 6.0 nM ^{3}H-R5020 or 6.0 nM ^{3}H-R5020 with 100-fold excess radioinert R5020. The reactions were incubated 4 hr at 0 to 4°C, then assayed by DCC to measure the amount of specific binding as described in Materials and Methods. The protein concentration for each dilution was determined by the method of Lowry et al. (15).

reliable regardless of protein concentration since the cytosol is rarely more concentrated than 10 mg protein per milliliter.

Time Dependence of Specific ^{3}H-R5020 Binding

Time studies of ^{3}H-R5020 uptake show a rapid increase in the amount of specific binding up to about 3 to 4 hr (Fig. 2). After saturation, the amount

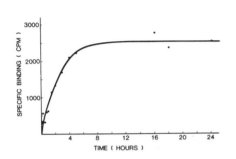

FIG. 2. Time dependence of specific ^{3}H-R5020 binding. Cytosol samples (6.0 ml) were added to incubation tubes containing either 6.0 nM ^{3}H-R5020 or 6.0 nM ^{3}H-R5020 with 100-fold molar excess radioinert R5020 which had just been dried down under nitrogen. The reactions were incubated at 0 to 4°C with periodic agitation. Duplicate 0.2-ml samples were removed from each tube at various times ranging from 3 min to 24 hr and assayed by DCC to determine the amount of specific binding as outlined in Materials and Methods.

of specifically bound ^{3}H-R5020 remains almost constant up to 24 hr at 0 to 4°C, thus enabling us to use either 3-hr or overnight incubation for receptor assay with essentially identical results.

Titration of Specific Binding Sites in Crude Cytosol with ^{3}H-R5020

Increasing concentrations of ^{3}H-R5020 were incubated with cytosol either with or without 1,000-fold molar excess radioinert R5020 for 18 hr at 0 to 4°C and then assayed by the DCC procedure to determine specific binding. Specific binding sites were saturated at 2 nM ^{3}H-R5020 (Fig. 3). Scatchard analysis (28) of these data gave a linear plot (Fig. 4). For this particular preparation, the specific R5020 binding protein displayed

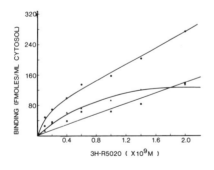

FIG. 3. Titration of specific binding sites in crude cytosol with ^{3}H-R5020. Cytosol (0.2 ml) was incubated in duplicate with 0.1 to −2.0 nM ^{3}H-R5020 (total binding) or ^{3}H-R5020 in the presence of 1,000-fold molar excess radioinert R5020 (nonspecific binding). After 18 hr at 0 to 4°C, the amounts of total binding (●——●), specific binding (▲——▲), and nonspecific binding (■——■) were determined by the DCC assay as described in Materials and Methods for each concentration of ^{3}H-R5020 used.

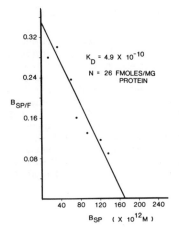

FIG. 4. Scatchard analysis of specific ^{3}H-R5020 binding proteins in crude cytosol. Cytosol was prepared, incubated, and treated as described in the legend for Fig. 3. Specific ^{3}H-R5020 binding for each point (B_{SP}) was calculated from the difference between total binding and nonspecific binding. The amount of free steroid (F) was determined from the difference between total counts and the total binding of ^{3}H-R5020 for each concentration used. The K_d was calculated from the slope of the line. The number of specific binding sites (N) was determined from the intercept on the abscissa.

an apparent dissociation constant (K_d) of 4.9×10^{-10} M and a binding capacity of 26 fmoles/mg cytosol protein as calculated from the intercept on the abscissa. The mean of three separate determinations of the K_d gave a value of $4.4(\pm 0.5) \times 10^{-10}$ M, and the average number of specific binding sites was $22.5(\pm 8)$ fmoles/mg of cytosol protein. The amount of specifically bound ^{3}H-R5020 as determined by SG analysis was in close agreement with the value calculated by DCC assay. Binding capacity determined by SG gave a value of $40(\pm 13)$ fmoles/mg cytosol protein (mean of six determinations).

Titration of Specific Binding Sites in Partially Purified Progestin Receptor with ^{3}H-R5020

Specific binding sites in partially purified (0 to 26% ammonium sulfate precipitate) progestin receptor were saturated at about 1.0 nM ^{3}H-R5020 (Fig. 5). Scatchard analysis of these data revealed a linear plot (Fig. 6). The

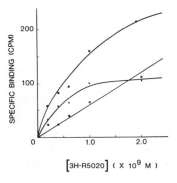

FIG. 5. Titration of specific binding sites in partially purified cytosol with ^{3}H-R5020. Cytosol was partially purified by ammonium sulfate fractionation according to procedures outlined in Materials and Methods. Aliquots (0.2 ml) were incubated in duplicate with 0.2 to 2.0 nM ^{3}H-R5020 (total binding) in the presence or absence of 1,000-fold molar excess radioinert R5020 (nonspecific binding). After 18 hr at 0 to 4°C, the amount of total binding (●——●), specific binding (▲——▲), and nonspecific binding (■——■) was determined by the DCC assay described in Materials and Methods for each concentration of ^{3}H-R5020 used.

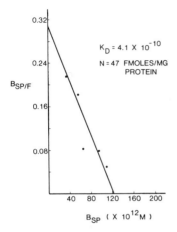

FIG. 6. Scatchard analysis of specific ³H-R5020 binding proteins in partially purified cytosol. Partially purified cytosol was incubated and treated as described in the legend for Fig. 5. Specific ³H-R5020 binding for each point (B_{SP}) was calculated from the difference between total binding and nonspecific binding. The amount of free steroid (F) was determined from the difference between total counts and the total binding of ³H-R5020 for each concentration used. The K_d was calculated from the slope of the line. The number of specific binding sites (N) was determined from the intercept on the abscissa.

apparent K_d for the partially purified ³H-R5020 receptor is 4.1×10^{-10} M. The binding capacity is 46 fmoles/mg cytosol protein as calculated from the intercept on the abscissa.

Temperature Stability of Specifically Bound ³H-R5020

The crude cytosol receptor was saturated with ³H-R5020, treated with DCC to remove free steroid, and incubated at 37°, 25°, and 0°C to follow dissociation from or degradation of ³H-R5020 binding sites with time. DCC analysis reveals that the ³H-R5020-receptor complex is relatively stable at 0°C. There is some initial dissociation, but after 2 to 3 hr the rate is much lower and the steroid-receptor complex is stable for up to 26 hr.

At 25°C the dissociation of specifically bound ³H-R5020 was more rapid with time, being essentially complete at 60 min. At 37°C the dissociation or degradation was even more rapid; only a very low amount of specifically bound ³H-R5020 remained after 5 min and after 10 min there was no specific binding of ³H-R5020 remaining.

Molecular Characteristics of Specific ³H-R5020 Binding Proteins as Determined by Sucrose Density Gradient Centrifugation

In low-salt sucrose gradients, the specific R5020 receptor has a sedimentation coefficient of 7.9S; however, there is also significant specific binding at 4.1S (Figs. 7 and 8). Analysis of many sucrose gradient profiles from different mammary cytosol preparations revealed the distribution of the two forms of specific receptor-steroid complex: the 8S binding component accounted for approximately $40(\pm 18)\%$ of the total specific receptor population whereas the 4S component accounted for $60(\pm 27)\%$. The same amount of specific binding was observed regardless of whether a 100- or

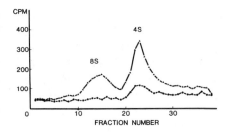

FIG. 7. Sucrose density gradient analysis of specific ³H-R5020 binding protein. ▲——▲, 6.0 nM ³H-R5020; ●——●, 6.0 nM ³H-R5020 plus 1000 × R5020. See Materials and Methods for specific details.

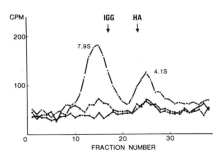

FIG. 8. Partial inhibition of specific ³H-R5020 binding with excess radioinert progesterone as determined by SG analysis. ▲——▲, 6.0 nM ³H-R5020; ●——●, 6.0 nM ³H-R5020 plus 100 × R5020; ■——■, 6.0 nM ³H-R5020 plus 25 × progesterone. See Materials and Methods for complete details.

1,000-fold molar excess unlabeled R5020 was used during incubation (Fig. 7). The specific binding component sediments only in the 4 to 5S region when the cytosol is prepared in TDK buffer and subsequently centrifuged through 10 to 30% sucrose gradients made in TDK buffer (*data not shown*).

Serum Binding of ³H-R5020

In view of the possible occurrence of specific binding components in goat serum that could complicate studies with the specific progestin receptor in mammary tissue, we investigated R5020 binding in goat serum. The serum (1:4 dilution) was incubated with ³H-R5020 either with or without 1,000-fold

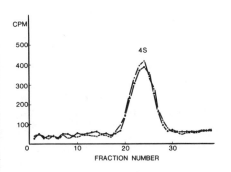

FIG. 9. Serum binding of ³H-R5020. Serum from a lactating goat was diluted with 3 volumes of TED buffer. The diluted serum was added to a pellet from 1.5 ml of DCC, vortexed, and incubated with periodic agitation at 0 to 4°C to remove endogenous steroids. After 10 min, the reaction was centrifuged at 10,000 × *g* twice for 5 min each. Several 0.4-ml portions of the diluted serum were incubated with 6.0 nM ³H-R5020 (▲——▲) either with or without the presence of 1,000-fold molar excess radioinert R5020 (●——●). After a 3-hr incubation at 0 to 4°C, the reactions were analyzed by SG as described in Materials and Methods.

molar excess radioinert R5020. After incubation the reaction was analyzed directly by SG with or without prior DCC treatment. In serum not treated with DCC prior to SG, a large amount of ^{3}H-R5020 was bound by serum proteins in the 4 to 5S region, but no binding was observed in the 8S region. A portion of the reaction mixture was also treated with DCC to remove free steroids and steroids from low-affinity binding sites, then analyzed by SG (Fig. 9). Treatment with DCC removed greater than 96% of ^{3}H-R5020 bound in the 4 to 5S region, and the remaining ^{3}H-R5020 bound was of noninhibitable nature (not inhibitable by excess radioinert R5020).

Ligand Specificity of ^{3}H-R5020 Binding Site

Lactating goat mammary cytosol was incubated with various radioinert steroid hormones and synthetic steroids in the presence or absence of 6 nM ^{3}H-R5020 to determine which were competitors for specific ^{3}H-R5020

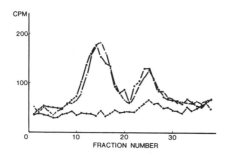

FIG. 10. Lack of inhibition of specific ^{3}H-R5020 binding with excess radioinert 17β-estradiol as described in Materials and Methods. ▲——▲, 6.0 nM ^{3}H-R5020; ●——●, 6.0 nM ^{3}H-R5020 plus 100 × R5020; ■——■, 6.0 nM ^{3}H-R5020 plus 100 × 17β-estradiol. See Materials and Methods for complete details.

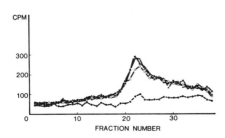

FIG. 11. Lack of inhibition of specific ^{3}H-R5020 binding with excess radioinert dexamethasone and partial inhibition with excess radioinert dihydrotestosterone as determined by SG analysis. ▲——▲, 6.0 nM ^{3}H-R5020; ●——●, 6.0 nM ^{3}H-R5020 plus 1,000 × R5020; ■——■, 6.0 nM ^{3}H-R5020 plus 500 × dexamethasone; ○---○, 6.0 nM ^{3}H-R5020 plus 500 × dihydrotestosterone. See Materials and Methods for details.

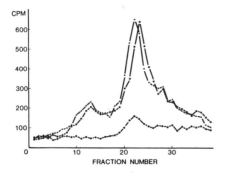

FIG. 12. Lack of inhibition of specific ^{3}H-R5020 binding with excess radioinert cortisol as determined by SG analysis. ▲——▲, 6.0 nM ^{3}H-R5020; ●——●, 6.0 nM ^{3}H-R5020 plus 1,000 × R5020; ■——■, 6.0 nM ^{3}H-R5020 plus 1,000 × cortisol. See Materials and Methods for details.

TABLE 1. *Ligand specificity of ^{3}H-R5020 binding by cytosol from normal mammary tissue*

^{3}H-R5020 + 500- or 1,000-fold molar excess competitor	^{3}H-R5020 specifically bound (%)
None	100
R5020	0
Progesterone	0
Triamcinolone acetonide	0
Dexamethasone	100
17β-Estradiol	100
5α-Dihydrotestosterone	87

binding sites. Sucrose gradient analysis revealed lack of inhibition of specific ^{3}H-R5020 binding in either the 8S or 4S region by 100- to 1,000-fold molar excess 17β-estradiol (Fig. 10), dexamethasone (Fig. 11), or cortisol (Fig. 12). A 500- to 1,000-fold molar excess 5α-dihydrotestosterone resulted in minor inhibition of specific ^{3}H-R5020 binding (Fig. 11). These data are summarized in Table 1. Unlabeled triamcinolone acetonide and progesterone (Pg) compete effectively with specific ^{3}H-R5020 binding to the progestin receptor. Incubation with 40- to 50-fold molar excess TA (Fig. 13) or 15- to 20-fold molar excess Pg (Fig. 8) was required for 50% inhibition of specific ^{3}H-R5020 binding. A 1- to 3-fold molar excess unlabeled R5020 was required for 50% inhibition (Fig. 14). From the concentration of competitor to inhibit 50% of specific ^{3}H-R5020 binding, the

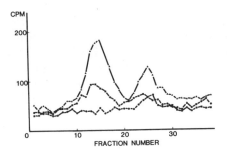

FIG. 13. Partial inhibition of specific ^{3}H-R5020 binding with excess radioinert triamcinolone acetonide as determined by SG analysis. ▲——▲, 6.0 nM ^{3}H-R5020; ●——●, 6.0 nM ^{3}H-R5020 plus 100 × R5020; ■——■, 6.0 nM ^{3}H-R5020 plus 50 × triamcinolone acetonide. See Materials and Methods for details.

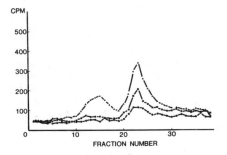

FIG. 14. Partial inhibition of specific ^{3}H-R5020 binding with excess radioinert R5020 as determined by SG analysis. ▲——▲, 6.0 nM ^{3}H-R5020; ●——●, 6.0 nM ^{3}H-R5020 plus 100 × R5020; ■——■, 6.0 nM ^{3}H-R5020 plus 5 × R5020. See Materials and Methods for details.

TABLE 2. *Dissociation constant of competitive inhibitors of specific ^{3}H-R5020 binding*

Competitor	Molar excess for 50% inhibition of specific ^{3}H-R5020 binding	Apparent K_d
R5020	1–3-fold	4.4×10^{-10} M
Progesterone	15–20-fold	3.9×10^{-9} M
Triamcinolone acetonide	40–50-fold	1.0×10^{-8} M

relationship suggested by Rodbard (27) is used to estimate the apparent K_d of the competitors; these values are given in Table 2.

Estrogen and Glucocorticoid Receptors

Sucrose gradient analysis of cytosol prepared from lactating goat mammary tissue and labeled with ^{3}H-17β-estradiol (E_2) demonstrates the presence of substantial amounts of a specific estrogen-binding protein that sediments at 7.9S and 4.1S in low-salt buffer (*data not shown*) (17). Separate competitive inhibition studies indicate that the binding site of the estrogen receptor is highly specific. Excess radioinert estradiol, estriol, and estrone compete effectively with ^{3}H-E_2 binding to the specific estrogen receptor. However, 5α-dihydrotestosterone, cortisol, dexamethasone, TA, and R5020 did not inhibit ^{3}H-E_2 binding at molar concentrations up to 1,000-fold that of ^{3}H-E_2.

Lactating goat tissue also contains substantial amounts of a glucocorticoid receptor. When cytosol is labeled with ^{3}H-TA and analyzed by SG, the specific ^{3}H-TA binding protein sediments primarily in the 8S region but also in the 4S region (Fig. 15). In the presence of 100-fold molar excess radioinert R5020, the specific binding of ^{3}H-TA to the glucocorticoid receptor is inhibited approximately 65% (Fig. 15). Even with 1,000- or 5,000-fold molar excess radioinert R5020, specific ^{3}H-TA binding was still not completely suppressed (*data not shown*). A 100-fold molar excess radioinert dexamethasone completely inhibits ^{3}H-TA binding to the glucocorticoid receptor (Fig. 15).

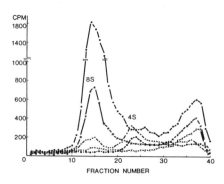

FIG. 15. Sucrose gradient analysis of specific ^{3}H-R5020 and ^{3}H-TA binding. Partial inhibition of ^{3}H-TA binding with excess radioinert R5020. Complete inhibition of ^{3}H-TA binding and ^{3}H-R5020 binding with excess radioinert TA and R5020, respectively. ▲——▲, 6.0 nM ^{3}H-R5020; ●---●, 6.0 nM ^{3}H-R5020 plus 100 × R5020; ■——■, 6.0 nM ^{3}H-TA; X——X, 6.0 nM ^{3}H-TA plus 100 × TA; ★——★, 6.0 nM ^{3}H-TA plus 100 × R5020. See Materials and Methods for details.

Androgen Receptors

Sucrose gradient analysis of cytosol prepared from lactating goat mammary tissue and labeled with ^{3}H-DHT demonstrates the complete absence of any specific androgen receptor.

DISCUSSION

We have demonstrated the presence of a specific progestin binding protein in the lactating goat mammary gland (11). The R5020 binding protein satisfies three major physiochemical and kinetic criteria for classification as a specific progestin receptor. First, ^{3}H-R5020 was bound with high affinity (Figs. 4 and 6). By Scatchard analysis, the specific receptor has a K_d of approximately 4×10^{-10} M. This K_d value is in close agreement with those reported for the progestin receptor in other tissues and cells (9,10,24). Second, ^{3}H-R5020 binding was saturable and of low capacity. Corroboration between SG analysis and our DCC analysis indicates that there are only 20 to 55 fmoles of receptor per milligram of cytosol protein, with saturation occurring at approximately 1 to 2 nM ^{3}H-R5020 (Figs. 3 and 5). Finally, the specific R5020 receptor sediments consistently at 7.9S (and at 4.1S as well), as determined by SG in low salt. These findings distinguish the R5020 receptor from corticosteroid binding globulin and other nonspecific binding entities in plasma which sediment in the 4S region. Analysis by SG has also revealed the complete absence of specific ^{3}H-R5020 binding by lactating goat serum, although there is binding of noninhibitable nature in the 4S region (Fig. 9).

Competition studies indicate that the R5020 receptor behaves as other progestin receptors and is distinct from the estrogen receptor, the glucocorticoid receptor, and CBG. This was established by the lack of inhibition of specific ^{3}H-R5020 binding by excess radioinert 17β-estradiol, dexamethasone, and cortisol, respectively. There was very slight inhibition of specific ^{3}H-R5020 binding by excess radioinert DHT. Horwitz et al. (9) report that excess radioinert DHT depresses specific ^{3}H-R5020 binding sites by more than 50% in the MCF-7 human breast cancer cell line.

The competition for ^{3}H-R5020 binding sites by excess radioinert TA is interesting in that we find a complete lack of inhibition by dexamethasone (Fig. 11). Mehta and Wittliff (19) have also demonstrated competition for specific ^{3}H-R5020 by TA in normal rat mammary tissue. Furthermore, Horwitz et al. reported no competition by dexamethasone for specific ^{3}H-R5020 binding in the MCF-7 human breast cancer cell line (9) and human breast cancer tissue (10). Asselin et al. (1) similarly reported little if any effect of dexamethasone on ^{3}H-R5020 binding in dimethylbenzanthracene-induced mammary tumors in rats.

It has been suggested by Wittliff (8,33,34) that glucocorticoids and progestins operate through a common site in mammary tissue. However, we have several lines of evidence which suggest that these two receptors are separate and distinct in lactating goat mammary tissue. First, SG analysis of specific ^{3}H-TA binding to an 8S receptor indicates that there are significant differences in the relative amounts of progestin and glucocorticoid receptors present (Fig. 15). Second, 100-fold molar excess radioinert dexamethasone (*data not shown*) and TA (Fig. 15) completely inhibit specific ^{3}H-TA binding as shown by SG. However, although 100-fold molar excess radioinert TA completely inhibits specific ^{3}H-R5020 binding, even 1,000-fold molar excess of radioinert dexamethasone does not inhibit ^{3}H-R5020 binding to the specific receptor. Finally, although 100-fold molar excess radioinert R5020 inhibits specific ^{3}H-R5020 binding completely, a similar excess of radioinert R5020 inhibits specific ^{3}H-TA binding by only 65% (Fig. 15). In fact, 1,000- or 5,000-fold molar excess R5020 still does not completely inhibit specific ^{3}H-TA binding to the glucocorticoid receptor (*data not shown*).

In preliminary studies, we have developed procedures for extensive purification of intact nuclei from frozen lactating mammary glands. Initial experiments designed to study transformation of the R5020-receptor complex and subsequent nuclear interaction have been hampered by the relative instability of the R5020-receptor complex at elevated temperatures because of rapid dissociation and/or proteolytic degradation as indicated earlier. We have failed thus far to observe transformation of the receptor to a 5 to 6S form after treatment with various combinations of high salt and/or elevated temperature. However, purified nuclei which were incubated with ^{3}H-R5020-labeled cytosol at 0° for 2 hr, then washed thoroughly and extracted with high salt, did produce a small but significant 4S peak upon SG analysis in high-salt gradients. Other investigators studying activation and nuclear incorporation of progesterone receptor (4,30) and glucocorticoid receptor (3) have similarly failed to detect specific sedimentation coefficient differences between the activated and nonactivated forms of the receptor. However, Schrader et al. (29), DeSombre et al. (5), and others (20) have demonstrated the requirement for transformation of the 4S form to either the 5S or 6S upon activation before nuclear uptake is observed. Our studies will be continued using fresh tissue and purified receptors to attempt to resolve the problems previously encountered.

SUMMARY

Goat lactating mammary glands contain a specific progestin receptor that has been identified by sucrose density gradient centrifugation (SG) of a cytosol preparation. The specific R5020 receptor has a sedimentation coefficient of 7.9S; however, there is also significant specific binding in the 4.1S region. By Scatchard analysis, the receptor has a dissociation constant of 4×10^{-10} M, and the number of specific binding sites per milligram cytosol protein is approximately 20 to 55 fmoles. Analysis of goat serum binding of ^{3}H-R5020 by SG revealed the complete absence of specific R5020 binding.

Competitive binding studies indicate lack of inhibition of specific R5020 binding by 500- or 1,000-fold molar excess radioinert dexamethasone or cortisol and 100-, 500- or 1,000-fold molar excess 17β-estradiol (E$_2$). With 500- or 1,000-fold molar excess radioinert 5α-dihydrotestosterone, there was minor inhibition of R5020 binding. Radioinert triamcinolone acetonide (TA) and progesterone (Pg) compete effectively with ^{3}H-R5020 binding: incubation with 40- to 50-fold molar excess TA or 15- to 20-fold molar excess Pg is required for 50% inhibition of specific R5020 binding, whereas 1- to 3-fold molar excess radioinert R5020 is required for 50% inhibition.

Separate studies using SG show that lactating goat mammary tissue also contains a specific estrogen receptor and a glucocorticoid receptor; no androgen receptor was detectable by SG. R5020 does not compete with ^{3}H-E$_2$ binding to the estrogen receptor; however, R5020 does compete with TA for binding to the glucocorticoid receptor.

ACKNOWLEDGMENTS

The authors are gratefully indebted to Mrs. Laura Lee and Miss Y. Shih for their expert technical assistance and to Mrs. Karen Chalmers and Miss Mary Jo McNally for their help in preparation of the manuscript and figures. This work was supported by a grant from the USPHS, National Cancer Institute, Grant CA 14089, to the LAC/USC Cancer Center.

REFERENCES

1. Asselin, J., Labrie, F., Kelly, P. A., Philibert, D., and Raynaud, J. P. (1976): Specific progesterone receptors in dimethylbenzanthracene (DMBA)-induced mammary tumors. *Steroids*, 27:395–404.
2. Atger, M., Baulieu, E. E., and Milgrom, E. (1974): An investigation of progesterone receptors in guinea pig vagina, uterine cervix, mammary glands, pituitary and hypothalamus. *Endocrinology*, 94:161–167.
3. Atger, M., and Milgrom, E. (1976): Mechanism and kinetics of the thermal activation of glucocorticoid hormone-receptor complex. *J. Biol. Chem.*, 251:4758–4762.
4. Buller, R. E., Toft, D. O., Schrader, W. T., and O'Malley, B. W. (1975): Progesterone-binding components of chick oviduct. VIII. Receptor activation and hormone-dependent binding to purified nuclei. *J. Biol. Chem.*, 250:801–808.
5. DeSombre, E. R., Mohla, S., and Jensen, E. V. (1975): Receptor transformation, the key to estrogen action. *J. Steroid Biochem.*, 6:469–473.
6. Feil, P. D., and Bardin, C. W. (1975): Cytoplasmic and nuclear progesterone receptors in the guinea pig uterus. *Endocrinology*, 97:1398–1407.
7. Goral, J. E., Turnell, R. W., and Wittliff, J. L. (1975): Properties of progesterone-binding proteins in mammary tissues. *Proc. Am. Assoc. Cancer Res.*, 16:154.
8. Goral, J. E., and Wittliff, J. L. (1976): Characteristics of progesterone-binding components in neoplastic mammary tissues of the rat. *Cancer Res.*, 36:1886–1893.
9. Horwitz, K. B., Costlow, M. E., and McGuire, W. L. (1975): MCF-7; a human breast cancer cell line with estrogen, androgen, progesterone and glucocorticoid receptors. *Steroids*, 26:785–795.

10. Horwitz, K. B., and McGuire, W. L. (1975): Specific receptors in human breast cancer. *Steroids,* 25:497–505.
11. Hutchens, T. W., and Markland, F. S. (1976): Progestin receptor from lactating mammary glands of the caprine species: identification and characterization. *Fed. Proc. (Abst.),* 35:1559.
12. Jensen, E. V., and DeSombre, E. R. (1972): Mechanism of action of the female sex hormones. *Annu. Rev. Biochem.,* 41:203–230.
13. Kontula, K., Jänne, O., Vihko, R., Jager, E., Visser, J., and Zeelen, F. (1975): Progesterone binding proteins: in vitro binding and biological activity of different steroidal ligands. *Acta Endocrinol.* (Kbh.), 78:574–592.
14. Kuhn, R. W., Schrader, W. T., Smith, R. G., and O'Malley, B. W. (1975): Progesterone binding components of chick oviduct. X. Purification by affinity chromatography. *J. Biol. Chem.,* 250:4220–4228.
15. Lowry, O. H., Rosebrough, N. J., Farr, A. L., and Randall, R. J. (1951): Protein measurement with the Folin phenol reagent. *J. Biol. Chem.,* 193:265–275.
16. MacLaughlin, D. T., and Richardson, G. S. (1976): Progesterone binding by normal and abnormal human endometrium. *J. Clin. Endocrinol. Metab.,* 42:667–678.
17. Markland, F. S., Shih, Y., and Lee, L. (1976): Properties of the estrogen receptor from lactating mammary glands. *Am. Chem. Soc. 31st Meeting Northwest Region (Abst.),* G-38.
18. Martin, R. G., and Ames, B. N. (1961): A method for determining the sedimentation behavior of enzymes: application to protein mixtures. *J. Biol. Chem.,* 236:1372–1379.
19. Mehta, R. G., and Wittliff, J. L. (1976): Specific association of a synthetic progestogen with binding sites in lactating mammary glands. *Fed. Proc. (Abst.),* 35:1559.
20. Notides, A., and Nielsen, S. (1975): A molecular and kinetic analysis of estrogen receptor transformation. *J. Steroid Biochem.,* 6:483–486.
21. O'Malley, B. W., and Schrader, W. T. (1976): The receptors of steroid hormones. *Sci. Am.,* 234:32–43.
22. Philibert, D., and Raynaud, J. P. (1973): Progesterone binding in the immature mouse and rat uterus. *Steroids,* 22:89–98.
23. Philibert, D., and Raynaud, J. P. (1974a): Binding of progesterone and R5020, a highly potent progestin, to human endometrium and myometrium. *Contraception,* 10:457–466.
24. Philibert, D., and Raynaud, J. P. (1974b): Progesterone binding in the immature rabbit and guinea pig uterus. *Endocrinology,* 94:627–632.
25. Rao, B. R., Wiest, W. G., and Allen, W. M. (1973): Progesterone "Receptor" in rabbit uterus. I. Characterization and estradiol-17β augmentation. *Endocrinology,* 92:1229–1240.
26. Rao, B. R., Wiest, W. G., and Allen, W. M. (1974): Progesterone "Receptor" in human endometrium. *Endocrinology,* 95:1275–1281.
27. Rodbard, D. (1973): Mathematics of hormone-receptor interaction. In: *Receptors for Reproductive Hormones,* edited by B. W. O'Malley and A. R. Means, pp. 289–326. Plenum Press, New York.
28. Scatchard, G. (1949): The attraction of proteins for small molecules and ions. *Ann. N.Y. Acad. Sci.,* 51:660–672.
29. Schrader, W. T., Heuer, S. S., and O'Malley, B. W. (1975): Progesterone receptors of chick oviduct: identification of 6S dimers. *Biol. Reprod.,* 12:134–142.
30. Smith, R. G., Iramain, C. A., Buttram, V. C., and O'Malley, B. W. (1975): Purification of human uterine progesterone receptor. *Nature,* 253:271–272.
31. Terenius, L. (1972): Specific progestogen binder in uterus of normal rats. *Steroids,* 19:787–794.
32. Terenius, L. (1973): Estrogen and progestogen binders in human and rat mammary carcinoma. *Eur. J. Cancer,* 9:291–294.
33. Wittliff, J. L. (1974): Specific receptors of the steroid hormones in breast cancer. *Semin. Oncol.,* 1:109–118.
34. Wittliff, J. L. (1975): Steroid binding proteins in normal and neoplastic mammary cells. In: *Methods in Cancer Research, Vol. 11,* edited by H. Busch, pp. 293–354. Academic Press, New York.

Progesterone Receptors in Normal and Neoplastic Tissues, edited by W. L. McGuire et al. Raven Press, New York © 1977.

Interaction of R5020 with Binding Sites in Normal and Neoplastic Mammary Tissues

*James L. Wittliff, *Rajendra G. Mehta, and *Timothy E. Kute

University of Rochester Cancer Center, and Department of Biochemistry, University of Rochester School of Medicine and Dentistry, Rochester, New York 14642

Development and differentiation of mammary gland are influenced by the coordinate action of both steroid and protein hormones. Association of steroid hormones with their specific binding components in the cytoplasm appears to be a prerequisite for initiation of biologic response. Binding sites for each of the steroid hormones (estrogens, progesterone, and glucocorticoids) involved in the development of mammary gland structure and function have been characterized extensively from both normal and neoplastic breast tissues by our laboratory (2,3,24–26) as well as by others (4–6,19).

A major problem in the determination of specific progesterone binding components in mammary tissues has been that [^{3}H]progesterone may associate with high affinity with certain serum proteins such as corticosteroid binding globulin (23), which may contaminate certain mammary tissues. Additionally, progesterone apparently dissociates rapidly from its specific binding sites in mammary tumors (3). Therefore the progestomimetic steroid R5020 (17,21-dimethyl-19-nor-pregna-4,9-diene-3,20-dione), reported to complex with progesterone receptors in uterus with high affinity (13), was used to characterize the progesterone binding components in the normal and neoplastic mammary gland. With the synthesis of [^{3}H]R5020, specific binding proteins have been described in uterus (13,14), mammary gland (7,10,26), and certain mammary tumors from rodents (1) and from humans (4–6,16). In this investigation we determined several of the kinetic and molecular characteristics of [^{3}H]R5020 association with specific binding proteins in cytosols from lactating mammary gland. These were compared to those of steroid binding components in the R3230AC and dimethylbenz-(α)anthracene (DMBA)-induced mammary tumors of the rat and in human breast carcinomas.

---

* Present address: Department of Biochemistry, University of Louisville School of Medicine, Health Sciences Center, Louisville, Kentucky 40201.

MATERIALS AND METHODS

Chemicals

All chemicals were reagent grade unless otherwise specified. 6,7-[^{3}H]-triamcinolone acetonide (9α-fluoro-11β,16α,17α,21-tetrahydroxypregna-1,4-diene-3,20-dione-16,17-acetonide, 33 Ci/mmole) and 1,2,6,7-[^{3}H]progesterone (96 to 105 Ci/mmole) were obtained from New England Nuclear Corp., Boston, Mass. 6,7-[^{3}H]R5020 (51 Ci/mmole) and unlabeled R5020 were supplied through the courtesy of Dr. J. P. Raynaud and Roussel-Uclaf, Paris. α_1-Acid glycoprotein was a gift from Dr. M. Wickerhauser of the American National Red Cross, Blood Research Laboratory, Washington D.C. DEAE-cellulose (type DE-52) was obtained from Whatman, Inc., Clifton, N.J., and unlabeled steroids were purchased from Steraloids, Inc., Wilton, N.H.

Tissues

Lactating Sprague-Dawley rats were supplied by Holtzman and Co., Madison, Wis., and housed in the vivarium of the University of Rochester Medical Center. Mammary glands from animals 17 to 21 days post-partum were used for receptor preparations. Female Fischer 344 rats were used as hosts for the transplantable R3230AC mammary carcinoma, whereas the other tumors studied were induced by the carcinogen DMBA in female Sprague-Dawley rats 49 to 50 days old. Collection and handling of human breast tumor specimens were undertaken as described earlier (25).

Preparation of Cytosols

Tissues were minced and homogenized in 10 mM Tris-HCl buffer, pH 7.4, containing 1.5 mM EDTA and 10 mM monothioglycerol (TET) or in the buffers noted in figure or in table legends. A Polytron homogenizer (Brinkman Instruments, Westbury, N.Y.) was used at setting 5 with several 5-sec bursts; lower settings were used also with no significant difference in results observed. Cytosols were prepared by centrifugation of the homogenates at 105,000 g for 30 min.

Quantitation of Steroid Binding Components

Constant volumes of cytosol were incubated with radioactive steroid either alone or in the presence of unlabeled substance at 0 to 3°C in quantities and for times given in the Results. After the incubation period, 0.6 ml of 1% dextran-coated charcoal suspension was added and incubated for 10 min at 0 to 3°C. Vials were centrifuged at 3,000 rpm for 10 min in a

Sorvall RC3B centrifuge, and aliquots of supernatant were removed and counted for radioactivity. The dextran-coated charcoal procedure used has been described in detail earlier (24). Protein was determined in cytosol by UV spectrophotometry (22).

Analysis of Binding of R5020 to Serum Proteins

Human serum was diluted with TET buffer to give seven concentrations of protein ranging from 1.3 to 16.6 mg/ml as determined by the Waddell method (22). Aliquots of 0.2 ml were incubated 5 hr at 0°C with either 80 nM [^{3}H]R5020 alone or in the presence of 20 μM unlabeled R5020. Reactions were terminated by adding 1 ml of 1% dextran-coated charcoal suspension as described above. Supernatants were analyzed for [^{3}H]R5020 binding.

The fluorescence quenching titration of α_1-acid glycoprotein–R5020 interaction was determined by the method of Stroupe et al. (20). Protein and R5020 concentrations were 1.4×10^{-5} M and 1.9×10^{-3} M, respectively. The binding capacity and affinity constants were determined as described previously (8,20).

Separation of R5020 Binding Components

The procedure for separation of R5020-binding proteins using sucrose density gradient centrifugation was essentially that described earlier (24). Portions of cytosol were incubated with [^{3}H]R5020 (40 nM) for 3 hr and subsequently treated for 10 min with a pellet of dextran-coated charcoal obtained from 0.4 ml of suspension prior to centrifugation. Supernatant (0.2 ml) was layered onto linear gradients of 5 to 40% or 5 to 20% sucrose in either 10 mM Tris-HCl buffer, pH 7.4, containing 1.5 mM EDTA or in the buffer under examination as described in the text. Parallel incubations were carried out in the presence of excess unlabeled R5020 as a measure of low-affinity binding. Gradients were centrifuged for 16 hr at 0 to 3°C using an SW60Ti rotor at 308,000 g.

DEAE-Cellulose Ion-Exchange Chromatography

Two DEAE-cellulose columns (0.9 × 30 cm) were prepared from microgranular Whatman DE-52 with a bed height of ~15 cm. Each was washed extensively with TET buffer until the conductivity and pH of the eluate were the same as those of the buffer. Both columns were placed in the cold room on identical fraction collectors.

Tissue from lactating mammary glands was homogenized in TET buffer (1 to 3 wt/vol) and the cytosol prepared as discussed above. Aliquots of 1 ml were incubated at 0 to 3°C with either 50 nM [^{3}H]progesterone alone (total binding) or with 40 μM unlabeled R5020 (low-affinity binding). In separate experiments 25 nM [³H]triamcinolone acetonide either with or without 6.3 μM unlabeled triamcinolone acetonide was incubated with cytosol under identical conditions. Incubations were terminated by adding 2 ml of 0.5% dextran-coated charcoal suspension and incubating at 0 to 3°C for 10 min prior to centrifugation. Supernatant (2.5 ml) from each of these reactions was layered simultaneously onto separate DEAE-cellulose columns and each washed with 36 ml of TET buffer. Elution of bound proteins was accomplished with a 200-ml gradient of 0 to 0.4 M KCl in TET buffer. Two-milliliter fractions from each column were collected and assayed for radioactivity, relative protein concentrations, and ionic strength.

RESULTS AND DISCUSSION

Reports from several laboratories including our own have demonstrated that progesterone significantly inhibited the binding of labeled glucocorticoids to receptors in mammary tissues (2,4,19,26). Similarly, glucocorticoids have been noted to be competitive inhibitors of progesterone binding components in cytosol from certain mammary tumors (3). Since these data suggested the possibility that both classes of steroids may interact with the same site (3,26), we undertook a study of the properties of [³H]R5020 binding as this ligand appears to be specific only for intracellular progesterone binding proteins (13-16). Characteristics of R5020 association with binding proteins in mammary tissues were compared with those of progesterone and triamcinolone acetonide binding components.

Time Course of Ligand Association

Association of [³H]progesterone, [³H]R5020, and [³H]triamcinolone acetonide with their respective binding sites was studied as a function of time after incubation at 0 to 3°C (Fig. 1). [³H]R5020 exhibited a rate constant of association (K_a) of 1 to 2×10^5 M^{-1} min^{-1}, which was comparable with the K_a of progesterone (1 to 3×10^5 M^{-1} min^{-1}) and that of the glucocorticoid binding site (2 to 4×10^5 M^{-1} min^{-1}). However, unlike progesterone, R5020 did not dissociate readily from its specific binding sites (Fig. 1C). Maximum binding occurred within 2 hr and remained virtually unchanged over a 24-hr period. A similar type of stability was seen for triamcinolone acetonide binding to cytosol from lactating mammary gland (Fig. 1B). Rapid association and maintenance of R5020 binding was considered advantageous for studies of progestin receptors in mammary tissues. Although most previous investigations have used 2 to 4 hr for incubations with [³H]R5020, time courses have not been reported. In our experiments 3- to 4-hr incubations were used routinely.

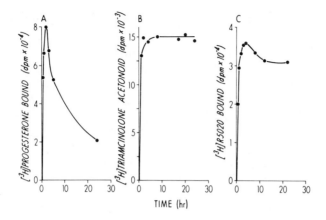

FIG. 1. Time course of specific progestin and glucocorticoid binding. Cytosols prepared from mammary glands of rats lactating 19 to 21 days were incubated in duplicate with either 50 nM [^{3}H]progesterone **(A)**, 25 nM [^{3}H]triamcinolone acetonide **(B)**, or 40 nM [^{3}H]R5020 **(C)** for various times at 0 to 3°C in the presence of appropriate inhibitors to correct for low-affinity binding. Unbound steroids were removed by the dextran-coated charcoal procedure. Only specific binding is presented.

Affinity of R5020 for Specific Binding Sites

Constant volumes of cytosol were incubated with increasing concentrations of [^{3}H]R5020 in either the presence or absence of 250-fold excess of unlabeled R5020 for 3 hr at 0 to 3°C. Following dextran-coated charcoal adsorption of unbound steroid, protein-bound radioactivity was measured. A representative titration curve is presented in Fig. 2. Saturation of specific binding sites was attained by approximately 60 nM [^{3}H]R5020 (Fig. 2A). Dissociation constants calculated from Scatchard plots ranged from 2 to 6×10^{-8} M (Fig. 2B). The number of binding sites in lactating mammary gland from rats 17 to 21 days post-partum ranged from 70 to 267 fmoles/mg cytosol protein. Binding affinities of [^{3}H]R5020 agreed well with the reported dissociation constants for progesterone binding components in mammary tissues (see ref. 24).

Titration of cytosol from three human breast tumors with [^{3}H]R5020 revealed K_d values of 1 to 4×10^{-8} M with specific binding capacities of 206 to 313 fmoles/mg protein. Earlier reports using preparations of R5020 binding proteins from either solid human breast carcinomas (5) or the MCF-7 cell line from a metastatic breast cancer (4) indicated dissociation constants of 1 to 2×10^{-9} M. The variation in these results may be due to differences in the incubation time (16 hr) used by these investigators compared to that (3 or 4 hr) used in our laboratory.

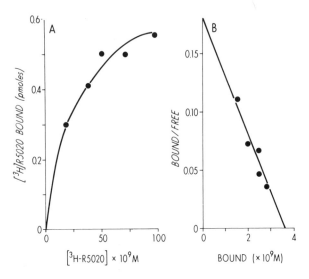

FIG. 2. Titration of R5020 binding sites in cytosol from lactating mammary gland. Portions of cytosols (0.2 ml) were incubated with increasing concentrations of [^{3}H]R5020 either alone or in the presence of 250-fold excess unlabeled R5020 for 3 hr at 0 to 3°C. Specific binding **(A)** was determined as the difference between total binding and binding in the presence of unlabeled steroid using the dextran-coated charcoal method. The plot given in **B** represents Scatchard analyses of the titration data shown in **A**. The dissociation constant estimated was 2×10^{-8} M, and the number of binding sites calculated was 221 fmoles/mg of cytosol protein for this preparation.

Specificity of R5020 Binding Sites

Various unlabeled steroids were incubated at a concentration of 1×10^{-5} M in the presence of 4×10^{-8} M [^{3}H]R5020 with cytosol from the mammary gland of lactating rats to determine ligand specificity. Inhibition was studied by competition analyses using the dextran-coated charcoal procedure (Table 1). [^{3}H]R5020 binding was inhibited significantly by progesterone as well as by the synthetic steroids, triamcinolone acetonide and dexamethasone. The latter two compounds contain a 9α-fluoro group thought to be important in the association of glucocorticoids with binding sites in mammary tissues (2,26). However, hydrocortisone was a weaker competitor, probably resulting from its rapid dissociation from the binding sites (see ref. 24). Similar specificity results for progesterone binding in cytosols of the R3230AC- and DMBA-induced mammary tumors have been reported from our laboratory (3,26). Likewise, in normal Caprine mammary gland R5020 binding sites were inhibited by unlabeled triamcinolone acetonide when present at 100-fold excess (7). However, dexamethasone at 100-fold excess did not compete for R5020 binding sites in cytosol from the MCF-7 tumor cell line, although unlabeled R5020 at this concentration inhibited over 80% of [^{3}H]dexamethasone binding (4).

TABLE 1. *Ligand specificity of [³H]R5020 and [³H]triamcinolone acetonide binding by cytosol from lactating rat mammary gland*

Competitive substance[a]	Inhibition of specific binding (%)	
	[³H]R5020 (40 nM)	[³H]Triamcinolone acetonide (25 nM)
R5020	100[b]	99[b]
Progesterone	90	91
Triamcinolone acetonide	85	100
Dexamethasone	77	92
Hydrocortisone	42	48
Dihydrotestosterone	36	2
Diethylstilbestrol	30	7
Estradiol-17β	18	0
Estriol	0	0

[a] Concentration was 250-fold that of the [³H]ligand.
[b] Mean ± SEM of 3 determinations.

In our study estrogens did not compete significantly for R5020 binding sites, which is consistent with findings using human breast cancers (16). However, in DMBA-induced mammary tumors, estradiol-17β at 100-fold excess inhibited approximately one-half of specific R5020 binding (1). Dihydrotestosterone was a weak competitor of R5020 binding sites in mammary gland (Table 1). Similar results have been observed using receptors from either DMBA-induced mammary tumors from rodents (1) or the MCF-7 cell line of human tumors (4).

In comparison, both unlabeled R5020 and progesterone were excellent competitors of [³H]triamcinolone acetonide binding (Table 1). As expected, hydrocortisone was also a good competitor although examples of androgens and estrogens were not (Table 1). Progesterone inhibition of glucocorticoid binding has been observed previously in mammary tissues (e.g., 2,19,26). However, since R5020 is purported to interact specifically with progesterone binding sites in other tissues (13–16), our finding that it inhibited significantly [³H]triamcinolone binding was unexpected.

To investigate further the nature of R5020 competition by glucocorticoids, we titrated cytosol preparations from lactating mammary glands of rats with increasing concentrations of [³H]R5020 (10 to 60 nM) either alone or in the presence of limiting concentrations of unlabeled R5020, progesterone, or triamcinolone acetonide for 3 hr at 0 to 3°C (Fig. 3A). From the titration data we constructed double reciprocal plots to determine the nature of inhibition by these steroids (Fig. 3B). It was observed that both progesterone and triamcinolone acetonide inhibited R5020 binding in a competitive fashion. From this analysis a K_m of 4×10^{-8} M was obtained for [³H]R5020 which agreed well with the dissociation constant calculated from Scatchard analysis of the same data. These data suggest that both

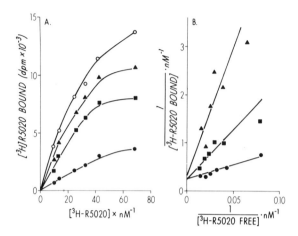

FIG. 3. Nature of inhibition by progesterone and triamcinolone acetonide of [^{3}H]R5020 binding sites in cytosol from lactating mammary gland. Constant volumes of cytosol were incubated with increasing concentrations of [^{3}H]R5020 (10 to 70 nM) either alone (○) or in the presence of 1.2×10^{-5} M unlabeled R5020 (●), 3×10^{-6} M unlabeled progesterone (■), or 3×10^{-6} M unlabeled triamcinolone acetonide (▲). **A:** Binding capacities were determined by the dextran-coated charcoal procedure. **B:** Double reciprocal plots were constructed from these results.

R5020 and glucocorticoid may interact with the same binding site as proposed earlier (3,17,26). However, recently we have shown that the triamcinolone binding site is exceedingly sensitive to the ionic strength of the incubation environment whereas R5020 binding sites are not (R. G. Mehta and J. L. Wittliff, *unpublished observations*).

Association of R5020 with Serum Proteins

The necessity of eliminating serum contamination of preparations of intracellular steroid binding proteins in studies of progesterone receptors in

TABLE 2. *Specific binding of [^{3}H]R5020 to serum proteins*

Serum protein (mg/ml)	Total [^{3}H]R5020 bound (dpm/ml)	[^{3}H]R5020 bound specifically (fmoles/mg protein)[a]
16.4	900	<1
9.0	862	<1
4.8	937	<1
2.6	900	<1
1.7	900	<1
1.3	825	<1

[a] Incubations were performed with 8×10^{-8} M [^{3}H]R5020 for 5 hr at 3°C in the presence and absence of 2×10^{-5} M unlabeled R5020.

target tissues is of obvious importance. The synthetic steroid R5020 has the apparent advantage of binding specifically only to intracellular components in uteri of a number of animal species and of humans (13–15). However, little is known of its association with serum proteins known to bind progestins (23).

In Table 2, the results of an experiment concerning [^{3}H]R5020 association with components in whole serum are given. At concentrations of serum protein ranging from 1.3 to 16.4 mg/ml, insignificant specific binding was observed. In spite of the low capacity, Stroupe and Westphal (*personal communication*) have determined an affinity constant of 8×10^5 M^{-1} for R5020 association with human serum using equilibrium dialysis. These same workers also have measured the affinity constants of R5020 binding to a number of purified plasma proteins (R5020, a Tag for the Progestin Receptor, *this volume*).

Finally, using the method of fluorescence quenching analysis (8,20), we analyzed a purified progesterone binding protein from serum, α_1-acid glycoprotein, for R5020 binding (Fig. 4). The results show that there were 1.2 binding sites per molecule and that the K_d was 2×10^{-6} M. These results are similar to those obtained for progesterone binding to α_1-acid glycoprotein (8). From the control in Fig. 4 it should be noted that the steroid and not the ethanol solvent gave the quenching effect. A circular dichroism profile of α_1-acid glycoprotein either alone or with saturating amounts of unlabeled R5020 showed no observable perturbation between 205 and 250 nm. These data indicate that the association of R5020 with this serum

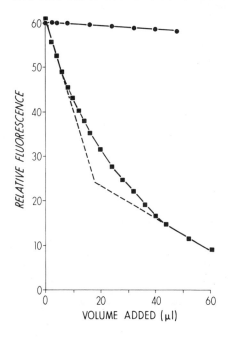

FIG. 4. Fluorescence quenching titration of α_1-acid glycoprotein by R5020. Small aliquots of 1.9×10^{-3} M unlabeled R5020 in 99% ethanol (■) or of 99% ethanol alone (●) were added to 2 ml of solution containing 1.4×10^{-5} M α_1-acid glycoprotein in 10 mM Tris-HCl buffer, pH 7.4, and the relative fluorescence of tryptophan residues measured. The equivalence point was determined by extrapolation of the two linear regions of the curve to determine the volume of R5020 needed to saturate the binding sites on the protein. The binding capacity and affinity constant were 1.2 sites per molecule and 2×10^{-6} M, respectively, as determined by method described earlier (8).

component does not result in a conformation change in the protein (*data not shown*).

Sedimentation Properties

Previously we demonstrated that cytosol prepared from normal mammary gland (26) as well as from two hormonally responsive mammary carcinomas of the rat (3) contains specific [^{3}H]progesterone binding components (Fig. 5B). To determine if [^{3}H]R5020 and [^{3}H]progesterone associate with specific binding components with similar sedimentation behavior, we reacted cytosol separately with each of the [^{3}H]ligands either alone or in the presence of unlabeled progesterone or R5020. After unbound steroid was removed with dextran-coated charcoal, binding components were separated by sucrose gradient centrifugation as described earlier (Fig. 5).

Regardless of the [^{3}H]progestin used, only a single component sedimenting at ~4S was observed in cytosol of lactating mammary gland. When [^{3}H]R5020 was employed as ligand, both unlabeled R5020 and proges-

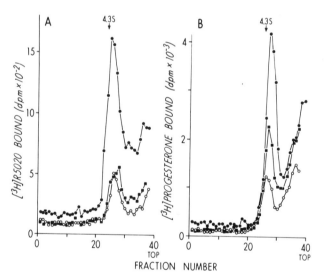

FIG. 5. Sedimentation properties of progestin binding components in cytosol. Tissues were homogenized in 10 mM Tris-HCl buffer, pH 7.4, containing 1.5 mM EDTA and 10 mM monothioglycerol. Cytosols were prepared as described in Materials and Methods. Aliquots (0.4 ml) of cytosol were incubated with 4×10^{-8} M [^{3}H]R5020 **(A)** either alone (●) or in the presence of 1×10^{-5} M unlabeled R5020 (○) or 1×10^{-5} M unlabeled progesterone (■) for 3 hr at 0 to 3°C. Parallel sets of incubations were performed with 5×10^{-8} M [^{3}H]progesterone **(B)** either alone (●) or in the presence of 1.25×10^{-5} M unlabeled progesterone (○) or 1.25×10^{-5} M unlabeled R5020 (■). After incubation, cytosols were treated with dextran-coated charcoal pellets for 10 min at 0°C as described earlier. Supernatants of each reaction were layered onto linear gradients of 5 to 40% sucrose and centrifuged for 16 hr at 3°C.

terone competed equally well for specific binding (Fig. 5A). However, using [³H]progesterone, we found that unlabeled progesterone in 250-fold excess inhibited binding to a greater extent than that observed with unlabeled R5020 at the same concentration. We suggest that this difference in competition is due principally to the fact that R5020 does not combine specifically with serum components likely to be present in preparations of mammary tissues. From more than 20 different analyses, we have not observed significant amounts of specific binding in the 8S region of sucrose gradients when cytosols of lactating mammary gland were examined using either [³H]progestin. In contrast, both 8S and 4S species as well as only the 4S form of R5020 binding components have been reported in goat mammary gland (7).

When cytosols prepared from DMBA-induced mammary tumors of rats were examined for [³H]R5020 binding components using sucrose gradient centrifugation, the majority exhibited only the 4S species (Fig. 6A). This result is similar to that we reported earlier using [³H]progesterone as ligand (3). Asselin et al. (1) also observed specific [³H]R5020 binding components sedimenting at ~8S in DMBA-induced tumors, although distribution of the various sedimentation profiles was not given.

Previously we reported that progesterone binding components in cytosol of the transplantable mammary adenocarcinoma of the rat termed the R3230AC tumor contained only 4S forms when sedimented on sucrose gradients of low ionic strength (3). Substituting [³H]R5020 under similar conditions, we again found the ~4S species as well as a lower molecular weight component sedimenting at ~3S (Fig. 6B). Both of these components associated with R5020 in a specific fashion.

In addition to measurements of estrogen receptors in human breast tumors (25), we are examining a number of these tissues for R5020 binding components as it has been suggested that the latter entities also may be predictive indices of therapeutic response (6). The majority (83%) of [³H]R5020 binding components in cytosols of human breast tumors sedimented at ~4S only (Fig. 7A) in sucrose gradients prepared with 10 mM Tris-HCl buffers containing 1.5 mM EDTA and 10 mM monothioglycerol. However, a few cytosols exhibited both the 8S and 4S forms of specific R5020 binding components (Fig. 6C) using the same conditions. To determine if our finding that most of the R5020 binding components sedimented at ~4S was due to different homogenization and gradient conditions, we compared our method (Tris-HCl buffer) with that previously reported with phosphate buffer containing glycerol (5). Regardless of the buffer used, the sedimentation behavior and specific binding capacity were unchanged (Fig. 6 C and D). The results shown are from a tumor containing both the 8S and 4S species, although similar data were obtained for tumors exhibiting only the ~4S species such as that shown in Fig. 7A. Thus we must conclude that either buffer is satisfactory for preparation of these binding

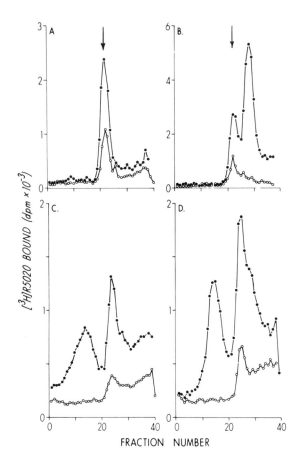

FIG. 6. Sedimentation of R5020 binding proteins in cytosol of certain mammary tumors. Cytosols were prepared by homogenizing specimens of DMBA-induced mammary tumors **(A)**, R3230AC transplantable mammary carcinomas **(B)**, and human breast tumors of the infiltrating ductal carcinoma type **(C)** in 10 mM Tris-HCl buffer, pH 7.4, containing 1.5 mM EDTA and 10 mM monothioglycerol. An additional cytosol was prepared from the same human mammary tumor using 5 mM sodium phosphate buffer, pH 7.4, containing 1 mM monothioglycerol and 10% glycerol **(D)**. Each of the cytosols was incubated with 4×10^{-8} M [^{3}H]R5020 either alone (●) or with 1×10^{-5} M unlabeled R5020 (○) for 3 hr at 0 to 3°C. After cytosol was treated with a dextran-coated charcoal pellet, 0.2 ml supernatant was layered onto either 5 to 40% **(A** and **B)** or 10 to 40% **(C** and **D)** linear gradients of sucrose and centrifuged overnight at 3°C in an SW60Ti swinging bucket rotor. The sedimentation of a 4.3S marker protein is indicated in **(A)** and **(B)** by the arrows.

components. Using the Tris-HCl buffer, we (26) have observed the 7 to 8S components in cytosol of rat uterus described earlier by Philibert and Raynaud (13).

Thus far we have examined 30 human tumors for both [^{3}H]R5020 and [^{3}H]estradiol-17β binding components in cytosol (Table 3). The hypothesis proposed earlier (6) suggests that one should *not* find progesterone receptors in a tumor lacking estrogen receptors. Contrary to this, we have observed

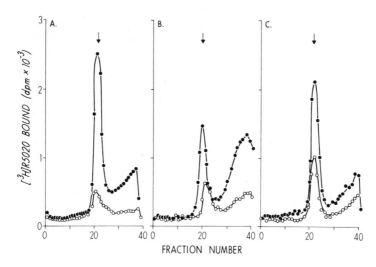

FIG. 7. Influence of proteolytic inhibitors on the sedimentation of R5020 binding components in cytosol. Human mammary tumors were homogenized either in 10 mM Tris·HCl buffer containing 1.5 mM EDTA and 10 mM monothioglycerol alone **(A)** or in the presence of 20 mM diisopropylfluorophosphate **(B)** or 20 mM phenylmethylsulfonylfluoride **(C)**. Cytosols were prepared as described in the text and incubated with 4×10^{-8} M [^{3}H]R5020 either alone (●) or in the presence of 1×10^{-5} M unlabeled R5020 (○) for 3 hr at 0 to 3°C. Sucrose density gradient analysis was performed as described in Materials and Methods. Sedimentation of a 4.3S marker protein is indicated by the arrows.

in 7 of 30 (23%) specimens of infiltrating ductal carcinoma of the human breast R5020 binding proteins in the absence of estrogen receptors (Table 3). Interestingly, the majority of tumors contained specific R5020 binding components (80%) regardless of the status of estrogen receptors. The finding that *most* of these tumors contained progesterone receptors is similar to that we reported previously for the R3230AC- and DMBA-induced mammary tumors of the rat (3,26).

Since human uterus is known to contain proteolytic activity resulting in the presence of ~4S estrogen binding components (12), we have examined the use of protease inhibitors in the preparation of R5020 binding compo-

TABLE 3. *Distribution of estradiol and R5020 receptors in cytosol of human breast carcinomas*

Receptor distribution[a]	No. of tumors
E⁻/P⁻	3 (10%)
E⁺/P⁻	3 (10%)
E⁺/P⁺	17 (57%)
E⁻/P⁺	7 (23%)
	Total 30 (100%)

[a] Binding capacities of <10 fmoles/mg protein were considered negative.

TABLE 4. *Influence of the composition of homogenizing buffer and of sucrose gradients on sedimentation of [³H]R5020 binding components in lactating mammary gland*

Homogenizing buffer	Sucrose gradients	Sedimentation velocity(s)
1. Tris-HCl, pH 7.4, containing 10 mM monothioglycerol and 1.5 mM EDTA	5 to 40% in Tris-HCl with 1.5 mM EDTA	~4
2. Same as 1	Same as above + 10 mM monothioglycerol	~4
3. Same as 1	5 to 20% in Tris-HCl with 1.5 mM EDTA and 10% glycerol	~4
4. Same as 1 + 10% glycerol	Same as above	~4
5. Same as 1 + 15 mM diisopropylfluorophosphate	Same as above	~4
6. Tris-HCl, pH 7.4, containing 1.5 mM EDTA and 10 mM monothioglycerol	5 to 20% in Tris-HCl with 1.5 mM EDTA and 0.4 M KCl	~4
7. 5 mM sodium phosphate, pH 7.4, containing 1 mM monothioglycerol and 10% glycerol	5 to 20% in 5 mM sodium phosphate, pH 7.4, containing 1 mM monothioglycerol and 10% glycerol	~4

nents. Using human mammary tumors, we have observed *no* difference in sedimentation behavior when cytosols were prepared with Tris-HCl buffer alone (Fig. 7A) or with buffer containing either diisopropylfluorophosphate (Fig. 7B) or phenylmethylsulfonylfluoride (Fig. 7C). Although R5020 binding components from each of these preparations sedimented at ~4S (Fig. 7), both of the protease inhibitors usually brought about a reduction in binding capacity. However, the quantity of their loss was variable among preparations.

Further investigation of alternative compositions of the homogenizing buffer and the sucrose gradients indicated that none of the combinations examined changed the sedimentation properties. In all cases [³H]R5020 binding components in cytosol of lactating mammary gland sedimented at 4S (Table 4). Thus the presence of the 4S R5020 binding protein in cytosol of mammary gland does *not* appear to result from a methodological problem. Milgrom et al. (11) have proposed that progesterone receptor size in uterus is dependent on the extent of estrogen stimulation. Similar findings have been reported for the progesterone receptor of the chick oviduct (21).

Distinction Between Specific Progestin and Glucocorticoid Binding Components

Previously we reported that certain glucocorticoids compete significantly for progesterone binding sites in mammary tissues and vice versa

(2,3,10,24,26). At that time we proposed that the action of these two classes of steroid hormones may be mediated by a common mechanism and possibly may share certain binding sites. In support it was observed that glucocorticoids translocated sites into nuclei of lactating mammary gland which were exchangeable with [^{3}H]progesterone (9).

A notable difference in the properties of binding by these two types of ligands includes sedimentation behavior. Under conditions of low ionic strength, the [^{3}H]triamcinolone acetonide binding components sedimented at ~7S (2), whereas those for either R5020 or progesterone sedimented at ~4S in low and high ionic strength sucrose gradients (3,10). Another difference is the affinity constant of the primary ligand for binding sites. Triamcinolone acetonide exhibited a K_d value of 10^{-9} M (2) whereas that for progestins was 10^{-8} M (3,10).

Further investigation of ionic strength revealed a marked difference in the sensitivity of R5020 and glucocorticoid binding sites (Fig. 8). If cytosol was incubated with various concentrations of KCl 1 hr prior to loading of sites with [^{3}H]ligand, triamcinolone acetonide binding was decreased by KCl in a concentration-dependent fashion resulting in *no* observable binding at 0.4 M salt. The decrease in binding capacity by either ligand was not accompanied by a change in the K_d value of the steroid-binding protein complexes (R. G. Mehta and J. L. Wittliff, *unpublished observations*). Inhibition of binding was observed only if cytosol was treated with KCl prior to addition of ligand.

The results described above suggested that R5020 and triamcinolone acetonide bound to separate sites. Additional characterization of the cytosolic components associating with these ligands in lactating mammary gland

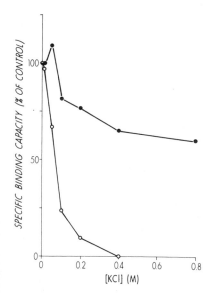

FIG. 8. Influence of ionic strength on the R5020 and triamcinolone acetonide binding capacities. Cytosols were prepared from lactating mammary glands of rats 19 to 20 days postpartum and incubated with KCl of the final concentrations given for 1 hr prior to the treatment with either 40 nM [^{3}H]R5020 (●) or 25 nM [^{3}H]triamcinolone acetonide (○) for 3 hr at 0 to 3°C using the appropriate unlabeled ligands to correct for low-affinity binding. Specific binding, obtained in the absence of KCl, was considered as 100%.

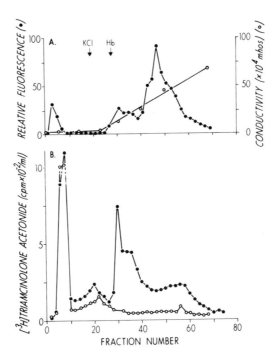

FIG. 9. DEAE-cellulose chromatography of glucocorticoid binding components in cytosol from lactating mammary gland. Cytosol prepared in TET buffer was incubated with 25 nM [^{3}H]triamcinolone acetonide either alone or with 250-fold excess of unlabeled triamcinolone acetonide overnight at 3°C, and the unbound steroid was removed by dextran-coated charcoal. Each reaction was layered onto identical DEAE-cellulose columns as described in Materials and Methods. After each column was washed with 36 ml of TET buffer, additional proteins estimated by relative fluorescence (●) were eluted with a gradient of 0.0 to 0.4 M KCl indicated by conductivity (○) in **A**. Components associating with [^{3}H]triamcinolone acetonide either when it was present alone (●) or when unlabeled steroid was included in the reaction (○) are represented in **B**.

was accomplished by DEAE-cellulose chromatography (Figs. 9 and 10). Following the procedures described in Materials and Methods, the profiles of [^{3}H]triamcinolone acetonide binding revealed at least two components associating specifically with the ligand. The proteins eluting between fractions 28 and 40 (Fig. 9B) were removed from the column at an average ionic strength of 20×10^{-4} mhos (~0.1 M KCl). Hemoglobin, contaminating the cytosol, eluted just prior to this peak (Fig. 9A). A second peak of bound radioactivity was eluted from the column at an average ionic strength of 50×10^{-4} mhos (~0.2 M KCl) between fractions 48 and 60 (Fig. 9B).

In contrast to our result with glucocorticoid binding proteins, DEAE-cellulose chromatography revealed that specific progesterone binding components were eluted as a single radioactive peak between fractions 46 and 56 (Fig. 10B) with an average ionic strength of 51×10^{-4} mhos (~0.2 M KCl).

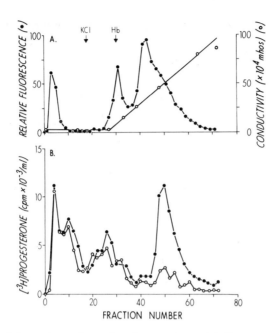

FIG. 10. DEAE-cellulose chromatography of progesterone binding components in cytosol from lactating mammary gland. Cytosol prepared in TET buffer was incubated with 50 nM [^{3}H]progesterone either alone or with 250-fold excess of unlabeled R5020 for 3 hr at 3°C, and the unbound steroid was removed by dextran-coated charcoal. Each reaction was layered onto identical DEAE-cellulose columns as described in Materials and Methods. After each column was washed with 36 ml of TET buffer, additional proteins indicated by relative fluorescence (●) were eluted with a gradient of 0.0 to 0.4 M KCl measured by conductivity (○) as given in **A**. **B:** Components associating with [^{3}H]progesterone either when it was present alone (●) or when unlabeled R5020 was included in the reaction (○).

These components eluted at an ionic strength similar to that of the smaller peak of glucocorticoid radioactivity (Fig. 9B). Another significant peak of [^{3}H]progesterone bound radioactivity, observed between fractions 24 and 30, was *not* displaced by unlabeled R5020 (Fig. 10B). This component may represent transcortin, which has been reported to elute at low ionic strength (18). In general our results suggested that [^{3}H]triamcinolone acetonide is accepted by either glucocorticoid or progesterone binding sites under the conditions described. However, use of either [^{3}H]R5020 or [^{3}H]progesterone with unlabeled R5020 as competitor permits one to study only the specific progesterone binding site.

In summary, we have shown that certain molecular and kinetic properties of [^{3}H]R5020 binding components of mammary tissues were similar to those exhibited by [^{3}H]progesterone. However, it appears that R5020 associates with specific sites in a more stable manner than progesterone itself, permitting more exact quantitation of binding capacities in mammary tissues. Finally, in the event that [^{3}H]R5020 is unavailable, [^{3}H]progesterone may be used as a suitable ligand under appropriate incubation conditions using unlabeled R5020 to correct for low-affinity (nonspecific) binding.

ACKNOWLEDGMENTS

This research was supported in part by U.S. Public Health Service Grants CA-11198 and CA-12836 and by a grant from the Monroe Community Cancer and Leukemia Association, Inc. Dr. Kute is a fellow in the Endocrinology Training Program (USPHS Grant AM-07092-01 from NIAMDD) of the University of Rochester Medical Center. The excellent technical assistance of Dewitt T. Baker, Jr., is greatly acknowledged. We also wish to thank Eve-Marie Cuthbert for her devotion to the preparation of this manuscript.

REFERENCES

1. Asselin, J., Labrie, F., Kelly, P. A., Philibert, D., and Raynaud, J. P. (1976): Specific progesterone receptors in DMBA induced mammary tumors. *Steroids,* 27:395–404.
2. Goral, J. E., and Wittliff, J. L. (1975): Comparison of glucocorticoid-binding proteins in normal and neoplastic mammary tissues of the rat. *Biochemistry,* 14:2944–2952.
3. Goral, J. E., and Wittliff, J. L. (1976): Characteristics of progesterone-binding components in neoplastic mammary tissues of the rat. *Cancer Res.,* 36:1886–1893.
4. Horwitz, K. B., Costlow, M. E., and McGuire, W. L. (1975): MCF-7: a human breast cancer cell line with estrogen, androgen, progesterone and glucocorticoid receptors. *Steroids,* 26:785–795.
5. Horwitz, K. B., and McGuire, W. L. (1975): Specific progesterone receptors in human breast cancer. *Steroids,* 25:497–505.
6. Horwitz, K. B., McGuire, W. L., Pearson, O. H., and Segaloff, A. (1975): Predicting response to endocrine therapy in human breast cancer: a hypothesis. *Science,* 189:726–727.
7. Hutchens, T. W., and Markland, F. S. (1976): Progestin receptors from lactating mammary gland from Caprine species: identification and characterization. *Fed. Proc.,* 35:1559.
8. Kute, T. E., and Westphal, U. (1976): Chemical modification of α_1-acid glycoprotein for characterization of the progesterone binding site. *Biochem. Biophys. Acta,* 420:195–213.
9. Mehta, R. G. (1975): Progesterone binding sites in nuclear fractions from mammary glands of lactating rats. *Fed. Proc.,* 34:627.
10. Mehta, R. G., and Wittliff, J. L. (1976): Specific association of a synthetic progestogen with binding sites in lactating mammary gland. *Fed. Proc.,* 35:1559.
11. Milgrom, E., Atger, M., Perrot, M., and Baulieu, E. E. (1972): Progesterone in uterus and plasma. VI. Uterine progesterone receptors during the estrus cycle and implantation in the guinea pig. *Endocrinology,* 90:1071–1078.
12. Notides, A. C., Hamilton, D. E., and Rudolph, J. H. (1972): Estrogen-binding proteins of the human uterus. *Biochim. Biophys. Acta,* 271:214–224.
13. Philibert, D., and Raynaud, J. P. (1973): Progesterone binding in the immature mouse and rat uterus. *Steroids,* 22:89–98.
14. Philibert, D., and Raynaud, J. P. (1974a): Progesterone binding in the immature rabbit and guinea pig uterus. *Endocrinology,* 94:627–632.
15. Philibert, D., and Raynaud, J. P. (1974b): Binding of progesterone and R5020, a highly potent progestin to human endometrium and myometrium. *Contraception,* 10:455–466.
16. Raynaud, J. P., Bouton, M. M., Philibert, D., Delarue, J. C., Guerenot, F., and Bohuon, C. (1976): Progesterone and estradiol binding sites. *Res. Steroids (in press).*
17. Rousseau, G. G., Baxter, J. D., and Tomkins, G. M. (1972): Glucocorticoid receptors: Relations between steroid binding and biological effects. *J. Mol. Biol.,* 67:99–115.

18. Schrader, W. T., and O'Malley, B. W. (1972): Progesterone-binding components in chick oviduct. *J. Biol. Chem.,* 247:51–59.
19. Shyamala, G. (1973): Specific cytoplasmic glucocorticoid hormone receptors in lactating mammary glands. *Biochemistry,* 12:3085–3090.
20. Stroupe, S., Cheng, S. L., and Westphal, U. (1975): Fluorescence quenching of progesterone-binding globulin and α_1-acid glycoprotein upon binding of steroids. *Arch. Biochem. Biophys.,* 168:473–482.
21. Toft, D. O., and O'Malley, B. W. (1972): Target tissue receptors for progesterone: the influence of estrogen treatment. *Endocrinology,* 90:1041–1045.
22. Waddell, W. J. (1956): A simple ultraviolet spectrophotometric method for the determination of proteins. *J. Lab. Clin. Med.,* 48:311–314.
23. Westphal, U. (1971): Steroid-protein interaction. *Monographs on Endocrinology, Vol. 4.* Springer-Verlag, Berlin.
24. Wittliff, J. L. (1975): Steroid binding proteins in normal and neoplastic mammary cells. *Meth. Cancer Res.,* 11:293–354.
25. Wittliff, J. L., and Savlov, E. D. (1975): Estrogen binding capacity of cytoplasmic forms of the estrogen receptors in human breast cancer. In: *Estrogen Receptors in Human Breast Cancer,* edited by W. L. McGuire, P. P. Carbone, and E. P. Vollmer. Raven Press, New York.
26. Wittliff, J. L., Mehta, R. G., Boyd, P. A., and Goral, J. E. (1976): Steroid binding proteins of the mammary gland and their clinical significance in breast cancer. *J. Toxicol. Environ. Health (Suppl.),* 1:231–247.

Determinations of High-Affinity Gestagen Receptors in Hormone-Responsive and Hormone-Independent GR Mouse Mammary Tumors by an Exchange Assay

J. L. Daehnfeldt and P. Briand

† *The Fibiger Laboratory, DK 2100 Copenhagen, Denmark*

Progress in tracer technology has increased the power of resolution of biochemical methods to such a degree that knowledge and understanding of hormonal effects on cell growth and differentiation are rapidly expanding. By using tritium-labeled steroids of high specific activity (100 Ci/mmole), it is possible to observe the high-affinity binding of the steroids to binding proteins in the cytoplasm of target cells, the transportation of the steroid binding protein complex into the nucleus, and the acceptance of the steroid binding protein complex to acceptor sites of the chromatin. Although the steroid may be found in the cytoplasm or nucleus of cells other than target cells, the presence of the hormone in these cells does not initiate or alter gene transcription because nontarget cells do not contain the cytoplasmic binding protein.

A group of malignant tumors are known as endocrine tumors. These tumors, which originate from tissues under hormonal regulation, show a pattern of progression from hormone-dependent growth to hormone-independent (autonomous) growth. This feature has been demonstrated in experimental animals (7) and is suggested in humans by the transitory effect of endocrine therapy.

The increased knowledge of hormone receptor proteins has brought about renewed interest in the endocrine treatment of human cancer, particularly in treatment of mammary cancer. Determinations of the estradiol receptor protein in tumor tissue have already improved the rational basis for choice of therapy in metastatic breast cancer. For some time, however, it has been realized that determinations of estradiol receptor protein alone give only incomplete information concerning the endocrine status of a tumor (8). Other hormones, such as progesterone, androgens, and prolactin, may

---

† Sponsored by the Danish Cancer Society.

be involved in growth regulation; and receptors for these hormones are under investigation at several centers.

The following deals with the determination of progesterone and estradiol receptors in GR mouse mammary tumors and the correlation between steroid hormone receptors and hormone-dependent growth.

MATERIALS AND METHODS

Tumor Tissue

Mammary tumors were induced in GRS/A/Fib mice (15) according to the method of van Nie (*personal communication,* 1970) as described previously (1,16,21). Ten- to twelve-week-old female mice were spayed and immediately treated with progesterone plus estrone. Progesterone pellets prepared from a paste of progesterone and olive oil were injected subcutaneously once a week at a dose of 5 to 10 mg. Estrone was dissolved in ethanol and added to the drinking water at a concentration of 0.5 μg/ml. The induced tumors were transplantable to spayed mice treated with progesterone and estrone. Hormone-responsive tumors were defined as tumors that did not grow as well (or at all) in spayed mice receiving no hormone treatment. Hormone-independent (HI) tumors, on the other hand, grew equally well in hormone-treated and untreated spayed mice.

Chemicals

The gestagen R5020 in tritiated (specific activity 50 Ci/mmole) and in nonradioactive form was generously provided by Roussel-Uclaf, Romainville, France. Tritiated estradiol-17β (specific activity 90 Ci/mmole) was obtained from Radiochemical Center, Amersham, U.K. The tritiated hormones were delivered in solvents that were evaporated under nitrogen to dryness. The solvents were replaced by equal volumes of 99% ethanol. This solution was stored at $-20°$C. Estradiol-17β, estrone, and progesterone were a gift from Leo Pharmaceuticals, Ballerup, Denmark. Dexamethasone was obtained from Merck, Sharp, and Dohme, Haarlem, Holland, and hydrocortisone from DAK Laboratories, Copenhagen. Tris (hydroxymethyl) aminomethane (Tris), dithiothreitol (DTT), and activated charcoal were obtained from Sigma. The charcoal was sifted through a nylon net with an 80-μm mesh. Dextran T 70 was supplied by Pharmacia, Uppsala, Sweden. Ethylendinitrilotetraacetate disodium salt (EDTA) was purchased from Merck, Germany. The remaining chemicals were of analytical grade.

The solution for liquid scintillation counting contained Triton X-100: toluene (1:2 vol/vol) and 2,5-diphenyloxazole (PPO) (5 g/liter).

Tissue Preparation and Storage

The animals were killed by cervical dislocation. Tumors weighing 0.5 to 2.0 g were excised and immediately placed in a Petri dish on crushed ice. Occasional necrosis was removed. Tumors were then either stored at −80°C for a maximum of 2 weeks or immediately homogenized in a microdismembrator (Braun, Melsungen, West Germany). After the Teflon homogenizing chamber and tungsten ball were precooled in liquid nitrogen, tumor tissue was introduced into the chamber and cooled. The chamber was then shaken for 30 sec at maximal setting. Frozen pulverized tissue was distributed into centrifuge tubes for a Beckman Ti 60 rotor and suspended in buffer. The suspension buffer (TE buffer) for estradiol receptor determinations consisted of Tris 10 mM, pH 7.4, EDTA 1.5 mM, and sodium azide 1.0 mM (6). Three parts of TE buffer were used for one part of tissue (vol/wt). For the progesterone receptor determinations, DTT 0.5 mM and glycerol 10% were added to the TE buffer (22).

After centrifugation at $100,000 \times g$ at 4°C for 1 hr, the resulting particle-free supernatant (PFS) was divided into aliquots and immediately frozen at −80°C in a Revco Freezer (Revco Inc., Columbia, S.C.). Repeated assays showed that supernatants could be stored at this temperature for up to 2 to 3 weeks without appreciable loss of binding capacity. The protein content of the PFS was 6 to 10 mg protein per milliliter determined by the method of Lowry et al. (14).

Binding Assay

Two different types of assays for the determination of high-affinity hormone binding capacity in cytosols were applied. The first is the dextran-coated charcoal (DCC) method originally described by Korenman (12), which has been widely used in several modified forms for various hormone receptor determinations. This assay method measures free receptor sites only.

The second assay method included a step that allowed endogenously bound hormone to exchange with ^{3}H-hormone, and thus the total as well as free receptor binding could be determined (11).

DCC Assay

Aliquots of PFS preparations were incubated at 0°C with five different concentrations of ^{3}H-ligand (0.3 to 7×10^{-9} M) as described in Daehnfeldt's article (4). Ligand was added in a volume of 5 µl 99% ethanol to maximally 50 µl of PFS. The volume was made up to 80 µl by addition of TKE buffer (Tris 10 mM, pH 7.4, KCl 50 mM, EDTA 1.0 mM, and NaN$_3$ 1.0 mM)

(6). After 2 hr of incubation at 0°C, 10 µl of dextran-coated charcoal suspension (25 mg charcoal + 25 µg dextran per milliliter TKE buffer) was added and adsorption carried out at 0°C for 30 min with occasional shaking. After centrifugation at $600 \times g$ at 4°C, the radioactivity in the supernatant was counted in a Beckman LS 255 scintillation counter. Quench correction was done by channels' ratio method. Controls with buffer plus bovine serum albumin 0.5 mg/ml were used to correct for deficient adsorption with charcoal.

Specificity of R5020 Binding

Binding specificity was investigated by competition experiments. To 100 µl of PFS, nonradioactive competing steroid was added together with ^{3}H-R5020 at 10^{-2}, 10^{-1}, 10^0, 10^1, and 10^2 times the concentration of ^{3}H-R5020 (30 nM) and was allowed to react for 1 hr at 0°C. After this, free ligands were removed by adsorption to a charcoal slurry (PFS:slurry, 10:1). The slurry consisted of 50 mg charcoal and 0.5 mg dextran T 70 per milliliter TKE buffer. The charcoal was removed and the radioactivity in 50 µl of the supernatant was counted. To determine the unspecific binding of R5020, we incubated a sample with ^{3}H-R5020 at a concentration of 30 nM plus nonradioactive R5020 at a concentration of 3,000 nM.

Hydrocortisone, dexamethasone, and progesterone were used as competing steroids.

Exchange Assay

This assay was performed as described by Katzenellenbogen et al. (11) for rat uterus.

Stability Determination

Preparations of PFS obtained as described in the section of tissue preparation were incubated with charcoal slurry (PFS:slurry, 10:1) for 10 min at 0°C to remove endogenous hormone. After centrifugation at 600 g for 10 min at 4°C, the resulting supernatant in a preliminary assay was subjected to the following steps. One part of the supernatant was incubated with ^{3}H-ligand at a concentration of 30 nM (5 µl ligand in 99% ethanol per 100 µl supernatant) for 1 hr at 0°C ("hot" incubation). An equal part was incubated under the same circumstances with ^{3}H-ligand at 30 nM and nonradioactive ligand in a concentration of 3,000 nM ("hot + cold" incubation). After 1 hr the two samples were subjected to charcoal slurry adsorption and centrifugation as already described. The resulting supernatants were again incubated at "hot" conditions and "hot + cold" conditions. These incubations were performed at 25°, 10°, 5°, and 0°C. Samples were withdrawn at suitable intervals. The aliquots were immediately cooled for 10 min at 0°C and again adsorbed with charcoal and centrifuged as described. In this way the influence of temperature and time on the binding capacity for estradiol and R5020 was determined.

Exchange Determination

An aliquot of PFS was saturated with nonradioactive ligand at a concentration of 3,000 nM for 1 hr at 0°C. After this incubation, free ligand was adsorbed to charcoal and centrifuged as described for stability determinations. To the resulting supernatant, ^{3}H-ligand was added to a concentration of 30 nM. The sample was then heated to the same temperatures as described for stability determinations. In this way the binding of ^{3}H-activity into the supernatant was followed in time.

Determination of Apparent Binding Capacity and Total Binding Capacity

The samples were adsorbed with charcoal slurry (PFS:slurry, 10:1) and centrifuged as described for stability and exchange determinations. The resulting supernatant was divided into four portions. The first two were incubated with ^{3}H-ligand at 30 nM hot, the other two with ^{3}H-ligand plus nonradioactive ligand 3,000 nM hot + cold.

The four samples were allowed to equilibrate for 1 hr at 0°C. At this time one hot and one hot + cold were withdrawn. Bound ^{3}H-ligand was determined. The specific binding values were obtained from the counting results of hot and hot + cold incubations by subtracting hot + cold values from hot values. This was named apparent binding capacity (ABC). The remaining samples were heated to the optimal temperatures determined in the preliminary assays of stability and exchange. After a suitable time, samples were again adsorbed by charcoal and centrifuged. The radioactivity of the resulting supernatants was counted as for ABC. The result was named total binding capacity (TBC).

RESULTS AND DISCUSSION

The technical problems of estrogen as well as of progesterone receptor protein determinations have recently been discussed by Clark et al. (3). Only a few sides of the problems of receptor protein determinations will be mentioned here.

Steroid hormone receptor proteins or receptors may be characterized by the high-affinity binding of the ligand, the stereospecificity of the binding, and the limited binding capacity at near physiological concentrations of the ligand. We have investigated whether the R5020 binding to mouse mammary

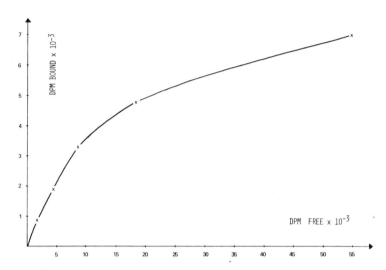

FIG. 1. ^{3}H-R5020 binding in PFS of a GR mouse mammary tumor determined by the DCC assay and plotted as bound versus free ^{3}H-R5020. Incubation was carried out at 0°C for 2 hr. Binding values were not corrected for unspecific binding.

tumor supernatants would fulfill these criteria for binding of the receptor type. In experiments using the conventional dextran-coated charcoal technique with five different concentrations of ^{3}H-R5020, we have obtained information on the binding capacity and K_d from a Scatchard plot $K_d \sim 2 \times 10^{-9}$ M. Rearranging the binding result by plotting bound versus free hormone (Fig. 1) or double reciprocal plots (*not included*), we can illustrate the saturable character of binding by leveling off of the curve at the higher concentrations of free hormone. Values are not corrected for unspecific binding. This may explain the nonhorizontal ending of the curve.

In the competition experiments of Fig. 2 it is shown that only progesterone significantly reduces the binding of R5020. No or little competition is found when incubation is carried out with dexamethasone or hydrocortisone. This evidence suggests that R5020 binds to a protein that also binds progesterone, and the use of the term receptor protein for the protein that binds R5020 in the supernatant seems justified.

The results of stability and exchange determinations for the R5020-receptor complex are shown in Figs. 3 and 4. The receptor appears to be highly unstable at 25°C as well as at 10°C. Even at 5° and 0°C there is an initial loss of R5020-receptor complexes. This loss remains unexplained. It may indicate the presence of two different binding proteins. After the initial loss, stability remains unaffected for 21 hr. Based on this we have chosen to determine R5020 TBC after incubation at 5°C for 20 hr.

In the GR strain of mice, mammary tumors appear early in breeding females. Most of these tumors are pregnancy responsive, that is, they grow during pregnancy and regress after parturition. Regrowth of regressed

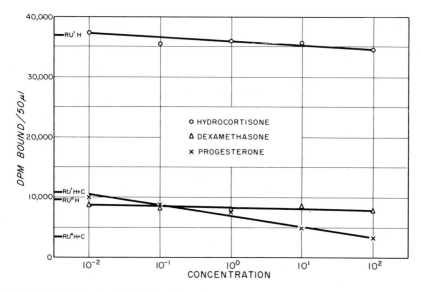

FIG. 2. ³H-R5020 binding in PFS in the presence of hydrocortisone, dexamethasone, and progesterone. Incubation was carried out at 0°C for 1 hr. RU' indicates binding of "hot" (H) and "hot + cold" (H + C) ³H-R5020 in the PFS used for hydrocortisone competition. RU" indicates binding of "hot" (H) and "hot + cold" (H + C) ³H-R5020 in the PFS used for dexamethasone and progesterone competition.

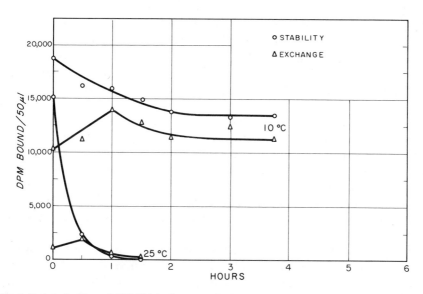

FIG. 3. Determination of R5020 binding protein stability and R5020 exchange rate in PFS of GR mouse mammary tumors at 25° and 10°C.

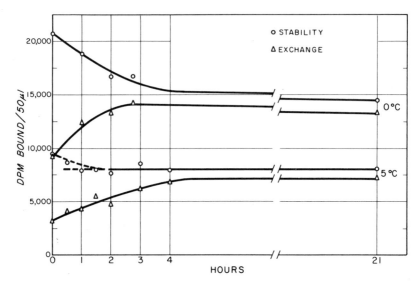

FIG. 4. Determination of R5020 binding protein stability and R5020 exchange rate in PFS of GR mouse mammary tumors at 5° and 0°C.

tumors occurs after combined treatment with estrone and progesterone but not after treatment with only one of the hormones (16; R. van Nie, *personal communication*). Mammary tumors can also be induced in these mice by treating spayed mice with estrone and progesterone (R. van Nie, *personal communication*). Such tumors are hormone responsive and in some cases regress completely within a week after discontinuation of estrone treatment (10,20). This regression occurs in spite of the continued presence of progesterone for up to 4 weeks after the last pellet implantation as demonstrated by anestrous vaginal smears (19). Thus, although it is apparent that estrone is essential for the growth of these tumors, the role that progesterone plays in tumor growth appears to be more subtle.

In previous experiments investigating the biochemical changes that occur during tumor induction (5), no progesterone binding could be demonstrated when using ^{3}H-progesterone as the ligand. At that time it was suggested that the failure to demonstrate progesterone binding might be due to endogenous blocking of the progesterone receptor protein because of the high doses of progesterone administered to the mice. Now it would appear that a more likely explanation might be the high lability of the receptor-hormone complex.

In order to clarify this problem we decided to determine the R5020 receptor protein using the method of Katzenellenbogen et al. (11), thus quantitating both the free binding sites (ABC) and, after an exchange step, the total binding sites (TBC). These results together with simultaneous determinations of estradiol receptor content are shown in Table 1. It is

TABLE 1. *Apparent binding capacity and total binding capacity for ^{3}H-estradiol and ^{3}H-R5020 in hormone-treated and untreated hormone-independent and hormone-responsive GR mouse mammary carcinomas*

	Tumor type	Passage	N[b]	ABC[a] Median	ABC[a] Range	TBC[a] Median	TBC[a] Range
Estradiol							
	HI−	1–7	7	2	2–3	2	1–3
	HI+	1–5	9	8(4)	2–10	11(4)	3–14
	HR	1	6	12	11–16	16	8–20
R5020							
	HI−	1–7	8	1	<1–2	<1	<1–1
	HI+	1–5	12	36(1)	<1–68	31(2)	1–61
	HR	1	9	60	14–87	58	13–92

Estradiol:
HI−, ABC compared to TBC $0.1 < p$.
HI+, ABC compared to TBC $0.1 < p$.
HR, ABC compared to TBC $0.1 < p$.
ABC, HI− compared to HI+ $p < 0.01$.
ABC, HI+ compared to HR $p < 0.05$.
ABC, HI− compared to HR $p < 0.01$.

R5020:
HI−, ABC compared to TBC $0.1 < p$.
HI+, ABC compared to TBC $0.1 < p$.
HR, ABC compared to TBC $0.1 < p$.
ABC, HI− compared to HI+ $p < 0.01$.
ABC, HI+ compared to HR $p = 0.02$.
ABC, HI− compared to HR $p < 0.01$.

[a] Dimension of ABC and TBC: ligand bound × 10^{14} moles/mg protein.
[b] N is the total number of tumors investigated. Except in 2 cases, each tumor line is represented by 2 tumors. From each tumor a PFS was obtained and investigated in 2 dilutions.
Figures in brackets are median values of determinations on tumors in more than 2 transplantation passages.

apparent that no significant differences were found between ABC and TBC for the R5020 receptor protein. Also, no difference between ABC and TBC for estradiol could be demonstrated.

The presence of hormone receptors for estradiol as well as for progesterone indicates that both of these hormones may exert a direct effect on the tumor cells. Estrogens may additionally affect the tumor cells through stimulation of the production of prolactin from the pituitary gland (13). Prolactin is known to be able to significantly stimulate tumor growth in these mice (2).

Estradiol binding was generally found to be low in hormone-independent and high in hormone-responsive tumors, which agrees with results found by Sluyser and van Nie (21). The R5020 binding followed a similar pattern. However, HI tumors in early transplant generations in animals treated with hormones had a high receptor binding (Table 1). This might indicate that these tumors consist of mixed populations of HR and HI tumor cells. However, if the receptor binding should be indicative of the proportion of HR tumor cells in a given tumor, this value is so high that the tumor would probably not behave as an HI tumor. Yet another explanation might be that the tumor cells are in an early stage of progression in which only growth and no other cellular functions have become hormone independent. In the later transplant generations, hormone independence is complete since no specific hormone action can occur without hormone receptor.

In conclusion, it might be stated that R5020 was found to bind to a cytoplasmic constituent with high affinity ($K_d = 2 \times 10^{-9}$). Binding capacity of R5020 concentration around the K_d value was demonstrated to be limited, and progesterone was found to depress the R5020 binding. The R5020 binding was not influenced by the presence of dexamethasone or hydrocortisone, which would indicate that R5020 was not cross-reacting with corticoid receptor protein or corticoid binding globulin.

On the basis of the results mentioned above, we suggest that R5020 binds to progesterone receptor protein, although the protein nature of the binding constituent has not been demonstrated. The investigation of the changes in receptor protein content following tumor progression clearly demonstrated that the occurrence of autonomy is followed by a decrease in both estradiol receptor protein and R5020 receptor protein content down to the limit of detection. In the group of HR tumors the ratio of estrogen binding to R5020 binding was fairly constant. From the results presented here it cannot be determined if an estrogen-inductive effect is involved in the synthesis of progesterone receptor protein. The correlation between the estradiol receptor content and the effect of endocrine therapy in human breast cancer fails in approximately 40% of the receptor-positive patients. We hope that additional information about progesterone receptor content can improve the prediction. Progesterone receptors have been determined in about one-third of human breast cancers (9), and another investigation (18) has shown that human breast cancers may contain estradiol receptors without progesterone receptors. Whether these are the cases that do not respond to endocrine therapy is not yet known.

On the basis of this information, we intend to include the R5020 determination in our current study on the effect of endocrine adjuvant treatment in primary human mammary cancer.

ACKNOWLEDGMENTS

We gratefully acknowledge the technical assistance of Lisbeth Huusom and Marianne Hansen. This work was supported by National Science Foundation Grants J. 512-5327 and 512-6558, the Daell Foundation, and Mrs. Agathe Neye.

REFERENCES

1. Briand, P., and Daehnfeldt, J. L. (1973): Enzyme patterns of glucose-catabolism in hormone-dependent and -independent mammary tumors of GR mice. *Eur. J. Cancer*, 9:763–770.

2. Briand, P., Thorpe, S. M., and Daehnfeldt, J. L.: The effect of prolactin and CB154 on the growth of transplanted hormone dependent GR mouse mammary tumors in vivo. (*In preparation*).
3. Clark, J. H., Peck, E. J., Jr., Schrader, W. T., and O'Malley, B. W. (1976): Estrogen and progesterone receptors: Methods for characterization, quantification, and purification. In: *Methods in Cancer Research, Vol. XII*, edited by H. Busch. Academic Press, New York.
4. Daehnfeldt, J. L. (1974): Endogenously blocked high affinity estradiol receptors in the immature and mature rat uterus. *Proc. Soc. Exp. Biol. Med.*, 146:159–162.
5. Daehnfeldt, J. L., Schülein, M., and Briand, P. (1975): Biochemical changes in GR mouse mammary tissue during hormonal tumor induction. *Eur. J. Cancer*, 11:509–515.
6. De Sombre, E. R., Puca, G. A., and Jensen, E. V. (1969): Purification of an estrophilic protein from calf uterus. *Proc. Natl. Acad. Sci. U.S.A.*, 64:148–154.
7. Foulds, L. (1969): *Neoplastic Development*, p. 51. Academic Press, New York.
8. Hilf, R., and Wittliff, J. L. (1974): Characterization of human breast cancer by examination of cytoplasmic enzyme activities and estrogen receptors. In: *Hormones and Cancer*, edited by K. W. McKerus. Academic Press, New York.
9. Horwitz, K. B., and McGuire, W. L. (1975): Specific progesterone receptors in human breast cancer. *Steroids*, 25:497–505.
10. Janik, P., Briand, P., and Hartman, N. R. (1975): The effect of estrone-progesterone treatment on cell proliferation kinetics of hormone-dependent GR mouse mammary tumors. *Cancer Res.*, 35:3698–3704.
11. Katzenellenbogen, J. A., Johnson, H. J., and Carlson, K. E. (1973): Studies on the uterine cytoplasmic estrogen binding protein. Thermal stability and ligand dissociation rate. An assay of empty and filled sites by exchange. *Biochemistry*, 12:4092–4099.
12. Korenman, S. G. (1968): Radio-ligand binding assay of specific estrogens using a soluble uterine macromolecule. *J. Clin. Endocrinol. Metab.*, 28:127–130.
13. Kwa, H. G., van der Gugten, A. A., Sala, M., and Verhofstad, F. (1972): Effects of pituitary tumours and grafts on plasma prolactin levels. *Eur. J. Cancer*, 8:39–54.
14. Lowry, O. H., Rosebrough, N. J., Farr, A. L., and Randall, R. J. (1951): Protein measurement with the folin phenol reagent. *J. Biol. Chem.*, 193:265–275.
15. Mühlbock, O. (1965): Note on a new inbred mouse-strain GR/A. *Eur. J. Cancer*, 1:123–124.
16. Nie, R. van, and Hilgers, J. (1976): Genetic analysis of mammary tumor induction and expression of mammary tumor virus antigen in hormone-treated ovariectomized GR mice. *J. Natl. Cancer Inst.*, 56:27–32.
17. Nie, R. van, and Thung, P. J. (1965): Responsiveness of mouse mammary tumours to pregnancy. *Eur. J. Cancer*, 1:41–50.
18. Raynaud, J. P., Bouton, M. M., Philibert, D., Delarue, J. C., Guerinot, F., and Bohuon, C. (1975): Progesterone and estradiol binding sites in human breast carcinoma. Abstract, International Study Group for Steroid Hormones and Receptor Proteins. Rome, December 1975.
19. Röpcke, G. (1975): Interaction of hypophyseal isografts and ovarian hormones in mammary tumour development in mice. Thesis, Mondeel, Amsterdam.
20. Schülein, M., Daehnfeldt, J. L., and Briand, P. (1976): Biochemical changes during regression and regrowth of hormone-dependent GR mouse mammary tumors. *Int. J. Cancer*, 17:120–128.
21. Sluyser, M., and Nie, R. van (1974): Estrogen receptor content and hormone-responsive growth of mouse mammary tumors. *Cancer Res.*, 34:3253–3257.
22. Wiest, W. G., and Rao, B. R. (1971): Progesterone binding proteins in rabbit uterus and human endometrium. In: *Advances in the Biosciences, Vol. 7*, edited by G. Raspé, pp. 251–264. Pergamon Press, New York.

Progesterone and Estradiol Receptors in DMBA-Induced Mammary Tumors Before and After Ovariectomy and After Subsequent Estradiol Administration

A. J. M. Koenders, A. Geurts-Moespot, S. J. Zolingen, and Th. J. Benraad

Department Medical Biology, University of Nijmegen, Nijmegen, The Netherlands

In many laboratories an estrogen receptor assay has been introduced to serve as a marker of hormonal dependency of human mammary carcinoma (5,8,13). It is disappointing that about 40% of the estrogen receptor–positive tumors fail to regress after ablative or additive endocrine treatment (12). It has been emphasized that the prediction of tumor responsiveness would become more reliable if these tumors additionally could be analyzed for other hormone receptors (16,18). In normal target tissues estradiol induces the synthesis of the progesterone receptor (7,14). Therefore, the presence of the progesterone receptor, a product of hormonal action, could give information about the intactness of this hormonal regulatory entity in neoplastic cells (1,6).

The objectives of the present investigation can be described as follows: (1) to examine whether in DMBA-induced rat mammary tumors any correlation exists between the levels of the receptors for progesterone and estradiol; (2) to study if mammary tumor responsiveness to ovariectomy is related to the levels of progesterone and estradiol receptors; (3) to analyze whether estradiol administration to ovariectomized animals induces progesterone receptor synthesis in these DMBA-induced rat mammary tumors.

MATERIALS AND METHODS

Materials

Estradiol-17β-2,4,6,7-^{3}H (105 Ci/mmole) was obtained from New England Nuclear, Boston, Mass. Unlabeled 17β-estradiol, progesterone, and 5α-dihydrotestosterone were purchased from Steraloids, Pawly, N.Y.

Dexamethasone, 9,10-dimethyl-1,2-benzanthracene (DMBA), α-monothioglycerol, bovine serum albumin (BSA), and triamcinolone acetonide were obtained from Sigma Chemicals Co. Cyproterone acetate was kindly provided by Schering A.G., Berlin, Germany, and medroxyprogesterone acetate and clomiphene by Farmitalia. Unlabeled 17,21-dimethyl-19-norpregna-4,9-diene-3,20 dione (R5020) and 6,7-^{3}H-R5020 (51.4 Ci/mmole) were kindly supplied by Roussel-Uclaf, France.

Dithiothreitol (DTT), dextran T 70, and Instagel were obtained from British Drug House Chemicals (BDH), Pharmacia, and Packard, respectively. All other products were reagent grade and purchased from Merck.

The following buffers were used:

1. Tris-HCl 10 mM, EDTA 1.5 mM, DTT 0.5 mM, pH 7.4 for determination of the estrogen receptor.
2. Tris-HCl 10 mM, EDTA 1.5 mM, α-monothioglycerol 10 mM, 10% glycerol (vol/vol) pH 7.4 for determination of the progesterone receptor.

The dextran-coated charcoal suspension contained 0.25% Norit and 0.025% dextran T 70 in buffer 1. The microtiterplates (v shape, 96 holes), the A.M. 69 Microshaker, and the centrifuge plate carriers were obtained from Cooke Instruments.

Animals

Mammary tumors were induced in 50-day-old female Sprague-Dawley rats (Zentralinstitut für Versuchtstierzucht, Hannover/Linden, Germany) by a single intragastric feeding of 20 mg DMBA in 1.0 ml of sesame oil. These rats were housed in a temperature- and light-controlled room and were fed a normal diet and tap water *ad libitum*. The first mammary tumors appeared in the seventh week after DMBA administration, and at the end of the eighth week each animal had one or more mammary tumors. Tumor areas (length × width) were calculated at least twice a week from measurements by caliper. Tumor biopsies were taken under light ether anesthesia. Bilateral ovariectomy was performed under light ether anesthesia at the time the tumor area had reached about 4 cm². Tumors were classified as hormone dependent when their tumor areas were reduced at least 25% following castration. Tumors not influenced by castration were considered to be hormone independent. Biopsies of hormone-dependent and of hormone-independent tumors were taken on day 9 ± 0.7 (mean $\pm$ SEM) and on day 13 ± 2.1 after ovariectomy, respectively. Animals with hormone-dependent tumors subsequently received 7.22 µg estradiol benzoate (s.c.) daily.

After the tumor growth response to this endocrine treatment was assessed, a third biopsy was taken approximately 24 hr after the last injection. All biopsies used for receptor studies were histologically proven to be adenocarcinomas.

Preparation of Tumor Cytosol

The excised tumors were freed of fat and connective tissue and stored until use at $-75°C$ in a Revco deep freezer. The stored tissue was first pulverized in a stainless steel mortar immersed in liquid nitrogen. Thereafter, the small pieces were vibrated to a fine powder by means of a microdismembrator (Braun, Melsungen, Germany). All subsequent steps were performed at 0 to 4°C. The powder was weighed and homogenized in cold buffer (wt/vol = 1:4) using a Ten Broeck homogenizer. The homogenate was centrifuged at 105,000 g for 60 min to prepare the cytosol, which was removed by pipetting. Care was taken to avoid contamination from the upper lipid layer. The cytosol was analyzed without delay.

Dextran-Coated Charcoal Assay of Receptor Content

All the receptor assays were performed in microtiterplates described for this purpose by Katzenellenbogen et al. (9). Table 1 summarizes the procedure for the assay of progesterone and estradiol receptors in DMBA-induced rat mammary tumors. The dextran-coated charcoal blank was assessed from the data obtained from holes 1 and 2. This blank varied between 0.5 and 1% of the radioactivity added. To determine the total binding (B_{tot}), we used at least six different concentrations of tritiated steroid ranging between 10^{-9} and 10^{-8} M for R5020 and between 10^{-9} and 8×10^{-9} M for E_2 (hole 5-6). The nonspecific binding (B_{ns}) for R5020 and E_2 was determined in the presence of a large excess of nonradioactive R5020 and E_2, respectively (hole 7-8). After equilibrium was reached, the unbound steroid was removed by adding 0.1 ml dextran-coated charcoal suspension. The plates were shaken for 10 min on a mechanical shaker and thereafter centrifuged for 20 min at 1,000 g. Aliquots (0.1 ml) of the supernatant were added to counting vials containing 10 ml scintillation liquid. The radioactiv-

TABLE 1. *Procedure for the assay of progesterone and estradiol receptors*

Hole no.	1-2	3-4	5-6	7-8
Buffer	55 µl	155 µl	5 µl	– µl
Unlabeled steroid excess (10^{-6} M)	–	–	–	5
Cytosol	–	–	50	50
Preincubation 15 min at 4°C				
³H-steroid*	5	5	5	5
Incubation (16-20 hr at 4°C)				
Dextran-coated charcoal	100	–	100	100

Vibrate the plates for 10 min at 4°C and centrifuge at 1,000 g for 20 min; take 100-µl aliquots for radioactivity counting.
[³H]E_2: 10^{-9}–8×10^{-9} M.
[³H]R5020: 10^{-9}–10^{-8} M.

ity was counted with an efficiency of 47%. For each tritiated steroid concentration the following corrections were carried out. The bound radioactivities, corrected for the charcoal blank, were subtracted from the total amount of radioactivity present (T, holes 3 and 4), giving the corresponding amounts of unbound steroid (U = T − B). The ratio B_{ns}/U_{ns} was found to be independent of the tritiated steroid concentration. This ratio was multiplied by $(T - B_{tot})$ giving the correct amount of nonspecific binding. The specific receptor binding (B_s) was calculated according to the following equation (3):

$$B_s = B_{tot} - (T - B_{tot}) \frac{B_{ns}}{U_{ns}}$$

The calculated B_ss were plotted against the corresponding $B_s/T - B_{tot}$ ratios giving the Scatchard plot. Scatchard plot analysis was performed to determine the concentration of receptor binding sites (R) and the equilibrium dissociation constant (K_d). The concentration of receptor binding sites was expressed as fmoles/mg protein. The protein content of each supernatant was determined by the method of Lowry et al. (10), using BSA as standard.

Statistical Methods

Wilcoxon rank test for the difference between two groups was used to test significance of the differences in receptor levels between the hormone-dependent and the hormone-independent tumors. The effect of ovariectomy or estradiol administration on receptor levels was tested for significance using Wilcoxon test for paired observations. Spearman rank correlation test was used to correlate estradiol and progesterone receptor levels.

RESULTS

Some Characteristics of Specific Progesterone and Estradiol Binding Sites in DMBA-Induced Rat Mammary Tumors

The Scatchard plot obtained by incubating increasing concentrations of the labeled synthetic progestin ^{3}H-R5020 with a tumor cytosol before and after correction for nonspecific binding is shown in Fig. 1. The uncorrected Scatchard plot was not linear representing binding of ^{3}H-R5020 to at least two different binding proteins. After correction for nonspecific binding, we obtained a linear Scatchard plot (R = 41 fmoles/mg protein; K_d = 1.2 nM). The mean ± 1 SD of K_ds measured in 37 experiments was 2.9 ± 1.6 nM. Experiments were performed by incubating mammary tumor cytosol with ^{3}H-R5020 (5 nM) and increasing concentrations of various nonradioactive compounds (0 to 4 μM) (Fig. 2). Some steroids were bound with high affinity (R5020 > progesterone, medroxyprogesterone acetate, cyproterone acetate, and triamcinolone acetonide), whereas 17β-estradiol and corticosterone

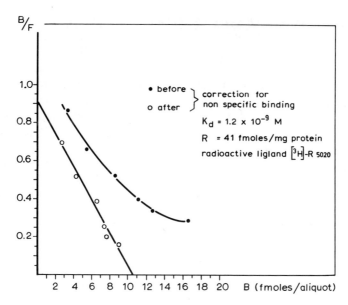

FIG. 1. Scatchard plot of mammary tumor cytosol incubated with 10^{-9} to 10^{-8} M ^{3}H-R5020. The Scatchard plot of total binding (●) was corrected for nonspecific binding to obtain the Scatchard plot of specific binding (○).

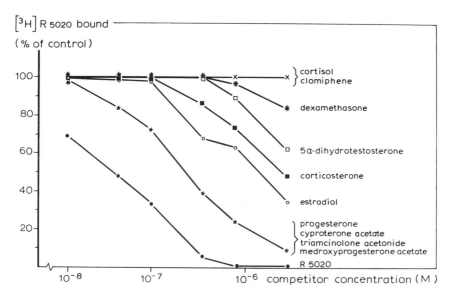

FIG. 2. Effect of increasing concentrations of unlabeled R5020, progesterone, cyproterone acetate, triamcinolone acetonide, medroxyprogesterone acetate, 17β-estradiol, corticosterone, 5α-dihydrotestosterone, dexamethasone, clomiphene, and cortisol on the binding of 5 nM ^{3}H-R5020 to mammary tumor cytosol.

competed only weakly. Dexamethasone and 5α-dihydrotestosterone showed minor competition for the R5020 binding sites, and cortisol and clomiphene did not interfere to any extent with the binding of R5020.

The Scatchard plot obtained by incubating increasing concentrations of ^{3}H-estradiol with mammary tumor cytosol was not linear before correction for nonspecific binding. However, after correction for nonspecific binding, linear Scatchard plots of estradiol binding were observed with limited binding capacity and high affinity for estradiol. The mean ± 1 SD of K_ds found in 52 experiments was 0.33 ± 0.15 nM.

Experiments in which plasma (diluted 1:6.5) of tumor-bearing rats was incubated with increasing concentrations of ^{3}H-R5020 or ^{3}H-E$_2$ showed no saturable binding (*data not shown*), excluding that contaminating plasma proteins, if present, can be responsible for the observed saturable binding in these mammary tumor cytosols.

Progesterone and Estradiol Receptor Levels Before Ovariectomy in Hormone-Dependent and Hormone-Independent Tumors

Tumor biopsies were taken from 29 animals before ovariectomy. Whenever possible, both the progesterone and estradiol receptor levels were determined in these biopsies. The effect of ovariectomy on the growth rate

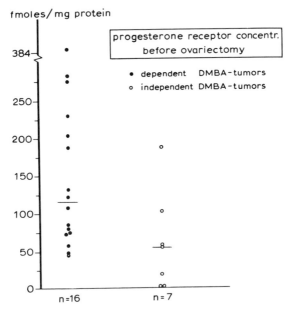

FIG. 3. Progesterone receptor levels before ovariectomy of hormone-dependent (●) and hormone-independent (○) DMBA-induced mammary tumors. Median values are indicated by horizontal lines.

PROGESTERONE AND ESTRADIOL RECEPTORS

of the remaining tissue of the tumor was assessed to distinguish between hormone-dependent and hormone-independent tumors. The progesterone and estradiol receptor levels are shown in Figs. 3 and 4, respectively.

It can be seen from Fig. 3 that all hormone-dependent and five out of seven of the hormone-independent tumors contained measurable amounts of progesterone receptors. The distributions of progesterone receptor levels of both groups of tumors were significantly different ($p < 0.05$). Although there was considerable overlap of the lower values of the hormone-dependent and the higher values of the hormone-independent tumors, hormone-dependent tumors had generally higher progesterone receptor levels than hormone-independent tumors.

As follows from Fig. 4, all mammary tumors contained estradiol receptors. The distributions of levels of hormone-dependent and hormone-independent tumors were significantly different ($p < 0.01$). All hormone-independent tumors except one had relatively low estradiol receptor levels, whereas almost all hormone-dependent tumors contained relatively high estradiol receptor levels.

From both Figs. 3 and 4 it follows that in two hormone-independent tumors estradiol receptors are present without measurable progesterone receptor activity.

The measured K_ds of the progesterone receptor in hormone-dependent and in hormone-independent tumors were 2.9 ± 0.4 and 2.4 ± 0.9 nM, respectively ($p > 0.1$). For the estradiol receptor the K_ds were 0.32 ± 0.03 and 0.40 ± 0.05 nM, respectively ($p > 0.1$). The ligand specificities of the progesterone receptor and the estradiol receptor did not differ for hormone-dependent and hormone-independent tumors.

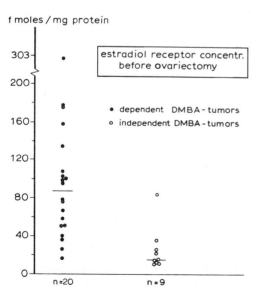

FIG. 4. Estradiol receptor levels before ovariectomy of hormone-dependent (●) and hormone-independent (○) DMBA-induced mammary tumors. Median values are indicated by horizontal lines.

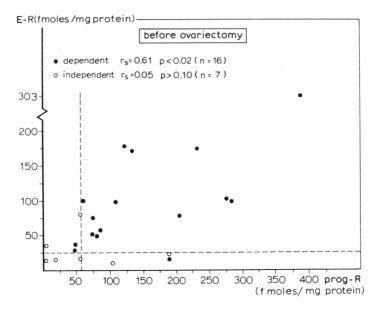

FIG. 5. Correlation of preovariectomy progesterone receptor and estradiol receptor levels of hormone-dependent (●) and hormone-independent (○) DMBA-induced mammary tumors. Broken lines indicate arbitrarily discriminatory receptor levels.

The significantly positive relation between the progesterone and estradiol receptor levels in hormone-dependent tumors is shown in Fig. 5. In addition, this figure shows that no such correlation was found in the hormone-independent tumors. Using discriminatory levels for the progesterone receptor above 57 fmoles/mg protein and for the estradiol receptor above 25 fmoles/mg protein, we found that all 13 tumors with both receptor levels above these discriminatory values regressed after ovariectomy. Ovariectomy had no effect on the growth of the three tumors, with both receptor levels below these discriminatory values. One of the tumors showing a progesterone receptor level above and an estradiol receptor level below these discriminatory levels regressed after ovariectomy. On the other hand, two tumors with an estradiol receptor level above and a progesterone receptor below these discriminatory levels regressed after endocrine ablation.

Effect of Ovariectomy on Progesterone and Estradiol Receptor Levels of Hormone-Dependent and Hormone-Independent Tumors

The progesterone receptor levels of hormone-dependent and hormone-independent tumors, which were analyzed before as well as after ovariectomy, are shown in Fig. 6. Ovariectomy strikingly decreased the progesterone receptor levels in both hormone-dependent and hormone-independent tumors. In contrast to preovariectomy values, after ovariectomy the proges-

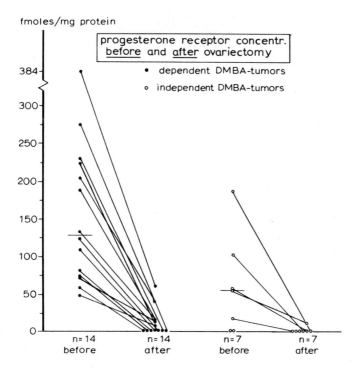

FIG. 6. Effect of ovariectomy on progesterone receptor levels of hormone-dependent (●) and hormone-independent (○) DMBA-induced mammary tumors. Median values are indicated by horizontal lines.

terone receptor values of hormone-dependent and hormone-independent tumors were not different. It seems worth mentioning that the highest three postovariectomy values of the hormone-dependent tumors were obtained in biopsies taken on days 3, 4, and 7 after ovariectomy.

Although ovariectomy significantly decreased ($p < 0.01$) the estradiol receptor levels of the hormone-dependent tumors, this effect was less consistent than the effect of ovariectomy on the progesterone receptor levels (Fig. 7). In six out of seven hormone-independent tumors, the estradiol receptor levels were slightly higher after ovariectomy. This number of observations was, however, too small to reach statistical significance. From the data of Fig. 7 it appears that postovariectomy estradiol receptor levels are definitely less discriminatory between hormone-dependent and hormone-independent tumors than the estradiol receptor levels before ovariectomy.

Progesterone and Estradiol Receptor Levels in Hormone-Dependent Tumors after Estradiol Administration to Ovariectomized Animals

After assessing hormone dependency by ovariectomy, we administered estradiol benzoate for varying periods. Figure 8 shows the progesterone

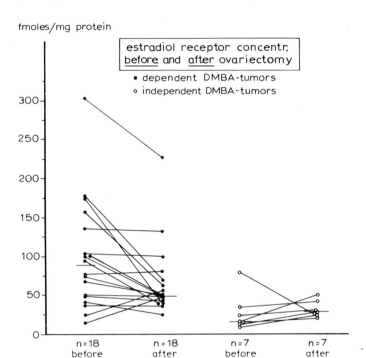

FIG. 7. Effect of ovariectomy on estradiol receptor levels of hormone-dependent (●) and hormone-independent (○) DMBA-induced mammary tumors. Median values are indicated by horizontal lines.

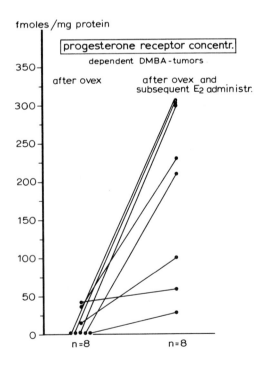

FIG. 8. Effect of estradiol administration to castrated animals on progesterone receptor levels of hormone-dependent DMBA-induced mammary tumors.

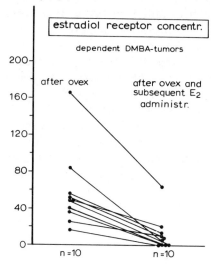

FIG. 9. Effect of estradiol administration to castrated animals on estradiol receptor levels of hormone-dependent DMBA-induced mammary tumors.

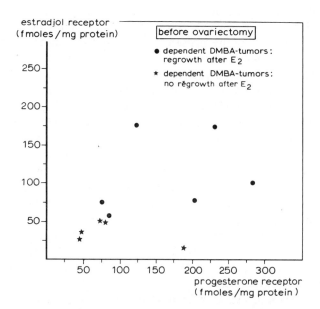

FIG. 10. Correlation of preovariectomy progesterone receptor and estradiol receptor levels with tumor response to estradiol administration to castrated animals. ●, tumors regressing after ovariectomy and regrowing after additional estradiol administration. ★, tumors regressing after ovariectomy and failing to regrow after additional estradiol administration.

receptor levels obtained before and after estradiol administration. The progesterone receptor levels all increased after estradiol treatment. It is worth mentioning that progesterone receptor concentrations increase in tumors with and without tumor regrowth after estradiol treatment.

In contrast to the progesterone receptor levels, this treatment decreased all estradiol receptor levels to very low or undetectable levels (Fig. 9). These diminished estradiol receptor levels might be explained by the relatively high dose of estradiol benzoate that was administered in this study. This could have caused occupation of cytoplasmic receptor sites by exogenous estradiol and/or translocation of estradiol receptor complexes to the nucleus. Interestingly, estradiol administration induced tumor regrowth only in those tumors which possessed high estradiol receptor levels before ovariectomy (Fig. 10). As is shown in Fig. 10, one tumor with a low estradiol receptor level (16 fmoles/mg protein) and a high progesterone receptor level (187 fmoles/mg protein) did not show regrowth of tumor tissue after estradiol administration.

DISCUSSION

Recently DeSombre et al. (4) reported that essentially all DMBA-induced rat mammary tumors contained some specific estrogen receptor, suggesting that previous considerations of hormone dependence on the basis of presence or absence of hormone receptors may be oversimplified. This observation was confirmed in the present investigation. Hormone-dependent as well as hormone-independent tumors showed estradiol receptor activity. The preovariectomy estradiol receptor levels of the hormone-dependent group, however, were significantly higher than the estradiol receptor levels of the hormone-independent group. With one exception, hormone-independent tumors contained markedly low estradiol receptor levels. Vignon and Rochefort (17) reported that the estrogen receptor concentrations decreased within the first 10 days after ovariectomy in "ovarian-dependent" rat mammary tumors. In the present study ovariectomy decreased the estradiol receptor levels in the hormone-dependent group, whereas the increase observed in the estradiol receptor in the hormone-independent group did not reach statistical significance. As a consequence, postovariectomy receptor levels are less suitable than preovariectomy receptor levels to discriminate between hormone-dependent and hormone-independent tumors. Among the hormone-dependent tumors a small number showed low preovariectomy receptor values indicating that there is no absolute correlation between estrogen binding capacity of a tumor and its growth response to ovariectomy as was also observed by Boylan and Wittliff (2) and Mobbs and Johnson (15).

Normal breast tissue is dependent on progesterone for growth and development, and it has been found that estradiol induces the synthesis of the receptor for progesterone in other target tissue. Therefore, a number of investigators (1,6) questioned whether measurement of the progesterone receptor could provide additional information for the prediction of the hormone dependency of mammary carcinoma. As appears from the results reported here, all hormone-dependent tumors contained specific progesterone receptor binding. Five out of seven hormone-independent tumors, however, also showed specific binding for progesterone, suggesting that even in these tumors the estradiol-mediated progesterone receptor synthesis is active. In accordance with this concept is the observation that not only in hormone-dependent but also in hormone-independent tumors ovariectomy caused a dramatic fall in progesterone receptor levels. This fall in receptor activity can be ascribed to the decrease of the plasma level of estradiol occurring after ovariectomy. This conclusion may be drawn from the fact that estradiol administration after ovariectomy induced a marked increase in the progesterone receptor levels in all hormone-dependent tumors analyzed. It is worth mentioning that this consistent increase in the progesterone receptor levels coincided in only a minority of the tumors with regrowth of tumor tissue. In other words, some tumors show estradiol-dependent progesterone receptor synthesis but nevertheless do not show tumor regrowth after estradiol administration. Some authors (11,15) consider a tumor as being unequivocally hormone dependent only when regression of the tumor occurs upon ovariectomy and a regrowth occurs when estradiol is administered subsequently. When such criteria were used in the present investigation, the unequivocal hormone-dependent tumors, which show regrowth, all had relatively high estradiol receptor levels (range 58 to 176 fmoles/mg protein) and relatively high progesterone receptor levels (range 74 to 283 fmoles/mg protein).

Interestingly, a significant positive correlation was found between preovariectomy progesterone receptor levels and preovariectomy estradiol receptor levels in the group of hormone-dependent tumors, whereas no such correlation was found in the group of hormone-independent tumors. From these observations one might argue that although the progesterone receptor–synthesizing system is active in these hormone-independent tumors, there is some disturbance in the regulation of this system. More extensive studies are needed to further evaluate this concept.

In the present study it was found that even when both estradiol as well as progesterone receptor levels were used as indicators for hormone dependency, hormone-dependent and hormone-independent tumors could not be discriminated completely. Nevertheless, in the hormone-independent group three out of seven tumors showed clearly lower progesterone receptor levels than those found in the hormone-dependent group. Thus, measuring progesterone receptor levels along with estradiol receptor levels provides additional although not definitive information with regard to hormonal dependency of the growth of DMBA-induced mammary carcinoma.

REFERENCES

1. Asselin, J., Labrie, F., Kelly, P. A., Philibert, D., and Raynaud, J. P. (1976): Specific progesterone receptors in dimethylbenzanthracene (DMBA)-induced mammary tumors. *Steroids*, 27:395-404.
2. Boylan, E. S., and Wittliff, J. L. (1975): Specific estrogen binding in rat mammary tumors induced by 7,12-dimethylbenz(a)anthracene. *Cancer Res.*, 35:506-511.
3. Chamness, G. C., and McGuire, W. L. (1975): Scatchard plots: common errors in correction and interpretation. *Steroids*, 26:538-542.
4. DeSombre, E. R., Kledzik, G., Marshall, S., and Meites, J. (1976): Estrogen and prolactin receptor concentrations in rat mammary tumors and response to endocrine ablation. *Cancer Res.*, 36:354-358.
5. Englesman, E., Korsten, C. B., Persijn, J. P., and Cleton, F. J. (1973): Human breast cancer and estrogen receptor. *Arch. Chir. Neerl.*, 25:393-397.
6. Horwitz, K. B., and McGuire, W. L. (1975): Specific progesterone receptors in human breast cancer. *Steroids*, 25:497-505.
7. Jänne, O., Kontula, K., Luukkainen, T., and Vihko, R. (1975): Oestrogen-induced progesterone receptor in human uterus. *J. Steroid Biochem.*, 6:501-509.
8. Jensen, E. V. (1975): Estrogen receptors in hormone-dependent breast cancers. *Cancer Res.*, 35:3362-3364.
9. Katzenellenbogen, J. A., Johnson, H. J., and Carlson, K. E. (1973): Studies on the uterine, cytoplasmic estrogen binding protein. Thermal stability and ligand dissociation rate. An assay of empty and filled sites by exchange. *Biochemistry*, 12:4092-4099.
10. Lowry, D. H., Rosebrough, N. J., Farr, A. L., and Randall, R. J. (1951): Protein measurement with the Folin phenol reagent. *J. Biol. Chem.*, 193:265-275.
11. McGuire, W. L., and Julian, J. A. (1971): Comparison of macromolecular binding of estradiol in hormone-dependent and hormone-independent rat mammary carcinoma. *Cancer Res.*, 31:1440-1445.
12. McGuire, W. L., Carbone, P. P., and Volmer, E. P. (Eds.) (1975): *Estrogen Receptors in Human Breast Cancer*. Raven Press, New York.
13. McGuire, W. L., Chamness, G. C., Costlow, M. E., and Shepherd, R. E. (1975): Hormone dependence in breast cancer. *Metabolism*, 23:75-100.
14. Milgrom, E., Luu Thi, M., Atger, M., and Baulieu, E. E. (1973): Mechanisms regulating the concentration and the conformation of progesterone receptor(s) in the uterus. *J. Biol. Chem.*, 248:6366-6374.
15. Mobbs, B. G., and Johnson, I. E. (1974): Estrogen-binding in vitro by DMBA-induced rat mammary tumors: its relationship to hormone responsiveness. *Eur. J. Cancer*, 10:757-763.
16. Persijn, J. P., Korsten, C. B., and Engelsman, E. (1975): Oestrogen and androgen receptors in breast cancer and response to endocrine therapy. *Br. Med. J.*, 4:503.
17. Vignon, F., and Rochefort, H. (1976): Regulation of estrogen receptors in ovarian-dependent rat mammary tumors. I. Effects of castration and prolactin. *Endocrinology*, 98:722-729.
18. Wagner, R. K., and Jungblut, P. W. (1976): Oestradiol- and dihydrotestosterone receptors in normal and neoplastic human mammary tissue. *Acta Endocrinol. (Kbh.)*, 82:105-120.

Progesterone Receptors in Normal and Neoplastic Tissues, edited by W. L. McGuire et al. Raven Press, New York © 1977.

Regulation of Hormone Receptor Levels and Growth of DMBA-Induced Mammary Tumors by RU16117 and Other Steroids in the Rat

P. A. Kelly, J. Asselin, F. Labrie, and *J. P. Raynaud

*Medical Research Council Group in Molecular Endocrinology, Le Centre Hospitalier de l'Université Laval, Quebec G1V 4G2, Canada; and *Centre de Recherches Roussel-Uclaf, Romainville, France 93230*

Approximately 35% of patients with breast cancer experience remission of their disease after endocrine ablative surgery or steroid therapy (16,26). It thus appears important to gain as much knowledge as possible about the mechanisms controlling the hormonal dependency of these cancers. The mammary carcinoma induced in the rat by dimethylbenzanthracene (14) has been the most widely accepted model of hormone-dependent breast cancer. Estrogens and prolactin have been shown to be important for the development and growth of these mammary tumors. In fact, procedures that reduce circulating levels of prolactin (hypophysectomy, ergot drugs) have been shown to reduce the number and size of these tumors (4,10,31). Moreover, tumor growth can be reinitiated in ovariectomized animals by treatment with estrogens or prolactin (25,27,33).

Progesterone has also been shown to stimulate DMBA tumor growth (15,15a). It is therefore important when evaluating responses in tumor model systems to measure hormone receptor levels for estradiol, progesterone, and prolactin.

In addition, we have recently found that RU16117 (11α-methoxyethinyl estradiol) has more potent antiestrogenic properties than the compounds previously available (8,32). It thus seemed of interest to study the effect of RU16117 on the development and growth of DMBA-induced mammary tumors in the rat. Finally, the hormonal mechanisms controlling estrogen, progesterone, and prolactin receptor levels in DMBA-induced tumors were examined.

SPECIFIC PROGESTERONE RECEPTORS

It became readily apparent that a simple and reliable method to predict the hormone responsiveness of mammary tumors was needed. A major advance in predicting hormonal dependency in experimental as well as human mammary carcinoma has been the measurement of estradiol receptor levels. In fact, in humans approximately two-thirds of the tumors containing a significant level of estrogen receptor regress after endocrine therapy, whereas only 5% of the tumors containing no measurable hormone receptors for estrogen have a positive response (16,26).

Receptors for prolactin have also been identified in rat and human tumors (6,7,11,20,35), and the growth response of DMBA tumors to injected prolactin has been shown to be correlated with the amount of prolactin receptors in the tumors (20).

The identification of progesterone receptors in target tissues had been difficult because [^{3}H]progesterone binds to corticosteroid binding globulin (CBG) and also to low-affinity, high-capacity binding sites. The synthetic progestin R5020 (17,21-dimethyl-19-nor-4,9-pregnadiene-3,20-dione) has been shown to bind to the progesterone receptor in the uterus with high affinity (28,32). Horwitz and McGuire (12) have identified a progesterone receptor in human breast cancer using R5020, and the hypothesis has been proposed that the simultaneous presence of receptors for estradiol and progesterone in tumor tissue increases the predictive value of receptor measurements (12).

As a preliminary step to our studies on hormonal control of the development and growth of DMBA-induced mammary carcinoma, we have studied the characteristics of binding of [^{3}H]R5020 to a specific progesterone receptor in DMBA-induced mammary tumors (2).

As illustrated in Fig. 1A, when the tumor cytosol incubated with [^{3}H]R5020 is analyzed by sucrose gradient without preliminary adsorption with dextran-coated charcoal (DCC), the radioactivity migrates with two components at 7 to 8S and 4S, respectively. Upon addition of an excess of the unlabeled steroid, the 7 to 8S peak disappears whereas the radioactivity associated with plasma proteins migrates at 4S. This binding to plasma proteins is, however, of relatively low affinity and is completely eliminated by treatment with DCC in the standard binding assay.

When the cytosol incubated with [^{3}H]R5020 is treated with DCC before sucrose gradient analysis, [^{3}H]R5020 is also bound to two components migrating at 7 to 8S and 4S. However, when a 100-fold molar excess of radioinert R5020 is added, the 7 to 8S peak always disappears completely whereas the 4S peak is usually decreased to 15 to 35% of the control level (Fig. 1B). [^{3}H]estradiol binding migrates at a slightly faster rate than [^{3}H]R5020 binding.

As illustrated in Fig. 2, R5020 is approximately four- to fivefold more potent than progesterone in displacing [^{3}H]R5020 binding (measured by the charcoal assay), whereas norgestrel is slightly more competitive than R5020. Estradiol-17β has approximately 10% the activity of progesterone, but 5α-dihydrotestosterone, testosterone, and diethylstilbestrol have appreciably lower activities.

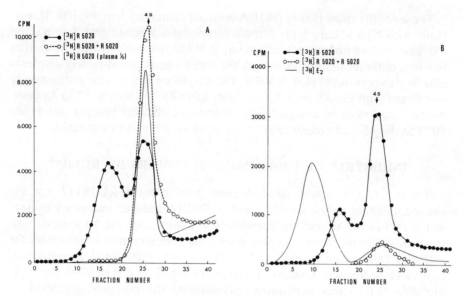

FIG. 1. Representative sedimentation patterns of radioactivity after incubation of mammary tumor cytosol with 1×10^{-8} M [^{3}H]R5020 in the presence (○--○) or absence (●——●) of 1×10^{-6} M unlabeled steroid. The cytosol from tumor illustrated in **B** was treated with dextran-coated charcoal before sucrose gradient, whereas in **A** no adsorption with DCC was performed. As reference, binding of [^{3}H]R5020 to plasma proteins (——) and [^{3}H]estradiol-17β (——) to tumor cytosol are indicated in **A** and **B**, respectively. The incubations were performed for 2 hr at 0 to 4°C (2).

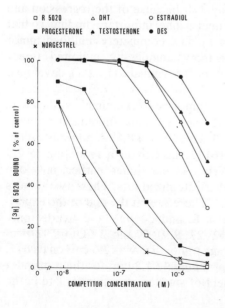

FIG. 2. Effect of increasing concentrations of unlabeled R5020, progesterone, norgestrel, estradiol-17β, testosterone, 5α-dihydrotestosterone, and diethylstilbestrol on the binding of 1×10^{-8} M [^{3}H]R5020 to mammary tumor cytosol. After incubation at 0 to 4°C for 16 hr, the bound fraction was separated by dextran-coated charcoal (2).

These results show that in DMBA-induced mammary tumors, [^{3}H]R5020 binds with high affinity with cytosol components migrating at 7 to 8S and 4S. The relative importance of binding to these two peaks varies somewhat between different tumors. Although the 7 to 8S peak is completely displaceable by excess unlabeled R5020, the displacement of the radioactivity associated with the 4S peak varies from 65 to 85% Thus, the 7 to 8S component appears to be completely progestin specific, but binding of [^{3}H]R-5020 to the 4S component appears to be more difficult to saturate.

INHIBITION OF DMBA-INDUCED TUMORS BY RU16117

Because of the potent antiestrogenic properties of RU16117, we examined its effect on the development of DMBA-induced mammary tumors in the rat. In an attempt to correlate the tumor response to antiestrogen treatment with hormone receptor levels, the concentration of receptors for estradiol-17β, progesterone, and prolactin was determined in individual tumors (19). Specific binding of [^{3}H]estradiol (E$_2$), [^{3}H]R5020, and ^{125}I-prolactin (PRL) was performed according to the methods described by Asselin et al. (2) and Kelly et al. (21).

As illustrated in Fig. 3A, tumors first appeared in control rats 53 days after DMBA administration, and the incidence of tumors increased to a maximum of 94.1% at 130 days. All treatments resulted in a delayed onset of tumor appearance. After a delay of 10 days, RU16117, at a dose of 0.5 µg/day, resulted in a curve similar to that of controls, although the incidence reached only 78.6%. When RU16117 was injected at a dose of 2 µg/day, the maximal incidence was 60% between days 99 and 106, after which the incidence fell to a value of 40% at day 130 because of the regression and disappearance of lesions in some animals. The important finding is that RU16117, at doses of either 8 or 24 µg/day, completely inhibited tumor development in all animals. Ovariectomy completely inhibited tumor appearance until day 95 when 2 out of 14 animals (14.2%) developed palpable tumors.

The average number of tumors per tumor-bearing animal is shown in Fig. 3B. Once again, control animals showed a gradual increment reaching a maximum of 3.3 tumors per rat 130 days after DMBA administration. At doses of 0.5 and 2 µg/day, RU16117 resulted in a reduction to 2.5 and 1.5 tumors per animal, respectively. As mentioned earlier, only one tumor developed in each of the two ovariectomized rats. These two tumors became, however, quite large (Fig. 3C). Therefore, at the end of the experiment in the control, RU16117 (0.5, 2, 8, and 24 µg), and ovariectomy (OVX) groups, there was a total of 52, 27, 9, 0, 0, and 2 tumors, respectively. In control rats, the largest average tumor size was 2.6 cm^2 on day 71, after which the values declined to approximately 1.2 cm^2 for the remainder of the study, reflecting the development of smaller, new tumors late in the

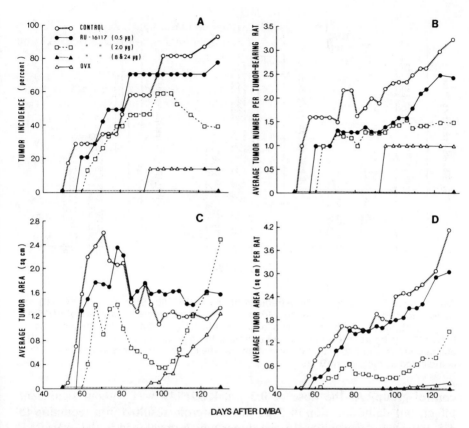

FIG. 3. Effect of treatment with increasing doses of RU16117 or ovariectomy (OVX) on the development of DMBA-induced mammary tumors. Injections began the day DMBA was administered and continued for the next 130 days. Ovariectomy was performed the day following DMBA administration. Animals were examined twice weekly for the presence of tumors, and when they were present, tumor area (length × width) was measured. **A:** Tumor incidence as a function of time after DMBA. **B:** Average tumor number per tumor-bearing rat. **C:** Average tumor area (cm²). **D:** Average tumor area (cm²) per rat (19).

study. Treatment with 0.5 µg RU16117 had no significant effect on tumor size, whereas rats receiving 2 µg RU16117 showed a more complex pattern of tumor growth. The overall effect of the treatment is most clearly shown in Fig. 3D in which there was a steady increase in the average tumor size per rat to 4.4 cm²/rat on day 130. In animals treated with 0.5 or 2 µg RU16117, the tumor averaged 0.8 cm²/rat in OVX animals.

Specific binding of [^{3}H]estradiol, [^{3}H]R5020, and [^{125}I]ovine PRL (oPRL) to DMBA-induced mammary tumors is shown in Fig. 4A, B, and C, respectively. Binding of [^{3}H]E$_2$ was 5.1 ± 1.0 pmoles/g tissue in tumors from control animals, whereas after daily treatment with 0.5 and 2.0 µg RU16117, reductions of binding to 2.7 ± 0.3 and 1.9 ± 0.6 pmoles/g tissue were observed. Binding of [^{3}H]R5020 was 8.9 ± 1.5 pmoles/g tissue in the

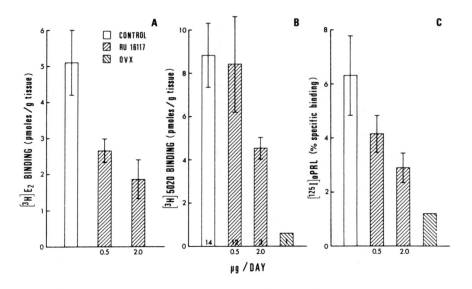

FIG. 4. Specific binding of [^{3}H]estradiol-17β **(A)** [^{3}H]R5020 **(B)** and [^{125}I]oPRL **(C)** to DMBA-induced rat mammary tumors from rats treated with increasing doses of RU16117 or ovariectomized. Rats were given a single injection of DMBA, and treatment with RU16117 was begun the same day. All tumors with an area larger than 1 cm^2 on the day animals were sacrificed (day 130) were removed. The number of tumors assayed per group is indicated in the bars in **B** (19).

control group. At the dose of 0.5 μg, RU16117 was without significant effect, but daily injection of 2 μg of the steroid resulted in a reduction to 4.5 ± 0.5 pmoles/g tissue. In one tumor which developed in the ovariectomized group, the level of progesterone receptors was very low at 0.6 pmoles/g tissue. RU16117 treatment caused a reduction of [^{125}I]oPRL to tumor plasma membranes from 6.3 ± 1.5% in control rats to 2.9 ± 0.6% in tumors from rats receiving 2 μg RU16117 per day (Fig. 4C). In the one tumor from an ovariectomized rat, there was low binding of [^{125}I]oPRL (1.2%). Specific binding of [^{125}I]ovine growth hormone was constant at 0.2 to 0.4% in all treatment groups (*data not shown*).

Plasma prolactin levels were slightly reduced by the 2- and 8-μg doses of RU16117 and increased by the highest dose (24 μg) of antiestrogen from 21 ± 7 to 41 ± 10 ng/ml. Ovariectomy decreased the plasma PRL concentration, whereas treatment with RU16117 led to a progressive decrease in plasma LH levels.

These data show that the new antiestrogenic compound RU16117 at the relatively low doses of 8 and 24 μg/day is capable of completely preventing the appearance of mammary tumors after DMBA administration. This compound has weak estrogenic activity in the mouse (Rubin test—1/100 of estradiol-17β) and castrated rat (Allen Doisy test—1/20 of estradiol-17β) and competes for uterine estradiol cytosol receptor in mouse (1/20 of estradiol-17β) and rat (1/10 of estradiol-17β) (32).

Although the two highest doses of RU16117 led to complete inhibition of tumor development, the two lower doses also had significant effects. The net effect is best illustrated in Fig. 3D where it can be seen that treatment with 0.5 or 2.0 µg RU16117 leads to inhibition of average tumor size at day 130 to respectively 75 and 37% of the control values. This important effect results from a lower incidence and smaller tumors after treatment with low doses of the antiestrogen.

The fact that estrogens can have a dual effect on mammary carcinoma induced by DMBA is well known. Administration of estrogens stimulates tumor development in ovariectomized animals (25), whereas in animals with developed tumors, large doses of estrogens can lead to tumor regression (13) or inhibition of development (23).

That the effect of RU16117 is not due to its low estrogenic activity is indicated by the finding that doses of estradiol-17β equivalent to the estrogenic activity of the doses of RU16117 used have no marked inhibitory effect on the development of DMBA-induced tumors (Fig. 5) and a stimu-

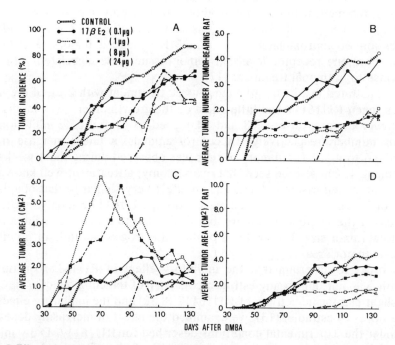

FIG. 5. Effect of treatment with increasing doses of E_2 (0.1, 1, 8, and 24 µg) on the development of DMBA-induced tumors. Daily injections beginning the day after DMBA administration were continued for 133 days. Animals were examined weekly for the presence of tumors. For description, see legend of Fig. 3.

latory effect on the size of tumors. Moreover, in this same system, high doses of estradiol-17β administered from the time of DMBA administration result in a delayed appearance of tumors with approximately one-half of the animals still developing tumors, and the tumors that are present are at least as large as those of the controls. RU16117, on the other hand, led to complete inhibition up to at least 130 days after DMBA administration (Fig. 3).

The present data clearly show that treatment with 2 μg RU16117 led to a 40 to 60% reduction of the levels of receptors for estradiol, progesterone, and PRL in the mammary tumors. Levels of progesterone and PRL receptors were reduced to 10 to 15% of control after ovariectomy, a finding confirmed on a larger scale in subsequent experiments (2).

REGRESSION OF DMBA-INDUCED TUMORS BY RU16117

Since the new antiestrogen RU16117, at relatively low doses (8 or 24 μg daily), completely prevented the development of rat mammary carcinoma when administered from the day after DMBA was administered (19), it was felt of interest to study the effect of this compound on the growth of DMBA tumors which were already developed and to compare the effect of such treatment with that of castration. Ovariectomy, a procedure leading to decreased levels of both estrogens and prolactin, is known to induce regression of approximately 90% of DMBA-induced tumors (31). Once again, hormone receptor levels in tumor tissue were correlated with the response to hormonal treatment (17).

As illustrated in Fig. 6, although 4-week treatment with 2 μg of the new antiestrogen RU16117 had little effect on the growth of already established mammary tumors, doses of 8 and 24 μg led to 45 and 65% inhibition of tumor number, respectively. In control animals, a linear increase from 3.2 ± 0.6 to 4.5 ± 0.7 tumors per rat was observed during the 4 weeks of treatment. It can also be seen that ovariectomy, a treatment well known to cause tumor regression (13,31), had an effect very similar to that of a daily dose of 24 μg RU16117. At the highest dose, RU16117 not only markedly decreased the number of tumors, but it also led to a marked reduction of the total tumor size. Lower doses of the antiestrogen had little or no effect on total tumor area.

In order to ascertain that the inhibitory effect of RU16117 on tumor growth was not due to any estrogenic activity of the compound in view of the slight estrogenic activity of RU16117 (8,32) and the inhibitory effect of large doses of estradiol (13), we examined the effect of increasing doses of E_2 under the experimental conditions described for RU16117. Daily injection of 0.1, 0.5, 2.5, or 12.5 μg E_2 had no significant effect on the number of tumors. Although the total tumor area was decreased from 4.0 ± 0.4 to 0.6 ± 0.2 cm²/rat after castration, the two low doses of E_2 induced some-

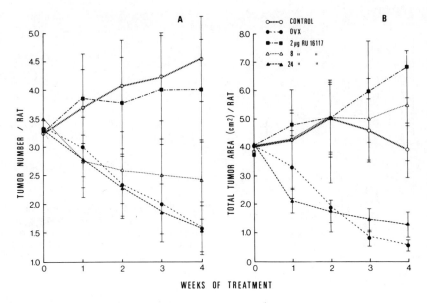

FIG. 6. Effect of 4-week treatment with 2, 8, or 24 μg RU16117 daily or ovariectomy on the number of DMBA-induced mammary tumors per rat **(A)** and total number area per animal **(B)**. Treatment was started approximately 4 months after DMBA administration (17).

what larger tumors, whereas the two larger doses resulted in similar or slightly smaller tumor size (*data not shown*).

The effect of 4-week treatment with E_2, RU16117, or ovariectomy on specific binding of $[^3H]E_2$, $[^3H]R5020$, and $[^{125}I]oPRL$ to mammary tumors is illustrated in Fig. 7. Binding of $[^3H]E_2$ and $[^{125}I]oPRL$ was lowered only with the highest dose of RU16117 and after ovariectomy.

The level of progestin receptors was not significantly affected at any of the doses of E_2 used. The two lower doses of RU16117 were without effect, but the 24-μg dose was also apparently ineffective. However, if two tumors having high values of 16.0 and 14.8 pmoles/g tissue were eliminated from the group, the level of $[^3H]R5020$ receptor would be reduced to 2.6 ± 1.5 pmoles/g tissue after treatment with 24 μg RU16117. The level of this receptor was reduced to 0.1 ± 0.1 pmole/g tissue after ovariectomy.

The data show that at the daily dose of 24 μg, RU16117 is as efficient as ovariectomy to inhibit tumor growth in rats bearing DMBA-induced mammary tumors. In fact, after 4 weeks of treatment (Fig. 6), the average number of tumors per rat and tumor size are reduced to approximately 30% of control.

These findings of low levels of hormone receptors after ovariectomy or treatment with 24 μg RU16117 may indicate that tumors unresponsive to hormonal treatment are those with low levels of receptors or that ovariec-

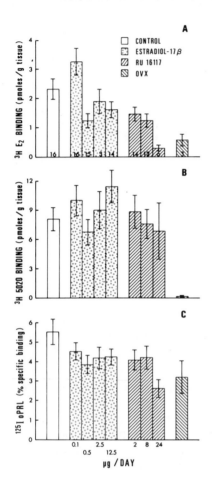

FIG. 7. Effect of 4-week treatment with 0.1, 0.5, 2.5, or 12.5 µg of estradiol-17β; 2, 8, or 24 µg of RU16117; or ovariectomy on the concentration of receptors for [^{3}H]estradiol-17β **(A)**, [^{3}H]R5020 **(B)**, or [^{125}I]oPRL **(C)** (17).

tomy or RU16117 treatment causes a reduction of receptor levels. A possible mechanism of action of RU16117 in the tumor tissue could be a decrease of the hormone receptor level leading to relative unresponsiveness of the tissue to its hormonal environment.

Since we have found that RU16117 inhibits LH secretion (8,19) and treatment with the 24-µg dose inhibits tumor growth in the presence of increased plasma prolactin levels, it is likely that RU16117 exerts its inhibitory activity through an action at both the hypothalamic-pituitary and tumor levels.

STIMULATION OF DMBA-INDUCED TUMORS BY PROGESTERONE

Progesterone has been reported to be both stimulatory (15,15a) and inhibitory (23) on the development of DMBA-induced mammary tumors. We therefore studied the effects on tumor development of progesterone (PROG)

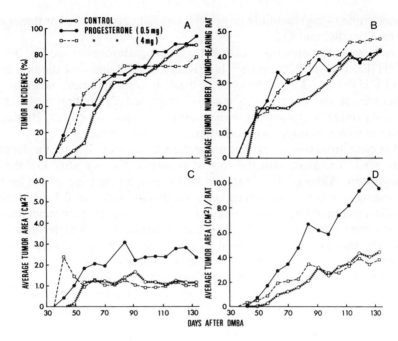

FIG. 8. Effect of treatment with 0.5 and 4 mg progesterone on the development of DMBA-induced mammary tumors (18).

at doses of 0.5 or 4 mg injected daily from the day after DMBA was administered (18).

As illustrated in Fig. 8A, treatment with both doses of progesterone resulted in an earlier appearance of tumors with tumors first appearing 42 days after DMBA injection. The incidence remained higher in the progesterone-treated groups during the first 90 days of the experiment to reach maximal values similar to those of controls.

Although progesterone increased the number of tumors per tumor-bearing animal most noticeably in the midportion of the experiment, maximum values similar to controls were seen at day 133. In fact, 4.3 ± 0.8 and 4.7 ± 1.0 tumors were seen for the two doses of progesterone, respectively (Fig. 8B).

In control rats, the largest average tumor size reached was 1.6 ± 0.2 cm^2 on day 91 (Fig. 8C). The value declined slightly thereafter but remained relatively stable for the rest of the study. Progesterone at a dose of 0.5 mg caused a marked increase in the average tumor area throughout the experiment with the final value at day 133 being 2.4 ± 0.5 cm^2. The stimulatory effect of progesterone is even more evident if the tumor area per rat is examined (Fig. 8D). The lower dose of progesterone caused a very striking increase in total tumor area per rat, reaching 9.6 ± 1.9 cm^2 compared to 4.5 ± 0.6 cm^2 in the control group 133 days after DMBA. The higher dose of progesterone (4 mg) had little or no effect on either tumor area or tumor area per rat (Fig. 8C and D).

There was little effect of treatment with progesterone on specific binding of $[^3H]E_2$, $[^3H]R5020$, and $[^{125}I]oPRL$ to DMBA-induced tumors. Binding of $[^3H]E_2$ was 2.9 ± 0.3 pmoles/g tissue in control tumors, whereas progesterone, at the dose of 0.5 mg/day, increased E_2 binding to 4.6 ± 0.3 pmoles/g tissue by treatment with 4 mg of the steroid per day. Binding of prolactin to the tumors was unaffected by the treatment.

It is clear from this study that progesterone can stimulate tumor development when treatment with the steroid is started the day after DMBA administration. Although it is possible that the small increase of the level of the estrogen receptor observed after treatment with the 0.5-mg dose of progesterone might be partly involved in the increased tumor incidence and growth, the concomitant inhibitory effect of this dose of protesterone on LH secretion with probable decreased ovarian function make it likely that progesterone exerts its stimulatory action by a direct interaction with its receptor in the tumor (2). It seems unlikely that progesterone is acting via an effect on prolactin secretion or on the level of prolactin receptor since both parameters remained unchanged after progesterone treatment.

CONTROL OF HORMONE RECEPTORS IN DMBA-INDUCED TUMORS

As already described in the foregoing sections, E_2, progesterone, and prolactin play an important role in the control of DMBA-induced mammary tumors. One of the difficulties with the DMBA model is that not all of the tumors are hormone dependent, and the hormone dependency may change with the age of the tumors (9). In addition, established tumors may undergo a spontaneous regression independent of the hormonal state of the animal (37). In fact, Bradley et al. (3) have proposed that although most DMBA tumors are dependent on E_2 and PRL, OVX or E_2 replacement alone may not accurately reflect estrogen dependency since prolactin secretion is also altered by estrogens.

Previous studies on the control of the various receptors in target tissues have shown that estrogen receptors are increased by estrogen or prolactin (24,36). Prolactin has also been shown to stimulate the estrogen receptor in rat liver (5), whereas the prolactin receptor in rat liver has been shown to be dependent on both estrogens (21,22,29) and prolactin (30). It has also been well demonstrated in the uterus that estrogens stimulate the progesterone receptor (34).

Experiments were undertaken to compare the effects of E_2, progesterone, and PRL given alone or in combination on the growth of hormone-dependent tumors (those which regressed after OVX) and on the levels of estrogen, progesterone, and prolactin receptors. Figure 9 shows the size of five to six

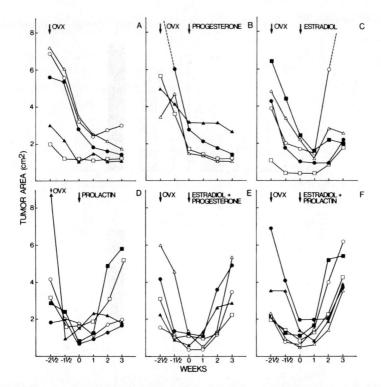

FIG. 9. Tumor growth (area, cm²) following ovariectomy (OVX). At 2.5 weeks after OVX animals received daily injections of PROG (0.5 mg), E_2 (0.5 µg), or PRL (2 mg), or combinations of E_2 + PROG or E_2 + PRL (1).

representative tumors following the mentioned treatments. Tumors from intact (not OVX) rats continued to grow (*not shown*). Following OVX, there was a rapid decline in the tumor area in all groups. Progesterone treatment alone had no effect on tumor growth (Fig. 9B). However, treatment with E_2 or PRL alone increased tumor size significantly (Fig. 9C and D). Combined E_2 and PROG, or E_2 and PRL increased tumor size markedly with some tumors reaching or exceeding the size measured before castration (Fig. 9E and F).

The levels of E_2, PROG, and PRL receptors are shown in Fig. 10. It shows that following castration, the levels of all three receptors are significantly reduced. Progesterone alone did not affect receptor levels, whereas estradiol injections increased significantly the level of all three receptors. Prolactin increased E_2 receptor concentration and caused a modest but nonsignificant increase in tumor prolactin receptors. Combined treatment of estradiol with either PROG or PRL increased the receptor levels for the three hormones to values similar to those observed with E_2 alone or in control tumors.

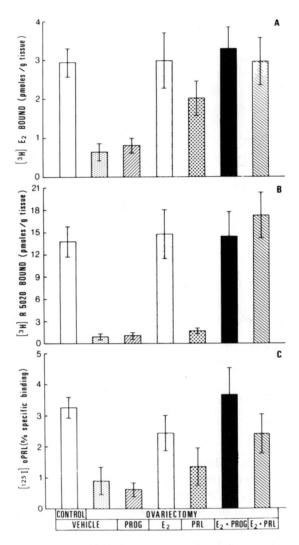

FIG. 10. Specific hormone receptors for estradiol, progesterone, and prolactin in the groups described in legend to Fig. 9 (1).

Recently, DeSombre et al. (7) have proposed that a better correlation of tumor response to endocrine ablation resulted from a combination of estrogen and prolactin receptor levels than from either receptor concentration alone in the DMBA mammary tumors of the rat. It was of particular interest to compare the effect of hormones on tumor growth and changes of receptor levels. In the present study, it was clearly demonstrated that the combined treatment of estradiol with progesterone or prolactin resulted in enhanced tumor growth. The levels of receptors for estrogen, progesterone, and prolactin were increased to control values by these treatments, thus suggesting a correlation between the effect of hormones on the level of hormone receptors and tumor growth (24,36).

These data confirm that the progesterone receptor in mammary tumors is dependent on estrogens. Moreover, a positive correlation between the estrogen and the progesterone receptor levels in control tumors was observed ($r = 0.65$, $p < 0.01$). In addition, estrogens increased progesterone receptor levels. This demonstrates that an estrogenic control of the progesterone receptor as observed previously in the uterus (34) is also existent in the DMBA mammary tumor. However, in the mammary tumors, the prolactin receptor is probably also implicated in the control of progesterone receptor since the estrogen receptor seems to be dependent on prolactin, whereas in the uterus an effect of prolactin on the estrogen receptor has not been demonstrated.

REFERENCES

1. Asselin, J., Kelly, P. A., Labrie, F., Auclair, G., and Leblanc, G. (1977): Control of estrogen, progesterone and prolactin receptor levels in dimethylbenzanthracene (DMBA)-induced mammary tumors of the rat. (*Submitted for publication*).
2. Asselin, J., Labrie, F., Kelly, P. A., Philibert, D., and Raynaud, J. P. (1976): Specific progesterone receptors in dimethylbenzanthracene (DMBA)-induced mammary tumors. *Steroids*, 27:395–404.
3. Bradley, C. L., Kledzick, G. S., and Meites, J. (1976): Prolactin and estrogen dependency of rat mammary cancers at early and late stages of development. *Cancer Res.*, 36:319–324.
4. Cassell, E., Meites, J., and Welsch, C. W. (1971): Effects of ergocornine and ergocryptine on growth of 7,12-dimethylbenzanthracene-induced mammary tumors in rats. *Cancer Res.*, 31:1051–1053.
5. Chamness, G. C., Costlow, M. E., and McGuire, W. L. (1975): Estrogen receptor in rat liver and its dependence on prolactin. *Steroids*, 26:363–371.
6. Costlow, M. E., Buschow, R. A., and McGuire, W. L. (1974): Prolactin receptors in an estrogen receptor-deficient mammary carcinoma. *Science*, 184:85–86.
7. De Sombre, E. R., Kledzik, G. S., Marshall, S., and Meites, J. (1976): Estrogen and prolactin receptor concentrations in rat mammary tumors and response to endocrine ablation. *Cancer Res.*, 36:354–358.
8. Ferland, L., Labrie, F., Hould, R., Bouton, M. M., Azadian-Boulanger, G., and Raynaud, J. P. (1975): Effects of RU16117 (11-methoxy ethinyl estradiol) on parameters of the estrous cycle in the rat. *Fed. Proc.*, 34:340.
9. Griswald, D. P., and Green, C. H. (1970): Observation on the hormone sensitivity of 7,12-dimethylbenz(a)anthracene-induced mammary tumors in the Sprague-Dawley rat. *Cancer Res.*, 30:819–826.
10. Heuson, J. C., Waelbroeck, C., Legros, N., Gallez, G., Robyn, C., and L'Hermite, M. (1971): Inhibition of DMBA-induced mammary carcinogenesis in the rat by 2-Br-α-ergocryptine (CB-154) an inhibitor of prolactin secretion and by nafoxidine (U-11,000A), an estrogen antagonist, hormones and antagonists. *Gynecol. Invest.*, 2:130–137.
11. Holdaway, I. M., and Worsley, I. (1976): Specific binding of human prolactin and insulin to human mammary carcinomas. *Endocrinology*, 96:160A.
12. Horwitz, K. B., and McGuire, W. L. (1975): Specific progesterone receptors in human breast cancer. *Steroids*, 25:497–505.
13. Huggins, C. (1965): Two principles in endocrine therapy of cancers: hormone deprival and hormone interference. *Cancer Res.*, 25:1163–1167.

14. Huggins, C., Grand, L. C., and Brillantes, F. P. (1961): Mammary cancer induced by a single feeding of polynuclear hydrocarbons and its suppression. *Nature,* 189:204–207.
15. Huggins, C. Moon, R. C., and Morh, S. (1962): Extinction of experimental mammary cancer. I. Estradiol and progesterone. *Proc. Natl. Acad. Sci. U.S.A.,* 48:379–386.
15a. Jabara, A. G. (1967): Effects of progesterone on 9,10 dimethyl-1,2-benzanthracene-induced mammary tumors in Sprague-Dawley rats. *Br. J. Cancer,* 21:418–429.
16. Jensen, E. V., Polley, T. Z., Smith, S., Block, G. E., Ferguson, D. J., and De Sombre, E. R. (1975): Production of hormone dependency in human breast cancer. In: *Estrogen Receptors in Breast Cancer,* edited by W. L. McGuire, P. P. Carbone, and E. P. Vollmer, pp. 37–56. Raven Press, New York.
17. Kelly, P. A., Asselin, J., Caron, M. G., Labrie, F., and Raynaud, J. P. (1977a): Potent inhibitory effect of a new antiestrogen (RU 16117) on the growth of DMBA induced mammary tumors. *J. Natl. Cancer Inst.* (*in press*).
18. Kelly, P. A., Asselin, J., Caron, M. G., Labrie, F., and Raynaud, J. P. (1977b): Effects of progesterone and R 2323, an antiprogestin on the development of dimethylbenz(a)anthracene-induced mammary tumors. (*Submitted for publication*).
19. Kelly, P. A., Asselin, J., Caron, M. G., Raynaud, J. P., and Labrie F. (1977): High inhibitory activity of a new antiestrogen, RU 16117 (11α-methoxyethinyl estradiol), on the development of dimethylbenz(a)anthracene-induced mammary tumors. *Cancer Res.* 37:76–81.
20. Kelly, P. A., Bradley, C., Shiu, R. P. C., Meites, J., and Friesen, H. G. (1974): Prolactin binding to rat mammary tumor tissue. *Proc. Soc. Exp. Biol. Med.,* 146:816–819.
21. Kelly, P. A., Posner, B. I., and Friesen, H. G. (1976): Effects of hypophysectomy, ovariectomy and cycloheximide on specific binding sites for lactogenic hormones in rat liver. *Endocrinology,* 97:1408–1415.
22. Kelly, P. A., Posner, B. I., Tsushima, T., and Friesen, H. G. (1974): Studies of insulin, growth hormone and prolactin binding: ontogenesis, effects of sex and pregnancy. *Endocrinology,* 95:532–539.
23. Kledzik, G. S., Bradley, C. J., and Meites, J. (1974): Reduction of carcinogen-induced mammary cancer incidence in rats by early treatment with hormones or drugs. *Cancer Res.,* 34:2953–2956.
24. Leung, B., and Sasaki, S. (1975): On the mechanism of prolactin and estrogen action in 7,12-dimethyl-benz(a)anthracene-induced mammary carcinoma in the rat. II. In vivo tumor responses and estrogen receptor. *Endocrinology,* 97:564–572.
25. Leung, B. S., Sasaki, G. H., and Leung, J. S. (1975): Estrogen-prolactin dependency in 7,12-dimethyl-benz(a)anthracene-induced tumors. *Cancer Res.,* 35:621–627.
26. McGuire, W. L., Carbone, P. P., Sears, M. E., and Escher, G. (1975): Estrogen receptors in human breast cancer: an overview. In: *Estrogen Receptors in Human Breast Cancer,* edited by W. L. McGuire, P. P. Carbone, and E. P. Vollmer, pp. 1–7. Raven Press, New York.
27. Meites, J. (1972): Relation of prolactin and estrogen to mammary tumorigenesis in the rat. *J. Natl. Cancer Inst.,* 48:1217–1224.
28. Philibert, D., and Raynaud, J. P. (1973): Progesterone binding in immature mouse and rat uterus. *Steroids,* 22:89–98.
29. Posner, B. I., Kelly, P. A., and Friesen, H. G. (1974): Induction of a lactogenic receptor in rat liver: influence of estrogen and the pituitary. *Proc. Natl. Acad. Sci. U.S.A.,* 71:2407–2410.
30. Posner, B. I., Kelly, P. A., and Friesen, H. G. (1975): Prolactin receptors in rat liver: possible induction by prolactin. *Science,* 187:57–59.
31. Quadri, S. K., Kledzik, G. S., and Meites, J. (1974): Enhanced regression of DMBA-induced mammary cancers in rats by a combination of ergocronine with ovariectomy or high doses of estrogen. *Cancer Res.,* 34:399–501.
32. Raynaud, J. P., Bonne, C., Bouton, M. M., Moguilewsky, M., Philibert, D., and Azadian-Boulanger, G. (1975): Screening for anti-hormones by receptor studies. *J. Steroid Biochem.,* 6:615–622.
33. Talwalker, P. K., Meites, J., and Mizumo, H. (1964): Mammary tumor induction by estrogen of anterior pituitary hormones in ovariectomized rats given 7,12-dimethyl-1,2-benzanthracene. *Proc. Soc. Exp. Biol. Med.,* 116:531–534.

34. Toft, D. O., and O'Malley, B. W. (1972): Target tissue receptors for progesterone: the influence of estrogen treatment. *Endocrinology,* 90:1041–1045.
35. Turkington, R. W. (1974): Prolactin receptors in mammary carcinoma cells. *Cancer Res.,* 34:758–763.
36. Vignon, F., and Rochefort, H. (1974): Régulation des "récepteurs" des oestrogènes dans les tumeurs mammaires; effet de la prolactine in vivo. *Comptes Rendus Acad. Sci. Paris,* 278:103–106.
37. Young, S., and Cowan, D. M. (1963): Spontaneous regression of induced mammary tumors in rats. *Br. J. Cancer,* 17:85–89.

Progesterone Receptors in Normal and Neoplastic Tissues, edited by W. L. McGuire et al. Raven Press, New York © 1977.

Estrogen and Progesterone: Their Relationship in Hormone-Dependent Breast Cancer

K. B. Horwitz and W. L. McGuire

University of Texas Health Science Center, San Antonio, Texas 78284

Because of the cyclic changes of blood estrogen and progesterone levels which occur in females and these hormones' interrelationships in regulating target tissue development and growth, it was inevitable that progesterone would be studied for its effect on breast cancer. Although progesterone itself is not a carcinogen, it may be a potent target-specific cocarcinogen for induction of mammary tumors by viral or chemical agents (1). As such, the hormone has been implicated in both tumor enhancement and tumor suppression.

In the reproductive tract progesterone-induced changes are generally superimposed on the major growth responses which accompany estrogen action, with few if any progesterone effects seen in tissues which have not first been prepared by estradiol (2). In confirmation of this long-established physiological principle, it has only recently been biochemically demonstrated that the cytoplasmic progesterone receptor (PgR) of the uterus is estrogen dependent (3).

Estrogen also produces major developmental changes in normal (4) or malignant (5) breast tissues, and it has now been conclusively demonstrated that the presence of estrogen receptors (ERs) in human breast tumor serves as an important marker for evaluating the response potential of tumors to endocrine therapy (6). Since progesterone similarly affects both normal and malignant breast tissue, we have been concerned with the interrelationships between estrogen and progesterone in controlling endocrine-responsive tumor growth, and with the possible modulation by estrogen of PgR in hormone-sensitive tumors.

CLINICAL EFFECTS OF PROGESTERONE IN BREAST CANCER

That progesterone plays a role in stimulating tumor growth is suggested by the pioneering studies of Huggins et al. (7–9). They showed that pregnancy promoted the growth of DMBA-induced rat mammary tumors. Administration of progesterone to intact rats accelerated the appearance of tumors, increased the number of tumors, and augmented the growth rate of established tumors.

If DMBA mammary tumor–bearing rats are ovariectomized and simultaneous lesions are placed in the median eminence to increase prolactin release, the tumors grow at an accelerated pace for only 10 to 12 days and then regress, even though prolactin levels remain elevated (10,11). The ovarian factor responsible for maintaining tumor growth under these circumstances has not been identified, but the following experiments suggest the importance of progesterone. Pregnancy stimulates the growth of DMBA mammary tumors, whereas parturition and weaning are followed by regression of a large number of these tumors (7,12,13). The tumor growth-promoting factor of pregnancy is probably placental lactogen (14), although prolactin has been implicated as responsible for the maintenance of tumor size and growth during lactation since tumors regress if suckling is prevented (15,16). Ovariectomy, however, blocks the stimulatory effects of endogenous or exogenous prolactin on tumor growth, and injection of progesterone removes this block (15). Either prolactin stimulation of tumors under these circumstances is dependent on progesterone or, alternatively, the high levels of circulating progesterone stimulated by prolactin in the lactating rat (17) are responsible for tumor growth. This does not mean that progesterone alone is responsible for maintaining rat mammary tumor growth, since in these experiments the animals had both high prolactin levels and intact adrenal glands. On the other hand, they do suggest that progesterone plays an important physiological role in stimulating tumor growth.

In contrast to the stimulatory effects of progesterone described above, progesterone can induce rat mammary tumor regression or prevent tumor appearance when combined with moderate-to-large doses of estrogen (7,18). In humans, the percentage of breast tumor regressions to a progesterone-estrogen combination is generally higher than with progesterone alone (19). The significance of these observations is somewhat obscured by the fact that moderate-to-large doses of estrogen alone can cause mammary tumor regression in rats (20–22) and humans (23). That progesterone is not superfluous is shown in patients whose tumors failed to regress following treatment with high-dose estrogen alone but responded to a combination of estrogen-progesterone (24–26). Postmenopausal patients with endogenous estrogen levels (presumably of adrenal origin) sufficient to cornify the vaginal mucosa have a 29% tumor remission rate with progesterone therapy, whereas patients with an atrophic vaginal smear experience only 6% remission rate with progesterone alone (27). These studies suggest that estrogen is required in progesterone-mediated tumor regression, perhaps for synthesis of progesterone receptor (see below). However, the mechanism by which progesterone promotes tumor regression remains unclear. Large doses of synthetic progestins can cause significant lowering of serum LH and cortisone levels, suggesting that alteration of pituitary function may be involved (28), but at least four previously hypophysectomized patients are reported to have had breast tumor regression following combinations of estrogen-progesterone (29,30). This is in contrast to the lack of tumor response to estrogens alone in hypophysectomized patients (31–33). The explanations of these perplexing, at times contradictory effects will come from our understanding of the biochemical mechanisms of hormone action. The first step in this mechanism is, in the case of steroid hormones, their binding to intracellular receptor proteins.

PROGESTERONE RECEPTOR REVIEW

Chick Oviduct

The most exhaustive studies of PgR have been done in the estrogen-primed chick oviduct. In this tissue, administration of a single dose of progesterone *in vivo* stimulates chromatin template capacity, DNA-dependent RNA polymerase activity, and synthesis of a specific messenger RNA, culminating with induction of the specific protein avidin (34). Unlabeled progesterone, its active metabolites, and testosterone (a good inducer of avidin) block ^{3}H-progesterone binding to the receptor (35). Progesterone receptor concentration is increased 20-fold by estradiol (36). The receptor has been purified to homogeneity and found to consist of two similar subunits which have distinctly different affinities for DNA and chromatin (37).

Mammalian Uterus

Extensive investigations using guinea pig uterus unequivocally demonstrated PgR in a mammal (38–51). The receptor migrates at 6 to 8S in sucrose gradients and does not bind glucocorticoids. Uterine levels of PgR are maximum at proestrus and fall progressively during estrus and metestrus to a 16-fold lower level at diestrus. Injection of estrogen causes an 8-fold rise in PgR within 24 hr, which can be prevented by inhibition of protein or RNA synthesis. The normal half-life of PgR is approximately 3 to 5 days, but an injection of progesterone will deplete the uterine cytoplasm of PgR within 3 hr. Evidence for nuclear translocation of PgR is available from direct biochemical and autoradiographic studies. Similar conclusions have been reached about PgR in the hamster (52,53), rabbit (54–59), mouse (60,61), rat (54–56,60–69), and human (70–77). In the latter two species PgR has been difficult to demonstrate reproducibly because the majority of radioactive progesterone binds in the 4S region of the sucrose gradient, making it difficult to distinguish PgR from corticosteroid binding globulin

(CBG). With incorporation of glycerol into buffers or gradients, the receptor complex is stabilized (77), and a specific, high-affinity (K_d 3 to 8 × 10^{-9} M) 7S binding component is seen occasionally in myometria from women treated with estrogen (72), in endometrium obtained from proliferative (estrogen-dominated) tissues (77,78), in hyperplastic endometrium (79), and in endometrial carcinoma (78).

The receptor is precipitated by ammonium sulfate (80), and this property has been used to purify it (81). The pure receptor (which migrates as a single band of molecular weight 110,000 on SDS polyacrylamide gels) sediments at 3.7S on sucrose gradients (after elution with hypertonic salt), has a K_d of 10^{-9} M, and does not bind hydrocortisone. The purified receptor-hormone complex will bind to nuclei; the nuclear bound form also sediments at 3.7S.

Mammary Gland PgR

Although the breast is also a target of progesterone action, virtually no information about receptor binding of the hormone exists. After injection, an apparent selective accumulation of progesterone by the normal human breast has been described (82). Terenius (83) used charcoal-resistant radioactivity of cytosols containing ^{3}H-progesterone and excess hydrocortisone to show a progesterone binder in human and rat mammary carcinomas. Attempts to demonstrate receptors in normal mammary tissue using progesterone have been unsuccessful, even though glucocorticoid receptors (which are often difficult to distinguish from PgR) are present (84,85).

Use of R5020

Recently the new synthetic progestin R5020 has been used to unequivocally demonstrate PgR in each of the species described above. In the rabbit and guinea pig, ^{3}H-progesterone and ^{3}H-R5020 bind at identical positions in sucrose gradients, with similar competitive effects and to approximately the same number of sites, indicating that the progesterone and progestin receptors are in all likelihood one and the same. However, the K_d of R5020 (0.7 × 10^{-9} M) is lower than that of progesterone (2.5 × 10^{-9} M) (51).

R5020 binds to a specific 7S cytosol component in both immature mouse and rat uteri. The K_d of R5020 binding (5 × 10^{-9} M) is again lower than that of progesterone (11 × 10^{-9} M), although for both it is higher in the rat than in the rabbit and guinea pig (61), explaining perhaps the relative ease with which PgR is measured in the latter two species.

In adult rat uterine cytosols, progesterone binds principally at 4S with a 6S shoulder. In contrast, R5020 binds to a heavy form of the receptor. Examples of these peaks in two different cytosol preparations are seen in Fig. 1. We have shown that the R5020 binding protein is sensitive to the action of estradiol. In chronically hypophysectomized, ovariectomized, and castrated

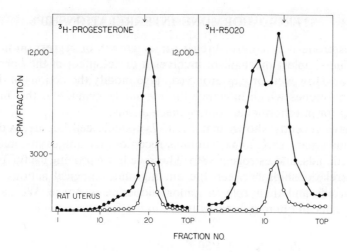

FIG. 1. Comparison of ³H-progesterone and ³H-R5020 binding to cytosols of estrogen-primed rat uteri. ³H-progesterone, cytosol 9.6 mg/ml protein; ³H-R5020, cytosol 15.7 mg/ml protein. (●) Radioactive hormone only or (○) with excess unlabeled hormone. Arrow indicates position of ¹⁴C-bovine serum albumin (BSA).

rats, uterine receptor levels regress to low basal values (Table 1). An 8- to 10-fold stimulation of PgR was obtained when estrogen alone was injected into rats. Prolactin alone or in combination with estrogen had no effect. Thus estrogen priming acts directly on the uterus to stimulate PgR synthesis and is not an indirect effect of increased pituitary prolactin secretion.

We have also used R5020 to demonstrate PgR in experimental rat mammary tumors and in human breast cancer, in which demonstration of this receptor using progesterone has been difficult (see Progesterone Receptors in DMBA Tumors and PgR in Human Breast Cancer).

TABLE 1. *Role of prolactin and estradiol in PgR priming*

	R5020 binding, fmoles/mg protein	
	8S	4S
Propylene glycol	88	76
	87	75
Estradiol	801	—
	696	
Prolactin	95	62
	73	28
Prolactin + estradiol	618	—
	710	

Chronically (>3 weeks) hypophysectomized, ovariectomized, and adrenalectomized rats injected 4 days with estradiol (1 µg/day), prolactin (2 mg/day), or both.

STEROID HORMONE INTERRELATIONSHIPS

Progesterone may control breast tumor growth or regression in several ways. The simplest mechanism involves a direct effect of the hormone on the tumor. However, progesterone can also modify the actions of the other steroid hormones which influence the mammary gland, and this may form the basis for interhormonal control mechanisms.

We have recently shown that MCF-7, a stable cell line derived from a human mammary carcinoma, contains receptors for progestins, androgens, glucocorticoids, and estrogens (86). These cells are proving useful for studying interrelationships between the binding and biological actions of these four steroids and their role in tumor endocrine response. We have used MCF-7 cells and human breast tumors in the following studies.

Progesterone, R5020, and Estrogen Receptors

The ability of progesterone to antagonize and/or modify the action of estrogen is well documented (87,88). Tamoxifen and nafoxidine, two widely used antiestrogens, exhibit progesterone-like effects (89,90). In this they resemble, R5020 which is also an antiestrogen. Hsueh et al. (87) have shown that after depletion of cytoplasmic ER by high-dose estrogen treatment, progesterone blocks the overshoot of ER seen during replenishment. They propose that this reduction of ER is correlated to reduced sensitivity of the uterus to estrogen. There is no evidence, however, that progesterone affects replenishment of ER after physiological estrogen treatments or alters basal ER levels. In short, estrogen and progesterone may exert feedback control on each other at the target tissue. Estradiol pretreatment enhances tissue sensitivity to progesterone correlated with increased PgR levels. Progesterone in turn may modify cytoplasmic ER and redirect the cells' ability to respond to estradiol.

To study the binding specificity of R5020, we have used MCF-7 cells and

TABLE 2. *Steroid hormone binding interrelationships*

	R5020	Pg	TA	Dex	DES	DHT	F	Cyp
^{3}H-R5020	++	++	+	−	−	±	−	++
^{3}H-TA	+	+	++	+	−	−	+	
^{3}H-Pg	++	++	±	−	−		−	
^{3}H-Dex	++	++	++	++				
^{3}H-DHT	+	+	−	−	−	++		++
^{3}H-E$_2$	−	−		−	++	−		

Competition for radioactive hormone binding to MCF-7 cell and human tumor cytosols by 100-fold excess unlabeled hormone as determined on sucrose density gradients. Competition: >85% (++); 40–85% (+); 20–40% (±); <20% (−). Pg, progesterone; TA, triamcinolone acetonide; Dex, dexamethasone; DES, diethylstilbestrol; DHT, dihydrotestosterone; F, hydrocortisone; Cyp, cyproterone acetate.

human breast tumors in binding and competition studies among four steroid classes and determined the percent displacement of 7 to 8S and 4S peaks by 100-fold excess unlabeled steroids. The results are summarized in Table 2. Since tamoxifen and nafoxidine competitively inhibit estradiol action by binding to estrogen receptor (90), the possibility existed that R5020 would similarly bind ER. We find, however, that ER is bound exclusively by estrogens; no significant competition is rendered by the other steroids. Thus, the antiestrogenic properties of R5020 are more than likely due to its ability as a progestin to antagonize the action of estrogen indirectly.

Progesterone, R5020, and Androgen Receptors

The androgenic properties of progestins are well known, and fetal virilization can result from their use in man (91). Progestins can masculinize the reproductive tract of rat fetuses (92) and can mimic androgen effects in several organs (92–96). Recently Bullock et al. (95) and Mowszowicz et al. (93) have demonstrated that progestins can be either synandrogenic (by potentiating androgen effects) or antiandrogenic (by inhibiting these effects) depending on steroid structure, dose, and tissue. If androgens have similar modifying effects on progesterone actions, it may be one reason why they are effective in treatment of hormone-dependent breast cancer. Although the mechanism of androgen-induced regression of breast tumors is not known, androgens cause regressions of fetal mammary buds (97) and may have similar effects on dedifferentiated malignant cells. It is possible that progestin-induced tumor regression is a reflection of the progestin's androgenic properties.

Table 2 shows that progestins, including cyproterone acetate, are potent inhibitors of ^{3}H-R5020 binding as expected. Cyproterone (*data not shown*), a nonprogestational antiandrogen, and dihydrotestosterone (DHT) were much weaker competitors, although not altogether ineffectual. We have not determined whether the latter were competing for binding of R5020 to PgR, or possibly, its binding to androgen receptor. Both progesterone and R5020 as well as DHT are good competitors of ^{3}H-DHT binding in human breast tumor cells. Therefore, it appears that androgens can inhibit R5020 binding. Since neither R5020 nor progesterone is metabolized under our incubation conditions (*data not shown*), their conversion to C19 androgens does not explain these findings, and we conclude that progestin inhibition of DHT binding is due to their direct interaction with the androgen receptor. Based on these data we would predict that R5020 may be an antiandrogen.

Progesterone, R5020, and Glucocorticoid Receptors

By far the most familiar model for the interaction of two steroids is that proposed by Rousseau et al. (98) to explain the inhibitory effects of progestins and the stimulatory effects of glucocorticoids on tyrosine aminotransferase production in rat hepatoma tissue culture (HTC) cells. Competition by progestins for glucocorticoid binding has also been demonstrated in mammary carcinomas (99,100) and lactating mammary glands (84,85). Since glucocorticoids are involved in mammary gland maturation, it is possible that progestins may affect mammary tumors by modifying glucocorticoid action.

Although there seems little doubt that progestins may compete for binding to glucocorticoid receptors, less is known about the reverse effect, that is, the extent of competition by glucocorticoids for PgR binding. This may be due in part to the fact that until recently ^{3}H-progesterone binding ascribed to PgR may have been to CBG. In that case competition by glucocorticoids would hardly be surprising, but its significance would be doubtful. Dexamethasone prevents ^{3}H-progesterone binding to HTC cells (98), although hydrocortisone is a poor competitor in several systems tested, and neither dexamethasone nor hydrocortisone translocates PgR to rat uterine nuclei or exchanges with nuclear bound progesterone (67).

In MCF-7 cells and human tumors, when ^{3}H-dexamethasone is the ligand, both R5020 and progesterone as well as dexamethasone itself are effective competitors (Table 2). This agrees with the established efficacy of progestins to compete for glucocorticoid receptor binding, a property shared by R5020. ^{3}H-triamcinolone acetonide (TA) was similarly inhibited by progestins. When ^{3}H-progestins were the ligands, neither dexamethasone nor hydrocortisone displaced ^{3}H-progesterone or ^{3}H-R5020. However, triamcinolone displaced both progesterone and R5020 binding, and cytoplasmic PgR in MCF-7 cells is depleted by triamcinolone or R5020 pretreatment but not by dexamethasone treatment (*data not shown*). We would predict on the basis of this that TA may be a progestin, and in fact, this compound, unlike the natural glucocorticoids, has been reported to have progesterone-like antiovulatory properties (101,102).

PROGESTERONE RECEPTORS IN DMBA TUMORS

R5020 has been used to demonstrate a PgR in DMBA mammary tumors (103). We have shown that the receptor resembles PgR of other tissues in that it binds hormone with high affinity (K_d approximately 0.8 nM) (Fig. 2) and migrates at least in part at 7S on sucrose density gradients (Fig. 3). Neither of these properties has been demonstrated when native progesterone is the ligand (104).

Rats bearing growing tumors were ovariectomized-adrenalectomized at proestrus, tumors were biopsied 24 hr later, and PgR was determined by a single saturating dose dextran-coated charcoal (DCC) assay. Tumor growth was reestablished by daily injections of estradiol (E, 1 μg) plus progesterone (Pg, 4 mg). When tumors regained prebiopsy size they were entered into

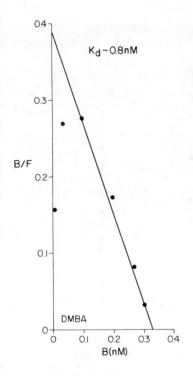

FIG. 2. Scatchard plot of progesterone receptor in DMBA tumors. Cytosol incubated with 4×10^{-11} to 1.5×10^{-8} M ^{3}H-R5020 and assayed by dextran-coated charcoal. Nonspecific binding has been subtracted.

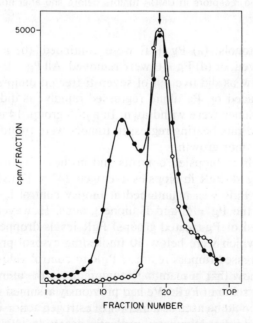

FIG. 3. Progesterone receptor in DMBA tumor. (●) ^{3}H-R5020 only or (○) with 100-fold excess unlabeled R5020. Arrow indicates position of ^{14}C-BSA.

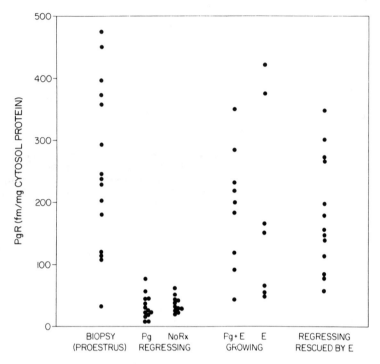

FIG. 4. Progesterone receptors in DMBA tumors before and after hormone treatments as described in text.

one of four protocols: (a) Pg + E were continued, (b) E was removed, (c) Pg was removed, or (d) Pg + E were removed. All Pg + E–treated tumors continued to grow as did five out of seven E-treated tumors. However, 11 of 13 tumors placed on Pg alone regressed rapidly, as did 11 of 11 from which both hormones were withdrawn. In a fifth group, 14 ovariectomized-adrenalectomized rats bearing regressed tumors were treated with E alone, which restored tumor growth.

The PgR levels in biopsied controls and in these five treatment groups are shown in Fig. 4. PgR in biopsies averaged 247 ± 27 (SEM) fmoles/mg cytosol protein. PgR was maintained at biopsy control levels in growing tumors of both the Pg + E and E alone groups. However, in regressing tumors (untreated or Pg-treated groups) PgR levels dropped precipitously to basal levels which were below 50 fmoles/mg cytosol protein. Estrogen treatment of regressed tumors restored PgR to control values.

Our studies show that in mammary tumors, as in the uterus (3), estrogen exerts acute control over PgR. We had previously assumed this in hypothesizing that PgR could be used as a marker of estrogen action in human breast tumors. Since estradiol stimulates prolactin secretion (105), we have not ruled out the possibility that endogenous prolactin may be stimulating PgR. Although we are testing this possibility in current studies, we believe that prolactin is not involved since estrogen alone, but not prolactin, can stimulate synthesis of PgR in the uterus of triple-operated rats (see above).

Although progesterone clearly enhances induction of DMBA tumors (7–9), its role in growth of established tumors is unclear, as is the mechanism by which progestins induce tumor regression (24–26). Our data show that in the absence of estradiol, physiological doses of progesterone do not sustain growth of established tumors despite the initial presence of PgR. It may be that cytoplasmic PgR is depleted by progesterone injection and no further PgR synthesis occurs in the absence of estradiol. This would render the cell refractory to further stimulation by progesterone. The basal PgR seen in all tumors appears to be inadequate in mediating a potential progesterone effect on tumor growth. Whether it is an effective mediator of other progesterone functions in mammary tissues remains to be seen.

In considering the role of progesterone and PgR in mammary tumors, it is important to distinguish between two points: first, the direct role of progesterone in hormone-dependent growth and regression; and second, its use as a marker of estrogen action and ER integrity. We have been measuring PgR in human breast cancer (106,107) for the purpose of demonstrating an intact estrogen response system. In rat tumors we have shown that a relationship exists between presence of estrogen and presence of PgR. This is not to say, however, that estrogen-responsive tumor growth and estrogen-induced PgR are inexorably linked. At least two tumors grew in the absence of estrogen so that their growth may be considered autonomous, whereas their PgR declined and could therefore be considered estrogen dependent. This suggests some dissociation between estrogen-controlled tumor growth and estrogen-controlled PgR synthesis. Similarly, from the persistent basal level of PgR seen in estrogen-deprived tumors and in the two tumors which failed to respond to estrogen treatment, we must conclude that at least some PgR appears to be estrogen independent.

PgR IN HUMAN BREAST CANCER

It is now generally appreciated that human breast tumors containing estrogen receptor (ER) often regress following endocrine therapy; tumors lacking ER usually fail to respond (6). The fact that not all ER-containing tumors respond has led to the concept that ER is a necessary but not sufficient marker of hormone dependence. Since ER is only an early step in the pathway from hormone binding to ultimate cellular response, endocrine-resistant ER-positive tumors may have lesions distal to the binding step. In that case a product of hormone action would be a better marker of endocrine responsiveness than simple presence of the receptor. Since in the uterus the synthesis of PgR was known to be dependent on the action of estrogen (3), we conjectured that PgR might be a useful marker of ER integrity in mammary tissue.

Using ^{3}H-R5020 we have been able to demonstrate in human breast

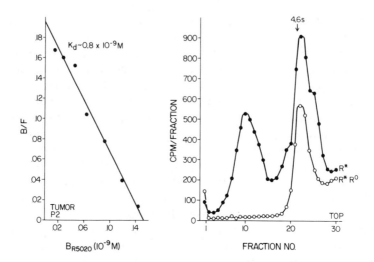

FIG. 5. Progesterone receptor in a frozen human tumor biopsy specimen. **Left:** Scatchard plot of tumor cytosol incubated with 1×10^{-10} M to 1.5×10^{-8} M ^{3}H-R5020 and assayed by protamine sulfate precipitation. Nonspecific binding has been subtracted. **Right:** Sucrose density gradient centrifugation of cytosol containing ^{3}H-R5020 only (R*) or with excess unlabeled R5020 (R* R°).

cancer a PgR that sediments at 7 to 8S on sucrose density gradients and binds hormones with high affinity (Fig. 5). Our earliest attempts to determine whether saturable, specific binding could be demonstrated in human breast cancer cytosol were with ^{3}H-progesterone. The sucrose gradient pattern of a representative tumor, F6, is seen in Fig. 6A. The protein-bound hormone sedimented at 4S, making it difficult to differentiate binding of progesterone to its own receptor from binding to CBG, glucocorticoid receptor, or nonspecific components. We estimated the relative contribution of these by differential competition studies. Unlabeled progesterone competed efficiently (91%) so that generalized nonspecific binding cannot account for the 4S peak; however, the competition by R5020 showed that only 30% of this binding was to progesterone receptor. Hydrocortisone competition showed that 57% of binding was to either CBG or glucocorticoid receptor. The latter possibility was excluded by the minimal competition of dexamethasone, which binds to glucocorticoid receptor but not to CBG.

Since competition studies evaluate receptor levels only indirectly, we also used ^{3}H-R5020 in the same tumor cytosol (Fig. 6B). Bound R5020 sedimented in the 8S and 4S regions of the gradient. Excess unlabeled R5020 or progesterone completely inhibited binding to the 8S peak. Hydrocortisone did not compete at all, and dexamethasone competed only minimally (15%). Estradiol competed only weakly, and when unlabeled R5020 was used as competitor of ^{3}H-estradiol binding in another tumor, the displacement of 8S estrogen receptor was less than 10% (*not shown*). Thus, R5020 was

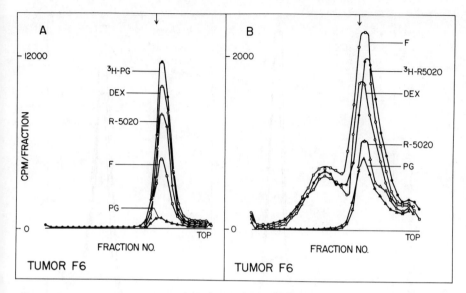

FIG. 6. Comparison of ³H-progesterone (PG) and ³H-R5020 binding in human mammary tumor cytosol and inhibition by unlabeled competitors. DEX, dexamethasone; F, hydrocortisone.

apparently binding to a specific 8S receptor; a considerable portion of the 4S binding in this tumor also appeared to be PgR.

³H-Progesterone and ³H-R5020 Binding to Human Serum, and Nature of 4S Peak

The original name for CBG, transcortin, implied that it bound only glucocorticoid to a significant extent. This is clearly not the case, for CBG binds progesterone with high affinity, and, in fact, all the steroids that bind to CBG compete for a single binding site (108). We showed that, in contrast to progesterone, R5020 did not bind CBG, by comparing their binding to human serum. Diluted serum stripped of endogenous hormone by preincubation with dextran-coated charcoal was incubated with ³H-R5020 or ³H-progesterone in the absence or presence of radioinert steroids (Fig. 7). As expected, progesterone bound to a specific 4S component, characteristic of CBG. Progesterone and hydrocortisone were effective competitors, but R5020 even when present in large excess failed to interfere with progesterone binding to CBG. R5020 also bound to a 4S component which was not decreased by any competitors tested, reflecting an inability to reach saturating levels. This is characteristic of nonspecific binding, as, for instance, to albumin. We found, therefore, that although R5020 can bind to a 4S serum component of high capacity as do a range of other steroids, it unlike progesterone does not bind to CBG.

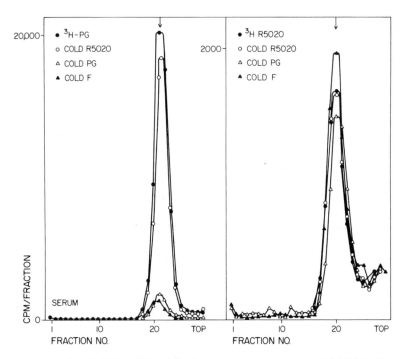

FIG. 7. Binding of ³H-R5020 and ³H-progesterone to human serum and inhibition by competitors. Arrows indicate sedimentation of ¹⁴C-BSA.

Thus, the 4S binding of R5020 in cytosols represents binding in two different components: the noncompetable binding is nonspecific; the competable 4S is not CBG and may represent specific cytoplasmic receptor. That the competable 4S binding could be to intracellular receptor is also shown by the following.

First, one tumor studied (Table 3) contained appreciable receptor levels, 68 fmoles/mg 8S and 31 fmoles/mg competable 4S. Both peaks were quantitatively removed from cytosol by protamine sulfate, whereas the noncompetable 4S fraction remained in solution. Second, when cytosol was centrifuged through a gradient of high ionic strength (0.4 M KCl), almost all the competable 8S binding was recovered in 4S form. The total competable 4S was the same as the sum of 8S plus 4S in low-salt gradients, suggesting that the two molecular forms were interrelated. Third, tumors which had 8S peaks almost always had 4S peaks, whereas most tumors with no 8S also had no competable 4S (see below). Thus the presence of competable 4S appeared to be related to the presence of 8S and may represent specific receptor molecules dissociated by the harsh conditions of tissue storage and assay.

TABLE 3. *Effect of protamine and KCl on PgR binding in human tumor cytosol*

	fmoles/mg Protein		
	8S	4S	Total
1. Specific ^{3}H-R5020 bound	68	31	99
2. Specific ^{3}H-Pg bound	–	84	84
3. Amount removed by protamine	70	24	94
4. Specific ^{3}H-R5020 bound in 0.4 M KCl	–	93	93

Definition of Positive and Negative PgR Assay

To arrive at a definition of positive and negative receptor, we analyzed the gradient profiles of 223 receptor assays. Table 4 contrasts the presence and absence of an 8S peak with presence or absence of a 4S peak in the same tumor. Note that suppressible 4S binding is almost always associated with presence of an 8S peak. A single suppressible 4S peak was seen in less than 12% of cases. We are uncertain whether these should be classified positive

TABLE 4. *Relationship between 8S and 4S PgR peaks in 223 human tumors*

	8S+	8S–
4S+	101	26
4S–	8	88

or negative, but our preliminary correlation data suggest that 4S greater than 8 fmoles/mg binding may represent receptor. This will have to be clarified when more data are available. In 8S binding, when more than 2 fmoles/mg binding is present, the assay is unequivocally positive.

Binding kinetics were determined in 23 breast tumors which were PgR positive by sucrose density gradients. The average K_d was 0.83 ± 0.38 (SD) $\times 10^{-9}$ M.

Estrogen receptors in breast tumors have been assayed in our laboratory for several years using the DCC assay and sucrose gradients. The level of sensitivity of the two methods is such that a value of less than 3 fmoles/mg is essentially equivalent to zero and is considered negative.

Distribution of ER and PgR in Breast Tumors and Clinical Correlations

We have now determined ER and PgR in over 500 human mammary tumors. We also have data in 44 cases involving a single biopsy and response to a single trial of endocrine therapy. The results are summarized in Table 5. Note first that if tumors are considered as a random set, 75% are ER positive. However, when primary versus metastatic tumors are considered separately, it is apparent that primary tumors are more likely (77%) than metastatic tumors (66%) to be ER positive.

In considering the distribution of PgR relative to ER, we find that in unselected tumors, when ER is negative, only 9% of tumors have PgR. When ER is positive, 74% of tumors also have PgR. This strongly suggests that, as we have shown in rat mammary tumors, in human breast cancer PgR is under control of estrogen acting through its receptor. In general we find that the likelihood of positive PgR in a tumor increases with increasing ER content (Fig. 8). As is the case for ER, primary tumors are more likely than metastatic tumors to have PgR. Although 77% of ER-positive primary tumors have PgR, only 59% of ER-positive metastatic tumors have PgR, suggesting perhaps that in dedifferentiating tumors, ER and estrogen responses become attentuated.

We have proposed (107) that if a tumor is capable of synthesizing a biological end-product under estrogen regulation, it may mean that endocrine responses are preserved. Conversely, the prospect of successful endocrine therapy would be low in tumors with ER but no PgR. Table 5 shows response data in 44 cases. When considering ER alone in unselected tumors,

TABLE 5. *Receptor distribution and response to endocrine therapy*

	Unselected	Primary	Metastatic
Distribution of ER[a]			
ER+	$\frac{392}{521}$ (75%)	$\frac{326}{421}$ (77%)	$\frac{66}{100}$ (66%)
Distribution of PgR			
ER−	$\frac{12}{129}$ (9%)	$\frac{11}{95}$ (12%)	$\frac{1}{34}$ (3%)
ER+	$\frac{289}{392}$ (74%)	$\frac{250}{326}$ (77%)	$\frac{39}{66}$ (59%)
Response to endocrine therapy			
ER−, PgR−	$\frac{0}{8}$ (0%)	—	$\frac{0}{8}$ (0%)
ER−, PgR+	—	—	—
ER+, PgR−	$\frac{6}{16}$ (37%)	$\frac{0}{1}$	$\frac{6}{15}$ (40%)
ER+, PgR+	$\frac{14}{20}$ (70%)	$\frac{3}{8}$ (38%)	$\frac{11}{12}$ (92%)

[a] ER+ total 8S + 4S > 3 fmoles.
PgR+ total 8S > 2 fmoles.

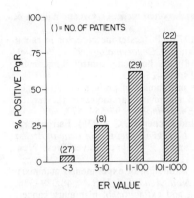

FIG. 8. Comparison of estrogen receptor content and number of PgR-positive tumors.

we find that when ER is absent no responsive tumors are seen. Of 36 ER-positive tumors, 20 or 55% responded to treatment, a value in close agreement with larger series of tumors from our (109) and other laboratories (6). If these tumors are subclassified by PgR, 37% of ER-positive, PgR-negative tumors responded, whereas 70% responded when PgR was present. Thus, PgR measurements appear to improve selection of ER-positive, responsive tumors. This is particularly striking when metastatic tumors are considered as a separate set. There, 92% of tumors responded to therapy when both receptors were present.

It may well be that during the extended period involving primary development, dissemination, and resumption of growth, tumors which have retained both receptors are likely to have remained hormone dependent. Although the case for use of ER to predict the hormone potential of tumors is now firmly established, there remains a large group of ER-positive but nonresponsive tumors. The information obtained from studies of the biological responses of tumors, not only to estrogens but also to androgens, glucocorticoids, and progestins, may permit precise description of the biochemical lesion in a tumor and the specific treatment to which it will respond.

ACKNOWLEDGMENTS

These studies were supported in part from the National Institutes of Health (CA 11378, CB 23862) and the American Cancer Society (BC 23). The clinical response data were provided by our collaborators Drs. O. H. Pearson and A. Segaloff. We thank Dr. J. P. Raynaud and Roussel-Uclaf for providing the R5020.

REFERENCES

1. Poel, W. E. (1968): Progesterone enhancement of mammary tumor development as a model of co-carcinogenesis. *Br. J. Cancer,* 33:867–873.
2. Steinetz, B. G. (1973): Secretion and function of ovarian estrogens. In: *Handbook of*

Physiology. II. Female Reproductive System, Part I, edited by R. O. Greep. William & Wilkins Co., Baltimore.
3. Rao, B. R., Wiest, W. G., and Allen, W. M. (1973): Progesterone receptor in rabbit uterus. Characterization and 17β-estradiol augmentation. *Endocrinology,* 92:1229–1240.
4. Lyons, W. R., Li, C. H., and Johnson, R. E. (1958): The hormonal control of mammary growth and lactation. *Recent Prog. Horm. Res.,* 14:219–254.
5. Leung, B. S., and Sasaki, G. H. (1975): On the mechanism of prolactin and estrogen action in 7,12-dimethylbenz(a)anthracene-induced mammary carcinoma in the rat. II. In vivo tumor responses and estrogen receptors. *Endocrinology,* 97:564–572.
6. McGuire, W. L., Carbone, P. P., Sears, M. E., and Escher, G. C. (1975): Estrogen receptors in human breast cancer: An overview. In: *Estrogen Receptors in Human Breast Cancer,* edited by W. L. McGuire, P. P. Carbone, and E. P. Vollmer. Raven Press, New York.
7. Huggins, C., Moon, R. C., and Morii, S. (1962): Extinction of experimental mammary cancer. I. Estradiol-17β and progesterone. *Proc. Natl. Acad. Sci. U.S.A.,* 48:379–386.
8. Huggins, C., and Yang, N. C. (1962): Induction and extinction of mammary cancer. *Science,* 137:257–262.
9. Huggins, C. (1965): Two principles in endocrine therapy of cancers: Hormone deprival and hormone interference. *Cancer Res.,* 25:1163–1175.
10. Clemens, J. A., Welsch, C. W., and Meites, J. (1968): Effects of hypothalamic lesions on incidence and growth of mammary tumors in carcinogen-treated rats. *Proc. Soc. Exp. Biol. Med.,* 127:969–972.
11. Sinha, D., Cooper, D., and Dao, T. L. (1973): The nature of estrogen and prolactin effect on mammary tumorigenesis. *Cancer Res.,* 33:411–414.
12. Dao, T. L., and Sunderland, H. (1959): Mammary carcinogenesis by 3-methylcholanthrene. 1. Hormonal aspects in tumor induction and growth. *J. Natl. Cancer Inst.,* 23:567–586.
13. McCormick, G. M., and Moon, R. C. (1965): Effect of pregnancy and lactation on growth of mammary tumors induced by 7,12-dimethylbenz(a)anthracene (DMBA). *Br. J. Cancer,* 19:160–166.
14. Nagasawa, H., and Yanai, R. (1973): Effect of human placental lactogen on growth of carcinogen-induced mammary tumors in rats. *Int. J. Cancer,* 11:131–137.
15. McCormick, G. M., and Moon, R. C. (1967): Hormones influencing postpartum growth of 7,12-dimethylbenz(a)anthracene-induced rat mammary tumors. *Cancer Res.,* 27:626–631.
16. McCormick, G. M. (1972): The effect of varying the length of the nursing period on the postpartum growth of chemically induced rat mammary tumors. *Cancer Res.,* 32:1574–1576.
17. Tomogane, H., Ota, K., and Yokoyama, A. (1975): Suppression of progesterone secretion in lactating rats by administration of ergocornine and the effect of prolactin replacement. *J. Endocrinol.,* 65:155–161.
18. McCormick, G. M., and Moon, R. C. (1973): Effect of increasing doses of estrogen and progesterone on mammary carcinogenesis in the rat. *Eur. J. Cancer,* 9:483–486.
19. Muggia, F. M., Cassileth, P. A., Ochoa, M., Flatow, F. A., Gellhorn, A., and Hyman, G. A. (1968): Treatment of breast cancer with medroxyprogesterone. *Ann. Intern. Med.,* 68:328–337.
20. Pearson, O. H., and Nasr, H. (1970): Hormonal steroids. *Excerpta Medica International Congress Series,* 219:602.
21. Nagasawa, H., and Yanai, R. (1971): Reduction by pituitary isograft of inhibitory effect of large dose of estrogen on incidence of mammary tumors induced by carcinogen in ovariectomized rats. *Int. J. Cancer,* 8:463–467.
22. Meites, J., Cassell, E. E., and Clark, J. H. (1971): Estrogen inhibition of mammary tumor growth in rats: counteraction by prolactin. *Proc. Soc. Exp. Biol. Med.,* 137:1225–1227.
23. Council on Drugs (1960): Androgens and estrogens in the treatment of disseminated mammary carcinoma. *J.A.M.A.,* 172:1271–1283.
24. Growley, L. G., and MacDonald, I. (1965): Delalutin and estrogens for the treatment of advanced mammary carcinoma in the postmenopausal woman. *Cancer,* 18:436–446.
25. Stoll, B. A. (1967): Progestin therapy of breast cancer: Comparison of agents. *Br. Med. J.,* 3:338–341.

26. Stoll, B. A. (1967): Effect of Lyndiol, an oral contraceptive, on breast cancer. *Br. Med. J.,* 1:150–153.
27. Stoll, B. A. (1967): Vaginal cytologyas, an aid to hormone therapy in postmenopausal cancer of the breast. *Cancer,* 20:1807–1813.
28. Sadoff, L., and Lusk, W. (1974): The effect of large doses of medroxyprogesterone acetate (MPA) on urinary estrogen levels and serum levels of cortisol T^4 LH and testosterone in patients with advanced cancer. *Obstet. Gynecol.,* 43:262–266.
29. Landau, R. L., Ehrlich, E. N., and Huggins, C. (1962): Estradiol benzoate and progesterone in advanced human breast cancer. *J.A.M.A.,* 182:632–636.
30. Kennedy, B. J. (1965): Hormone therapy for advanced breast cancer. *Cancer,* 18:1551–1557.
31. Pearson, O. H., and Ray, B. S. (1959): Results of hypophysectomy in the treatment of metastatic mammary carcinoma. *Cancer,* 12:85–92.
32. Lipsett, M. B., and Bergenstal, D. M. (1960): Lack of effect of human growth hormone and ovine prolactin on cancer in man. *Cancer Res.,* 20:1172–1178.
33. Kennedy, B. J., and French, L. (1965): Hypophysectomy in advanced breast cancer. *Am. J. Surg.,* 110:411–414.
34. O'Malley, B. W., McGuire, W. L., Kohler, P. O., and Korenman, S. G. (1969): Studies on the mechanism of steroid hormone regulation of synthesis of specific proteins. *Recent Prog. Horm. Res.,* 25:105–160.
35. Sherman, M. R., Corvol, P. L., and O'Malley, B. W. (1970): Progesterone binding components of chick oviduct. I. Preliminary characterization of cytoplasmic components. *J. Biol. Chem.,* 245:6085–6096.
36. Toft, D. O., and O'Malley, B. W. (1972): Target tissue receptors for progesterone: The influence of estrogen. *Endocrinology,* 90:1041–1045.
37. Schrader, W. T., Toft, D. O., and O'Malley, B. W. (1972): Progesterone-binding protein of chick oviduct. *J. Biol. Chem.,* 247:2401–2407.
38. Milgrom, E., Atger, M., and Baulieu, E.-E. (1970): Progesterone in uterus and plasma. IV. Progesterone receptor(s) in guinea pig uterus cytosol. *Steroids,* 16:741–754.
39. Falk, R. J., and Bardin, C. W. (1970): Uptake of tritiated progesterone by the uterus of the ovariectomized guinea pig. *Endocrinology,* 86:1059–1063.
40. Milgrom, E., Atger, M., Perrot, M., and Baulieu, E.-E. (1972): Progesterone in uterus and plasma. VI. Uterine progesterone receptors during the estrus cycle and implantation in the guinea pig. *Endocrinology,* 90:1071–1078.
41. Milgrom, E., Perrot, M., Atger, M., and Baulieu, E.-E. (1972): Progesterone in uterus and plasma. V. An assay of the progesterone cytosol receptor of the guinea pig uterus. *Endocrinology,* 90:1064–1070.
42. Corvol, P., Falk, R. J., Freifeld, M. L., and Bardin, C. W. (1972): *In vitro* studies of progesterone binding proteins in guinea pig uterus. *Endocrinology,* 90:1464–1469.
43. Faber, L. E., Sandmann, M. L., and Stavely, H. E. (1972): Progesterone binding in uterine cytosols of the guinea pig. *J. Biol. Chem.,* 247:8000–8004.
44. Milgrom, E., Thi, L., Atger, M., and Baulieu, E.-E. (1973): Mechanisms of regulating the concentration and the conformation of progesterone receptor(s) in the uterus. *J. Biol. Chem.,* 248:6366–6374.
45. Atger, M., Baulieu, E.-E., and Milgrom, E. (1974): An investigation of progesterone receptors in guinea pig vagina, uterine cervix, mammary glands, pituitary and hypothalamus. *Endocrinology,* 94:161–167.
46. Warembourg, M. (1974): Radioautographic study of the guinea pig uterus after injection and incubation with ^{3}H-progesterone. *Endocrinology,* 94:665–670.
47. Sar, M., and Stumpf, W. E. (1974): Cellular and subcellular localization of ^{3}H-progesterone or its metabolites in the oviduct, uterus, vagina and liver of the guinea pig. *Endocrinology,* 94:1116–1125.
48. MacLaughlin, D. T., and Westphal, U. (1974): Steroid protein interactions. XXX. A progesterone binding protein in the uterine cytosol of the pregnant guinea pig. *Biochim. Biophys. Acta,* 365:372–388.
49. Kontula, K., Janne, O., Rajakoski, E., Tanhuapaa, E., and Vihko, R. (1974): Ligand specificity of progesterone binding proteins in guinea pig and sheep. *J. Steroid Biochem.,* 5:39–44.
50. Freifeld, M. L., Feil, P. D., and Bardin, C. W. (1974): The *in vivo* regulation of the pro-

gesterone "receptor" in guinea pig uterus: Dependence on estrogen and progesterone. *Steroids*, 23:93-103.
51. Philibert, D., and Raynaud, J.-P. (1974): Progesterone binding in the immature rabbit and guinea pig. *Endocrinology*, 94:627-632.
52. Leavitt, W. W., and Grossman, C. J. (1974): Characterization of binding components for progesterone and 5α-pregnane-3,20-dione in the hamster uterus. *Proc. Natl. Acad. Sci. U.S.A.*, 71:4341-4345.
53. Leavitt, W. W., Toft, D. O., Strott, C. A., and O'Malley, B. W. (1974): A specific progesterone receptor in the hamster uterus: Physiologic properties and regulation during the estrous cycle. *Endocrinology*, 95:1041-1053.
54. Reel, J. R., VanDeward, S. K., Shih, Y., and Callantine, M. R. (1971): Macromolecular binding and metabolism of progesterone in the decidual and pseudopregnant rat and rabbit uterus. *Steroids*, 18:441-461.
55. McGuire, J. L., and Bariso, C. D. (1972): Isolation and preliminary characterization of a progestigen specific binding macromolecule from the 273,000 g supernatant of rat and rabbit uteri. *Endocrinology*, 90:496-506.
56. Faber, L. E., Sandmann, M. L., and Stavely, H. E. (1972): Progesterone binding proteins of the rat and rabbit uterus. *J. Biol. Chem.*, 247:5648-5649.
57. Faber, L., Sandmann, M. L., and Stavely, H. (1973): Progesterone and corticosterone binding in rabbit uterine cytosols. *Endocrinology*, 93:74-80.
58. Rao, B. R., Wiest, W. G., and Allen, W. M. (1973): Progesterone "receptor" in rabbit uterus. I. Characterization and estradiol-17β augmentation. *Endocrinology*, 92:1229-1240.
59. Davies, J., Challis, J. R. G., and Ryan, K. J. (1974): Progesterone receptors in the myometrium of pregnant rabbits. *Endocrinology*, 95:164-173.
60. Feil, P. D., Glasser, S. R., Toft, D. O., and O'Malley, B. W. (1972): Progesterone binding in the mouse and rat uterus. *Endocrinology*, 91:738-746.
61. Philibert, D., and Raynaud, J.-P. (1973): Progesterone binding in the immature mouse and rat uterus. *Steroids*, 22:89-99.
62. McGuire, J. L., and DeBella, C. (1971): *In vitro* evidence of progesterone receptor in the rat and rabbit uterus. *Endocrinology*, 88:1099-1103.
63. Milgrom, E., and Baulieu, E.-E. (1970): Progesterone in uterus and plasma. Binding in rat uterus 105,000 g supernatant. *Endocrinology*, 87:276-287.
64. Davies, J., and Ryan, K. J. (1972): The uptake of progesterone by the uterus of the pregnant rat *in vivo* and its relationship to cytoplasmic progesterone binding protein. *Endocrinology*, 90:507-515.
65. Davies, I. J., and Ryan, K. J. (1973): The modulation of progesterone concentration in the myometrium of the pregnant rat by changes in cytoplasmic "receptor" protein activity. *Endocrinology*, 92:394-401.
66. Saffran, J., Loeser, B. K., Haas, B., and Stavely, H. E. (1973): Binding of progesterone by rat uterus in vitro. *Biochem. Biophys. Res. Commun.*, 53:202-209.
67. Hsueh, A. J. W., Peck, E. J., and Clark, J. H. (1974): Receptor progesterone complex in the nuclear fraction of the rat uterus demonstrated by tritiated progesterone exchange. *Steroids*, 24:599-611.
68. John, P. N., and Rogers, A. W. (1972): The distribution of tritiated progesterone and tritiated megestrol acetate in the uterus of the ovariectomized rat. *J. Endocrinol.*, 53:375-386.
69. Stumpf, W. E., and Sar, M. (1973): Cellular and subcellular localization of tritiated progesterone and its metabolites in rat uterus studied by autoradiography. *J. Steroid Biochem.*, 4:477-481.
70. Verma, V., and Laumas, K. R. (1973): *In vitro* binding of progesterone to receptors in the human endometrium and the myometrium. *Biochem. Biophys. Acta*, 317:403-419.
71. Batra, S. (1973): Binding of progesterone *in vitro* by the cytoplasmic components of the human myometrium. *J. Endocrinol.*, 57:561-562.
72. Kontula, K., Janne, O., Luukkainen, T., and Vihko, R. (1973): Progesterone binding protein in human myometrium. Ligand specificity and some physiochemical characteristics. *Biochim. Biophys. Acta*, 328:145-153.
73. Haukkamaa, M. (1974): Binding of progesterone by rat myometrium during pregnancy and by human myometrium in late pregnancy. *J. Steroid Biochem.*, 5:73-79.

74. Smith, H. E., Smith, R. G., Toft, D. O., Neergaard, J. R., Burrows, E., and O'Malley, B. W. (1974): Binding of steroids to progesterone receptor proteins in chick oviduct and human uterus. *J. Biol. Chem.*, 249:5924–5932.
75. Guerigvian, J. L., Sawyer, M. E., and Pearlman, W. H. (1974): Comparative study of progesterone and cortisol binding activity in the uterus and serum of pregnant and nonpregnant women. *J. Endocrinol.*, 61:331–345.
76. Rao, B. R., Wiest, W. G., and Allen, W. M. (1974): Progesterone receptor in human endometrium. *Endocrinology*, 95:1275–1281.
77. Young, P. C. M., and Cleary, R. E. (1974): Characterization and properties of progesterone binding components in human endometrium. *J. Clin. Endocrinol. Metab.*, 39:425–439.
78. Pollow, K., Lubbert, H., Boquoi, E., Kruezer, G., and Pollow, B. (1974): Characterization and comparison of receptors for 17β-estradiol and progesterone in human proliferative endometrium and endometrial carcinoma. *Endocrinology*, 96:319–328.
79. Haukkamaa, M., and Luukkainen, T. (1974): The cytoplasmic progesterone receptor of human endometrium during the menstrual cycle. *J. Steroid Biochem.*, 5:447–452.
80. Kontula, K., Janne, O., Luukkainen, T., and Vihko, R. (1974): Progesterone binding protein in human myometrium. Influence of metal ions on binding. *J. Clin. Endocrinol. Metab.*, 38:500–503.
81. Smith, R. G., Iramain, C. A., Buttram, V. C., and O'Malley, B. W. (1975): Purification of human uterine progesterone receptor. *Nature*, 253:271–272.
82. Deshpande, N., Bulbrook, R. D., and Belzer, F. O. (1966): An apparent selective accumulation of progesterone by the human breast. *Excerpta Medica Int. Congr. Series*, 132:750–753.
83. Terenius, L. (1973): Estrogen and progesterone binders in human and rat mammary carcinoma. *Eur. J. Cancer*, 9:291–294.
84. Shyamala, G. (1973): Specific cytoplasmic glucocorticoid hormone receptors in lactating mammary glands. *Biochemistry*, 12:3085–3090.
85. Gardner, D. G., and Wittliff, J. L. (1973): Characterization of a distinct glucocorticoid-binding protein in the lactating mammary gland of the rat. *Biochim. Biophys. Acta*, 320:617–627.
86. Horwitz, K. B., Costlow, M. E., and McGuire, W. L. (1975): MCF-7: A human breast cancer cell line with estrogen, androgen, progesterone and glucocorticoid receptors. *Steroids*, 26:785–795.
87. Hsueh, A. J. W., Peck, E. J., and Clark, J. H. (1975): Progesterone antagonism of estrogen receptor and estradiol-induced uterine growth. *Nature*, 254:337–338.
88. Bullock, D. W., and Wellen, G. F. (1974): Regulation of a specific uterine protein by estradiol and progesterone in ovariectomized rabbits. *Proc. Soc. Exp. Biol. Med.*, 146:294–298.
89. Heuson, J. C., Waelbroeck, C., Legros, N., Gallez, G., Robyn, C., and L'Hermite, M. (1972): Inhibition of DMBA-induced mammary carcinogenesis in the rat by 2-Br-α-ergocryptine (CB154), an inhibitor of prolactin secretion, and by nafoxidine (U-11,100A) an estrogen antagonist. *Gynecol. Invest.*, 2:130–137.
90. Terenius, L., and Ljungkvist, I. (1972): Aspects on the mode of action of antiestrogens and antiprogesterones. *Gynecol. Invest.*, 3:96–107.
91. Voorhess, M. L. (1967): Masculinization of the female fetus associated with norethindrone-mestranol therapy during pregnancy. *J. Pediatr.*, 71:128–131.
92. Suchowsky, G. K., and Junkmann, K. (1961): A study of the virilizing effect of progestogens on the female rat fetus. *Endocrinology*, 68:341–349.
93. Mowszowicz, I., Bieber, D., Chung, K., Bullock, L. P., and Bardin, C. W. (1974): Synandrogenic and antiandrogenic effect of progestins: Comparison with nonprogestational antiandrogens. *Endocrinology*, 95:1589–1599.
94. Fahim, M. S., and Hall, D. G. (1970): Effect of ovarian steroids on hepatic metabolism. I. Progesterone. *Am. J. Obstet. Gynecol.*, 106:183–186.
95. Bullock, L. P., Barthe, P. L., Mowszowicz, I., Orth, D. N., and Bardin, C. W. (1975): The effect of progestins on submaxillary gland epidermal growth factor: demonstrating of androgenic, synandrogenic and antiandrogenic actions. *Endocrinology*, 97:189–195.
96. Naqvi, E., Zarrow, M. X., and Dennberg, V. H. (1969): Inhibition of androgen-induced precocious puberty by progesterone. *Endocrinology*, 84:669–670.

97. Kratochwil, K. (1971): *In vitro* analysis of the hormonal basis for the sexual dimorphism in the embryonic development of the mouse mammary gland. *J. Embryol. Exp. Morphol.*, 25:141–153.
98. Rousseau, G. G., Baxter, J. D., and Tomkins, G. M. (1972): Glucocorticoid receptors: Relations between steroid binding and biological effects. *J. Mol. Biol.*, 67:99–115.
99. Gardner, D. G., and Wittliff, J. L. (1973): Demonstration of a glucocorticoid hormone-receptor complex in the cytoplasm of a hormone-responsive tumor. *Br. J. Cancer*, 27: 441–444.
100. Shyamala, G. (1974): Glucocorticoid receptors in mouse mammary tumors. *J. Biol. Chem.*, 249:2160–2163.
101. Hagino, N. (1972): The effect of synthetic corticosteroids on ovarian function in the baboon. *J. Clin. Endocrinol. Metab.*, 35:716–721.
102. Cunningham, G. R., Capterton, E. H., and Goldzieher, J. W. (1975): Antiovulatory activity of synthetic corticoids. *J. Clin. Endocrinol. Metab.*, 49:265–267.
103. Asselin, J., Labrie, F., Kelly, P. A., Philibert, D., and Raynaud, J.-P. (1976): Specific progesterone receptors in dimethylbenzanthracene (DMBA)-induced mammary tumors. *Steroids*, 27:395–404.
104. Goral, J. E., and Wittliff, J. L. (1976): Characteristics of progesterone-binding components in neoplastic mammary tissues of the rat. *Cancer Res.*, 36:1886–1893.
105. Chen, C. L., and Meites, J. (1970): Effects of estrogen and progesterone on serum and pituitary prolactin levels in ovariectomized rats. *Endocrinology*, 86:503–505.
106. Horwitz, K. B., and McGuire, W. L. (1975): Specific progesterone receptors in human breast cancer. *Steroids*, 25:497–505.
107. Horwitz, K. B., McGuire, W. L., Pearson, O. H., and Segaloff, A. (1975): Predicting response to endocrine therapy in human breast cancer: A hypothesis. *Science*, 189:726–727.
108. King, R. J. B., and Mainwaring, W. I. P. (1974): Glucocorticoids. In: *Steroid-Cell Interactions.* University Park Press, Baltimore.
109. McGuire, W. L., Pearson, O. H., and Segaloff, A. (1976): Predicting hormone responsiveness in human breast cancer. In: *Estrogen Receptors in Human Breast Cancer*, edited by W. L. McGuire, P. P. Carbone, and E. P. Vollmer. Raven Press, New York.

Progesterone Receptors in Normal and Neoplastic Tissues, edited by W. L. McGuire et al. Raven Press, New York © 1977.

Clinical Correlations of Endocrine Ablation with Estrogen and Progesterone Receptors in Advanced Breast Cancer

Norman Bloom, Ellis Tobin, and George A. Degenshein

Department of Surgery, Maimonides Medical Center, Brooklyn, New York 11219

In 1896 Sir George Beatson (1) described the remission of advanced breast cancer in two women on whom he had performed bilateral oöphorectomy. In the 50 years subsequent to that publication, reports from numerous centers demonstrated that approximately one-third of all women with recurrent or metastatic breast cancer responded to either additive or ablative endocrine therapy. During the past decade the accurate selection of those patients most likely to respond has been improved to approximately 60%. The latter has been accomplished by selecting those individuals whose tumors demonstrate the capacity to take up and retain estradiol-17β. In an attempt to further improve the accuracy of this selection process, we undertook a clinical trial to evaluate the role progesterone receptors play in identifying hormone dependency.

MATERIALS AND METHODS

Tumor Specimens

Breast tumor specimens were received in the pathology department after surgical removal and were immediately placed in a cryostat maintained at −20°C. Portions of the tissue were trimmed of fat and necrotic areas and held for estrogen receptor (ER) and progesterone receptor (PgR) protein assay pending histological confirmation of malignancy. The tissues for receptor site assay were immersed in liquid nitrogen and stored at −70°C. The average lapsed time from surgical removal to immersion in liquid nitrogen was 30 min.

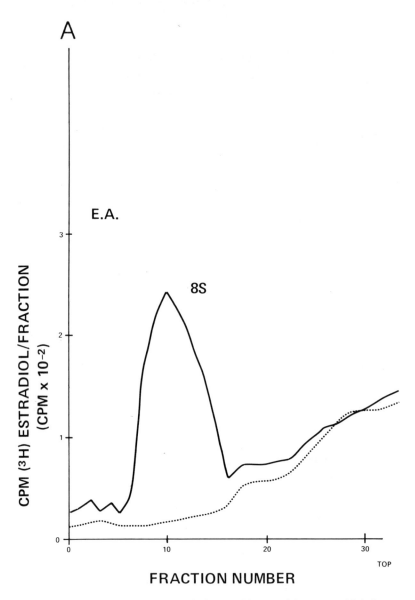

FIG. 1. A: Estrogen sucrose gradient analysis on a 34-year-old woman with inflammatory carcinoma, demonstrating a strongly positive receptor component (82 fmoles/mgP) sedimenting in the 8S region. Estradiol-17β binding to specific receptor protein (———). Estradiol-17β binding to nonspecific proteins (---).

Tissue Preparation

At the time of assay, the tissue was thawed to 4°C, weighed, and pulverized under liquid nitrogen to a powdered consistency (Thermovac Industries Corp., Copiaque, N.Y.). The pulverized tissue was placed in 4

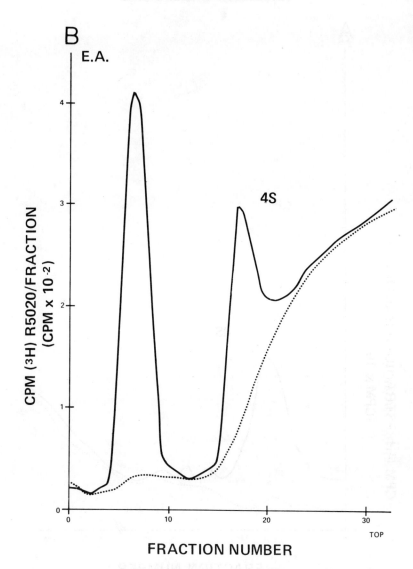

FIG. 1. B: Progesterone sucrose gradient analysis demonstrating a strongly positive receptor component sedimenting in both the 8S (147 fmoles/mgP) and 4S (122 fmoles/mgP) region. This patient responded to endocrine ablation. R5020 binding to specific receptor protein (——). R5020 binding to nonspecific proteins (---).

volumes of homogenization buffer, and maintained at 4°C throughout the assay.

Homogenization of tissue was accomplished on ice using a Teflon-glass homogenizer (A. H. Thomas Co., Philadelphia, Pa.) with three 15-sec pulses and a 45-sec cooling period between each pulse. The homogenate

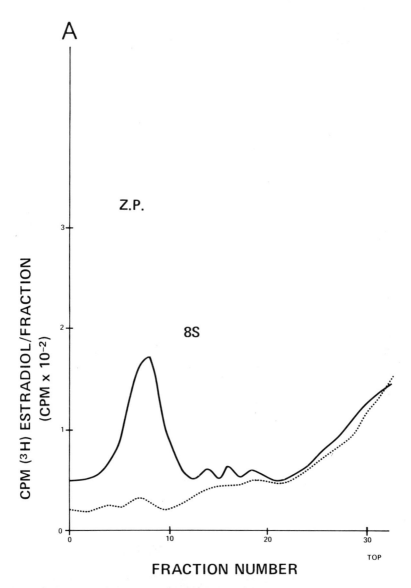

FIG. 2. A: Estrogen sucrose gradient analysis on a 46-year-old woman with infiltrating duct cell carcinoma metastatic to the bone and chest wall. This assay demonstrates the presence of a positive receptor component (22 fmoles/mgP) sedimenting in the 8S region. Estradiol-17β binding to specific receptor protein (———). Estradiol-17β binding to non-specific proteins (- - -).

was placed in a Beckman L Ultracentrifuge and spun on an SW 50.1 rotor (Beckman Instruments, Inc., Mountainside, N.J.) for 30 min at 240,000 × g to obtain the supernatant cytosol. Cytosol protein concentration was determined by the method of Waddell (8).

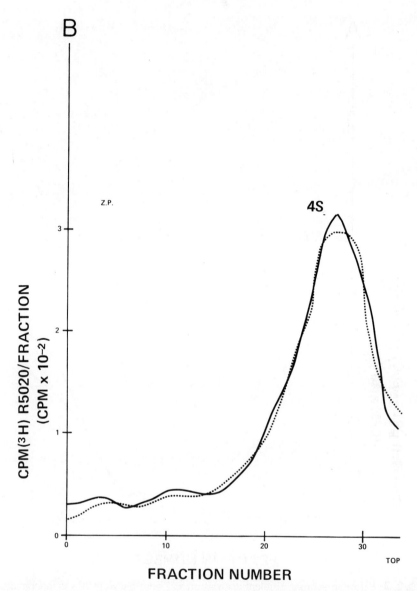

FIG. 2. B: Progesterone sucrose gradient analysis demonstrating nonspecific binding in the 4S region. Patient did not respond to endocrine ablation. R5020 binding to specific receptor protein (———). R5020 binding to nonspecific proteins (- - -).

ER Assay

The homogenization buffer for ER determination consisted of 10 mM Tris-HCl, pH 7.4, containing 0.5 mM dithiothreitol (Sigma Chemical Co., St. Louis, Mo.). Sucrose gradients (10 to 30%) consisting of ER homogeni-

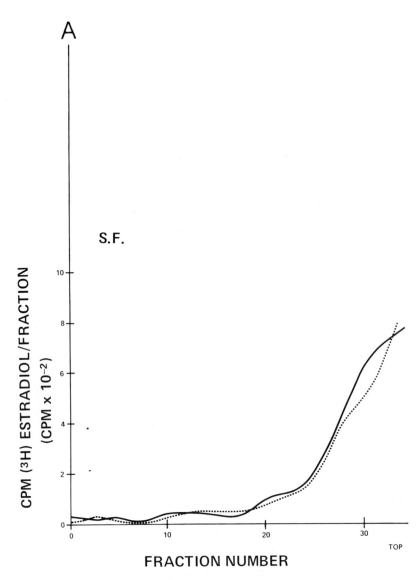

FIG. 3. A: Estrogen sucrose gradient analysis on a 27-year-old woman with inflammatory carcinoma of the breast with no demonstrable receptor component. Estradiol-17β binding to specific receptor protein (——). Estradiol-17β binding to nonspecific receptor protein (- - -).

zation buffer containing 10 mM KCl and 1 mM EDTA were prechilled overnight in polyallomar centrifuge tubes (Beckman).

Two 150-μl portions of cytosol were reacted with 50 μl of either buffer alone or with buffer containing 1 μM CI-628 (Parke-Davis and Co., Detroit,

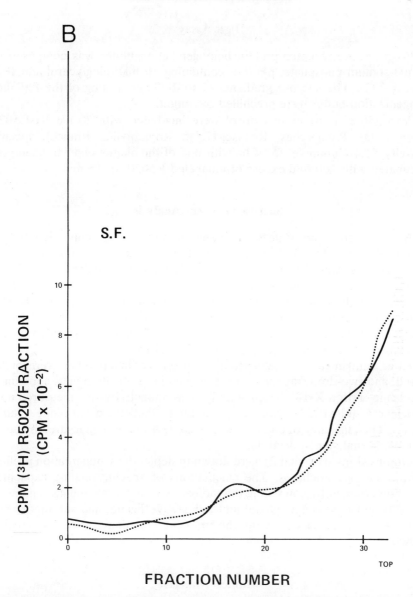

FIG. 3. B: Progesterone receptor analysis demonstrating no significant receptor component. Patient had a negative response to endocrine ablation. R5020 binding to specific receptor protein (——). R5020 binding to nonspecific proteins (---).

Mich.) for 15 min. A 50-µl aliquot of buffer containing 2.5 nM ^{3}H-estradiol-17β, specific activity 98.5 Ci/mmole (New England Nuclear Corp., Boston, Mass.), was added to each reaction mixture, agitated, and allowed to incubate for 4 hr.

PgR Assay

Progesterone receptor protein homogenization buffer was composed of 5 mM sodium phosphate, pH 7.4, containing 10 mM thioglycerol and 10% glycerol (2). The sucrose gradients (5 to 20%) consisting of the PgR homogenization buffer were prechilled overnight.

Two 250-μl portions of cytosol were incubated with 20 nM ^{3}H-R5020 (Centre De Recherches, Roussel-Uclaf, Romainville, France), specific activity 51.4 Ci/mmole, for 4 hr with one of the aliquots having been preincubated with 100-fold excess of unlabeled R5020 for 15 min.

Sucrose Gradient Analysis

A 1-ml suspension of dextran-coated charcoal (DCC) containing 0.25% Norit A and 0.0025% RIA-grade dextran (Sigma) in 10 mM Tris-HCl, pH 8.0, was centrifuged for 10 min at 2,000 × g to sediment the DCC pellet. The cytosol incubation mixtures were transferred onto the pellet, agitated for 10 min, and centrifuged for 10 min at 2,000 × g. Aliquots consisting of 200 μl of the resulting supernatant were layered on the sucrose gradients. Centrifugation at 240,000 × g for 17 hr was performed on the Beckman L Ultracentrifuge using the SW 50.1 rotor.

After centrifugation, the bottoms of the polyallomar tubes were punctured, and two-drop fractions were collected in scintillation vials containing a toluene–Triton X-100-based scintillant. Radioactivity was measured on a Nuclear Liquimat 220 scintillation counter (Picker Corp., White Plains, N.Y.). The efficiency of counting averaged 43% and was measured by automatic external standardization.

Graphs (Figs. 1, 2, and 3) were drawn to depict the sedimentation profiles of the sucrose gradients. The calculations of specific binding capacities were determined by computing the radioactivity in the 8S and 4S regions of the gradient containing cytosol and radioactive ligand, and subtracting the corresponding radioactivity in the gradient containing excess unlabeled ligand.

Clinical Material

The 37 patients included in this series ranged in age from 27 to 84 years with a mean of 56 years. Of these women, 23 were postmenopausal and 14 were premenopausal. The predominant tumor histology was infiltrating ductal carcinoma occurring in 69% of the patients. Additional histological types in order of frequency included: inflammatory carcinoma, 11%; infiltrating lobular carcinoma, 5%; medullary carcinoma, 5%; metastatic adenocarcinoma, 5%; and multiple independent tumors, 5% (Table 1).

TABLE 1. Clinical data of patients evaluated for estrogen and progesterone receptors

Patient	Age	Menopause	Clinical stage	Tumor type	Therapy[a]	ER	PgR	Response[b]
D. D.	53	Post	B	Infiltrating lobular	Modified radical	+	+	0
S. F.	27	Pre	D	Infiltrating medullary and inflammatory	Adrenalectomy, oöphorectomy	−	−	−
A. K.	65	Post	D	Infiltrating ductal	Adrenalectomy, oöphorectomy	+	+	+
E. R.	34	Pre	B	Infiltrating ductal	Radical oöphorectomy	+	+	−
R. S.	59	Post	D	Infiltrating ductal	Radiotherapy	+	+	−
S. S.	62	Post	D	Adenocarcinoma; bone metastasis	Adrenalectomy, oöphorectomy	−	−	−
I. W.	51	Post	D	Infiltrating ductal	Adrenalectomy, oöphorectomy	+	+	+
E. A.	36	Pre	D	Infiltrating ductal inflammatory	Adrenalectomy, oöphorectomy	+	+	+
M. B.	84	Post	A	Infiltrating medullary	Modified radical	−	−	0
A. B.	45	Post	A	Infiltrating lobular	Simple mastectomy	+	+	0
G. C.	44	Pre	A	Infiltrating ductal and lobular	Radical mastectomy	−	−	0
M. C.	49	Pre	D	Infiltrating ductal	Chemotherapy	+	+	+
B. D.	73	Post	D	Infiltrating ductal	Adrenalectomy, oöphorectomy	+	+	+
M. D.	71	Post	D	Infiltrating ductal	Adrenalectomy, oöphorectomy	+	+	+
I. G.	68	Post	B	Infiltrating ductal	Radical mastectomy	+	−	0
N. G.	58	Post	B	Infiltrating ductal	Radical mastectomy	+	+	0
J. K.	77	Post	D	Infiltrating ductal	Chemotherapy	+	−	−
M. L.	79	Post	C	Infiltrating ductal	Chemotherapy	+	+	+
T. L.	53	Post	D	Infiltrating ductal	Refused therapy	+	+	−

TABLE 1. (Continued)

Patient	Age	Menopause	Clinical stage	Tumor type	Therapy[a]	ER	PgR	Response[b]
S. M.	72	Post	A	Infiltrating ductal	Radical mastectomy	–	+	0
R. M.	58	Post	D	Infiltrating ductal	Adrenalectomy, oöphorectomy	+	–	–
D. M.	34	Pre	B	Infiltrating ductal comedocarcinoma	Radical oöphorectomy	+	+	0
R. M.	51	Post	B	Infiltrating ductal	Modified radical	+	+	0
L. N.	76	Post	D	Infiltrating ductal inflammatory	Radiotherapy, chemotherapy	+	+	–
B. N.	42	Pre	B	Infiltrating ductal	Radical mastectomy	+	+	0
Z. P.	46	Pre	D	Infiltrating ductal	Adrenalectomy, oöphorectomy	+	–	–
T. R.	35	Pre	B	Infiltrating ductal	Modified radical	+	+	0
E. S.	34	Pre	A	Infiltrating ductal	Radical mastectomy	–	–	0
F. S.	55	Post	A	Infiltrating ductal	Modified radical	–	–	0
B. S.	62	Post	D	Lung metastasis	Lobectomy, chemotherapy	–	+	+
T. W.	51	Post	B	Infiltrating ductal	Modified radical	+	+	0
A. Z.	79	Post	D	Infiltrating ductal	Adrenalectomy, chemotherapy	–	–	–
E. Z.	72	Post	D	Infiltrating ductal	Adrenalectomy, oöphorectomy	+	+	+
S. P.	50	Pre	D	Infiltrating ductal	Adrenalectomy, oöphorectomy	–	–	–
D. L.	45	Pre	A	Infiltrating ductal	Modified radical	+	+	0
R. D.	45	Pre	B	Infiltrating ductal	Modified radical	–	–	0
M. T.	84	Post	D	Inflammatory	Simple mastectomy, radiation therapy	–	–	–

[a] Patients receiving chemotherapy were given cyclophosphamide (Cytoxan®), 5-fluorouracil, and methotrexate.
[b] Response to therapy is indicated where applicable. Duration of remission exceeds 6 months in all cases indicated as positive.

RESULTS

Occurrence of ER and PgR in Human Breast Cancer

Sucrose gradient analysis was considered positive when the specific binding capacity, expressed as femtomoles per milligram protein (fmoles/mgP), was found to be greater than or equal to 7 fmoles/mgP. Twenty-one patients (57%) in this series demonstrated both ER and PgR in their tumors. Ten patients (27%) were ER and PgR negative. Four patients (11%) were ER positive and PgR negative, and two patients (5%) were ER negative and PgR positive (Table 2).

TABLE 2. *Occurrence of estrogen and progesterone receptors in human breast cancers*

Receptor profile	No. of patients	Percentage
ER^+PgR^+	21	57
ER^+PgR^-	4	11
ER^-PgR^-	10	27
ER^-PgR^+	2	5

Concentration of ER and PgR and Sucrose Gradient Sedimentation Patterns of the Hormone-Receptor Complex

It has been established that there are at least two sedimentation forms of the estrogen receptor protein: one appearing in the 8S region and the other appearing in the 4S region of the sucrose gradient (5). However, the significance of these two components remains in question; some investigators feel that the 8S form represents the component most associated with a hormone-dependent tumor. In this series, 12 of 37 tumors demonstrated only the 8S form, 1 of 37 tumors demonstrated only the 4S form, and 12 of 37 tumors demonstrated a combination of both the 8S and the 4S molecular forms of ER (Table 3). A range of from 0 to 542 fmoles/mgP could be determined in this group of 37 tumors.

TABLE 3. *Sedimentation forms of ER and PgR in human breast cancers*

	No. of patients with PgR component forms			
	8S	4S	8S and 4S	None
No. of patients with ER component forms				
8S	0	6	3	3
4S	0	1	0	0
8S and 4S	1	6	4	1
None	0	2	0	10

Both an 8S and 4S form of the progesterone receptor could be demonstrated in this series as well (Table 3). In our analysis, it appears, however, that the predominant receptor component was the 4S form of PgR. Only 1 of 37 tumors contained just the 8S component, 15 of 37 contained just the 4S component, and 7 of 37 contained both an 8S and 4S component with the latter predominating. Binding capacities ranged from 0 to 421 fmoles PgR/mgP.

A display of the different forms of the ER with the PgR can be seen in Table 3.

Clinical Correlation of Receptor Proteins with Endocrine Ablation

The criteria followed for objective remission include:

1. a 50% reduction in size of greater than 50% of all measurable lesions,
2. the appearance of no new lesions or progression of any lesion already present,
3. radiographic evidence of the sclerosing or disappearance of osseous lesions, and
4. a minimum period of remission of 6 months.

The clearance of a malignant pleural effusion or the disappearance of bone pain was not considered to be a criterion of remission.

Thirteen patients were subjected to surgical endocrine ablation (Table 4). The results obtained demonstrate 86% correlation of clinical remission with the presence of estrogen and progesterone receptors. Two patients whose tumors contained ER but in whom no PgR could be demonstrated did not respond to endocrine manipulation. The four patients in whom no receptor protein could be demonstrated also did not respond to endocrine ablative surgery.

The concentration of receptors as well as the sedimentation components in those patients who subsequently underwent endocrine ablation can be

TABLE 4. *Clinical correlation of objective remissions following ablative surgery with hormone receptors*

Receptor profile	No. of patients	No. of remissions	Percent remission
ER$^+$PgR$^+$	7	6	86[a]
ER$^+$PgR$^-$	2	0	0
ER$^-$PgR$^-$	4	0	0

[a] All patients in this clinical series were subjected to adrenalectomy and oöphorectomy with two exceptions. In the ER$^+$PgR$^+$ groups one patient had oöphorectomy only and was the one ablated treatment failure in this group. In the ER$^-$PgR$^-$ group one postmenopausal patient had an adrenalectomy only.

TABLE 5. Quantitation and sedimentation profile of estrogen and progesterone receptors in patients who subsequently underwent hormone ablation

| | Patient | ER | | | PgR | | | Months duration of response at present |
		8S	4S	Total[a]	8S	4S	Total[b]	
ER+PgR+	A. K.	94	7	101	6	53	59	13
	E. R.	73	0	73	8	80	88	N.R.[c]
	I. W.	124	18	142	143	5	148	11
	E. A.	82	0	82	147	122	269	8
	B. D.	36	12	48	0	23	23	7
	M. D.	26	8	34	16	12	28	7
	E. Z.	319	0	319	27	33	60	6[d]
ER+PgR−	R. M.	37	12	49	0	5	5	N.R.
	Z. P.	22	0	22	0	0	0	N.R.
ER−PgR−	S. P.	0	0	0	0	0	0	N.R.
	S. F.	0	0	0	0	6	6	N.R.
	S. S.	0	5	5	0	0	0	N.R.
	A. Z.	4	0	4	0	5	5	N.R.

[a] fmoles of ER bound/mg protein.
[b] fmoles of PgR bound/mg protein.
[c] No response.
[d] Patient died 6 months post-adrenalectomy. Metastatic supraclavicular node was ER+PgR+, malignant chest wall nodule ER−PgR−.

seen in Table 5. It should be noted that the 8S form of the ER and the 4S form of the PgR seem to be the predominating components in this group.

DISCUSSION

During the past decade, biochemical characteristics have been identified in breast tumors that have aided in predicting a patient's response to endocrine therapy. Folca et al. (3) first recognized in human breast cancers the selective uptake of tritiated hexestrol in the tumors of women who later experienced objective remissions to ablative surgery. Further investigations, based on the implications of this study, showed that approximately 60% of all primary breast cancers and 50% of metastatic breast tumors contained specific estrogenic binding proteins. Clinical trials evaluating the usefulness of this finding as a determinant of responsiveness to ablative or additive hormonal therapy yielded a 55% correlation in multiple centers.

Recently the identification by Philibert and Raynaud (6) of progesterone binding in the rat uterus and its identification and correlation with estrogen receptors in breast cancer tissues by Raynaud et al. (7) raised the possibility of a better discriminant of hormone responsiveness.

In a small clinical series, Horwitz et al. (4) reported 100% correlation in the response of advanced breast cancer patients to hormonal manipulation with the presence in their tumors of both estrogen and progesterone receptors. They postulated that the presence of the progesterone receptor reflected a functional expression of *in vivo* estrogen activity and based this hypothesis on animal studies which demonstrated the induction of progesterone receptor synthesis following estrogen stimulation.

In our clinical series, 86% of those patients whose tumors contained estrogen and progesterone receptors responded to endocrine ablation irrespective of the presence or absence of visceral metastasis. Surgical ablation was selected as the modality of therapy because it was felt by the authors to produce the most dramatic and prolonged remissions of all forms of hormonal manipulation.

The one failure in the estrogen- and progesterone-positive group occurred in a patient with diffuse liver metastasis. Fulminant hepatic failure developed immediately after surgery, and the patient died 4 weeks after oöphorectomy. All patients subjected to endocrine ablation who were PgR negative did not respond and showed advancement of their disease within a short time.

Based on our experience with estrogen and progesterone receptors, a protocol for the management of advanced breast cancer has been implemented at our institution (Fig. 4).

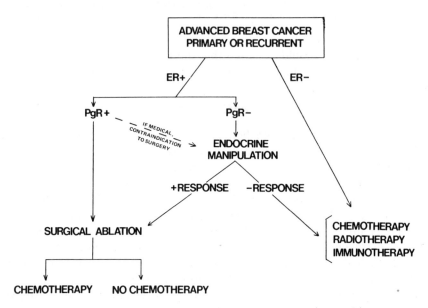

FIG. 4. Protocol for the management of advanced breast cancer at the Maimonides Medical Center.

ACKNOWLEDGMENTS

The authors wish to acknowledge the cooperation of the following individuals without whom this study could not have been undertaken: Drs.

Franco Ceccarelli, Burtan Herz, Sakda Suwan, Manuel Lagmay, Alan Kane, Rao Daluvoy, Qamar Amiruddin, and Ira Krafchin. The preparation of this manuscript would not have been possible without the help of Mrs. Noreen Katz and Mrs. Alice Bloom and the technical assistance of Mr. Peter Berman, Ms. Barbara Schreibman, Ms. Ellen Cutler, and Miss Christine Pagonis. This work was supported by the Jack R. Aron Research Foundation and the National Institute of Health Grant #5S01RR05497-B.

REFERENCES

1. Beatson, G. T. (1896): On the treatment of inoperable cases of carcinoma of the mamma: Suggestion for a new method of treatment with illustrative cases. *Lancet*, 2:796-798.
2. Faber, L. E., Sandmann, M. L., and Stavely, H. E. (1973): Effect of buffers and electrolytes on the stability of uterine progesterone receptors. *Fed. Proc.*, 22:229.
3. Folca, P. J., Glascock, R. F., and Irvine, W. T. (1961): Studies with tritium-labeled hexestrol in advanced breast cancer: Comparison of tissue accumulation of hexestrol with response to bilateral adrenalectomy and oophorectomy. *Lancet*, 2:796-798.
4. Horwitz, K. B., McGuire, W. L., Rearson, O. H., and Segaloff, A. (1975): Predicting response to endocrine therapy in human breast cancer: A hypothesis. *Science*, 189:726-727.
5. McGuire, W. L., Carbone, P. P., and Vollmer, E. P. (Eds.) (1975): *Estrogen Receptors in Human Breast Cancer*. Raven Press, New York.
6. Philibert, D., and Raynaud, J. P. (1973): Progesterone binding in the immature mouse and rat uterus. *Steroids*, 22:89-98.
7. Raynaud, J. P., Bouton, M. M., Philibert, D., Delarue, J. C., Guerinot, F., and Bohuon, C. (1975): Progesterone and estradiol binding sites on human breast carcinoma. Presented at the International Study Group for Steroid Hormones, Rome.
8. Waddell, W. J. (1956): A simple ultraviolet spectrophotometric method for the determination of proteins. *J. Lab. Clin. Med.*, 48:311-314.

Progesterone Receptors in Normal and Neoplastic Tissues, edited by W. L. McGuire et al. Raven Press, New York © 1977.

Estrogen and Progesterone Receptors in Human Breast Cancer

G. Leclercq, J. C. Heuson, M. C. Deboel, N. Legros, E. Longeval, and *W. H. Mattheiem

*Service de Médecine et Laboratoire d'Investigation Clinique†; and *Service de Chirurgie, Institut Jules Bordet, 1000 Brussels, Belgium*

It is now well established that most breast cancers contain cytoplasmic estrogen receptors (ERs). It is also agreed that the presence of ERs, at least in substantial concentration, is associated with a higher rate of response to endocrine treatments than their absence (7,14). A review of 436 trials revealed that 56% of patients with ER-positive tumors responded to endocrine treatments, whereas only 10% of ER-negative cases did (14). The question then arose why many ER-positive cases failed to be clinical responders.

The observation that ER-positive tumor cytosols are characterized by a large range of receptor concentrations (9,11) led us to investigate the potential predictive value of the quantitative assessment of receptors for guiding endocrine therapy. In a series of 34 patients, we found a highly significant relationship between ER concentrations and the probability of response to endocrine treatments (4). However, some patients with high ER levels appeared clinically refractory, and additional markers of hormone dependency were needed.

Because in target tissues the synthesis of progesterone receptors (PgRs) depends on the action of estrogens and requires an intact estrogen-receptive machinery, Horwitz et al. (6) proposed that PgR might be such a marker. This hypothesis found support in the observation that PgR was absent in almost all ER-negative tumors but present in approximately half of the ER-positive ones (6,19), a proportion close to the rate of responders in patients having tumors of this type. A small clinical study including nine patients provided further support insofar as all three responses occurred in cases with tumors containing both ER and PgR (6).

The purpose of the present study was to test Horwitz and co-workers'

---

† This service is a member of the European Organization for Research on Treatment of Cancer (E.O.R.T.C.).

theory on our own material. Tritiated estradiol-17β and ³H-R5020 (5,17,18) were used for assaying ER and PgR, respectively, in malignant and benign breast lesions.

METHODS

Biochemistry

TISSUES

We studied 209 samples from 148 primary (146 female, 2 male) and 61 metastatic (59 female, 2 male) breast cancers. They were carcinomas of various histological types in about the same proportion as in a previous study (10). We also analyzed 26 benign hyperplastic breast lesions including 10 dysplasias, 9 fibroadenomas, 6 gynecomastias, and 1 giant fibroadenoma. All specimens were obtained at operations and within minutes or, rarely, within 1 or 2 hr, were washed in cold Krebs-Ringer-Henselheit glucose buffer pH 7.3. They were then dipped into the homogenization buffer (10 mM HCl pH 7.5 containing 1.5 mM EDTA and 6 mM β-mercaptoethanol) before being frozen in liquid nitrogen.

ESTROGEN AND PROGESTERONE BINDING ASSAY

Estrogen and progesterone receptors in the tissue samples were assayed by measuring the binding affinity of their cytosol fraction for ³H-estradiol-17β (80 to 110 Ci/mmole, the Radiochemical Center, Amersham, England) or ³H-R5020 (51 Ci/mmole), respectively, using a dextran-coated method previously described (10); binding assays were carried out at 4°C according to the E.O.R.T.C. recommendations (1). Binding indices of the reactions were determined on Scatchard plots (20). In these coordinates, when linear functions were obtained, the dissociation constant of the reaction (K_d) was measured by the reverse of the slope; the concentration of binding sites (n) was read at the intersect of the line with the abscissa. When curved functions tending toward horizontal at high concentration of bound ³H-steroid were obtained, the indices of the reaction of higher binding affinity were evaluated by the extrapolating procedure of Mester et al. (15). The concentration of binding sites was expressed as fmoles ³H-E_2 or ³H-R5020 per milligram tissue protein. Tissue protein was calculated by the following equation (1):

$$\text{tissue protein} = \text{total protein} - \text{serum albumin} \times \frac{100}{60}$$

PROTEIN AND RADIOACTIVITY ASSAY

Total protein concentration was measured by the method of Lowry et al. (12) using serum albumin as standard. Human serum albumin was assayed by radial immunodiffusion on M-partigen plates (13). Radioactivity was determined in a PPO-POPOP-toluene-Triton X-100 scintillation medium (8).

Assessment of Clinical Response to Endocrine Treatment in Advanced Breast Cancer

PATIENTS

All women with inoperable, recurrent, or metastatic carcinoma of the breast whose response to an endocrine treatment could be assessed and who had a biopsy taken for ER and PgR determinations were included in the study. Any patient with a second cancer that could be confused with breast carcinoma was excluded. Any patient having received hormones or cytotoxic chemotherapy within 2 weeks of the biopsy (2 months in the case of depot preparations) or between biopsy and initiation of the treatment to be evaluated was also excluded. Another cause of exclusion was the presence of central nervous system involvement or massive liver involvement [i.e., more than one-third of the liver mass replaced by metastatic tissue (16)].

ASSESSMENT OF PATIENTS' RESPONSE TO TREATMENTS

The criteria for assessment of patients' response to treatment were basically those of the E.O.R.T.C. Breast Cancer Cooperative Group (2). The following modifications were, however, introduced. Premature interruption of treatment because of drug intolerance or patient loss to follow-up was interpreted as a cause of ineligibility instead of therapeutic failure because the purpose of the present study was to analyze response to treatments that were actually administered for a long enough period of time.

The medical records were presented in detail by one of the clinicians in charge and assessed for response by two independent senior clinicians who did not know the results of the receptor studies.

RESULTS

Presence of ER and PgR in Malignant and Benign Lesions of the Breast

High-affinity binding sites of ^{3}H-E$_2$ or ^{3}H-R5020 were found in a certain proportion of cytosol preparations from both malignant and benign lesions

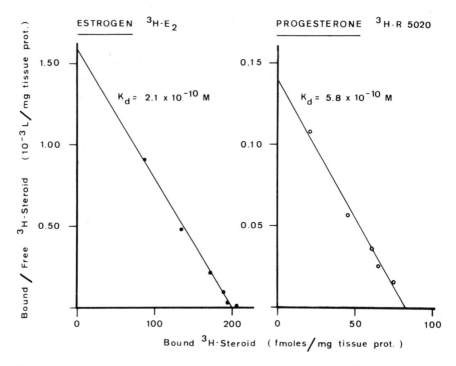

FIG. 1. Scatchard plots of the binding reactions of a cytosol preparation from a human breast cancer with ^{3}H-estradiol-17β (^{3}H-E$_2$) or ^{3}H-R5020.

of the breast. In most cases, Scatchard plots yielded straight lines demonstrating that only one class of binding sites was present (Fig. 1). Curved functions tending toward horizontal at high concentrations of bound labeled steroids were occasionally observed, then indicating the presence in such cytosols of several classes of sites. In the case of ^{3}H-R5020, horizontal lines corresponding to the binding of the steroid to nonspecific, unsaturable binding components were also rarely found. In the two first patterns, the estimated dissociation constants of the higher affinity binding reactions ranged from 0.3 to 27.4 × 10^{-10} M (mean: 4.0 × 10^{-10} M) for ^{3}H-E$_2$ and from 1.5 to 23.0 × 10^{-10} M (mean: 10.7 × 10^{-10} M) for ^{3}H-R5020. These values were similar to those previously reported for specific tissue steroid receptors (5,9,19). With regard to ^{3}H-R5020, the cellular origin of the receptors was further demonstrated by the fact that no binding components were detected in five diluted plasma samples (5.6 to 7.5 mg/ml) from two pre- and three postmenopausal women. In the case of ^{3}H-E$_2$, similar results were reported previously (9).

According to these criteria of positivity, ERs were present in 85% of primary breast cancers and 87% of metastases (Table 1). PgRs were present

TABLE 1. Proportion of ER- and PgR-positive breast cancers

	ER		PgR	
Primary				
Female	124/146		87/146	
Male	2/2		1/2	
Total	126/148	(85%)	88/148	(59%)
Metastatic				
Female	52/59		28/59	
Male	1/2		1/2	
Total	53/61	(87%)	29/61	(47%)

TABLE 2. Proportion of ER- and PgR-positive benign mammary lesions

	ER		PgR	
Benign dysplasia	6/10		3/10	
Fibroadenoma	7/9		4/9	
Giant fibroadenoma	0/1		0/1	
Gynecomastia	5/6		2/6	
Total	18/26	(69%)	9/26	(29%)

in a lower proportion of cases: 59% of primary tumors and 47% of metastases. In benign hyperplastic mammary lesions, ERs and PgRs were found in 69% and 29% of cases, respectively (Table 2). Remarkably, in malignant as well as in benign lesions, both receptors were observed in female and male patients (Tables 1 and 2).

ER and PgR Concentrations in Malignant and Benign Mammary Lesions

In primary and metastatic female breast cancers, ER and PgR concentrations were extremely variable from tumor to tumor (Fig. 2); ER concentrations ranged from 3 to 1,722 fmoles/mg tissue protein, PgR from 8 to 5,250 fmoles/mg tissue protein. Furthermore, for each receptor, the incidence of tumors increased progressively toward the lower values. In primary and metastatic male cancers, ER and PgR concentrations also appeared variable; ER varied from 129 to 442 fmoles/mg tissue protein, PgR from 92 to 171 fmoles/mg tissue protein. The scarcity of male breast cancers did not permit us to compare them with the female cancers in respect to range and distribution of receptor concentrations.

Benign mammary lesions contained small amounts of ER and PgR; ER concentrations varied from 8 to 174 fmoles/mg tissue protein (mean = 53), PgR from 35 to 156 fmoles/mg tissue protein (mean = 93). Similar amounts of both receptors were apparently present in female and male patients.

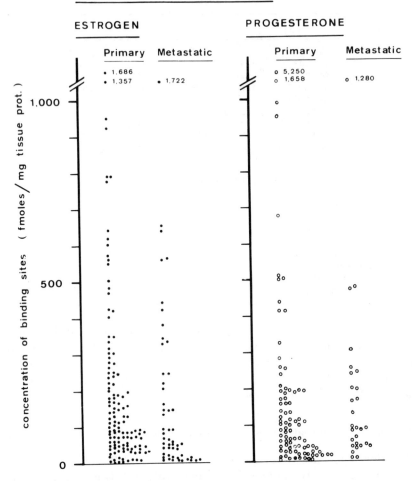

FIG. 2. ER and PgR concentrations in female primary and metastatic breast cancers.

Frequency of Presence and Concentrations of ER and PgR in Primary Breast Cancers of Pre- and Postmenopausal Patients

The possibility of influences of menopausal status on ER and PgR contents of breast cancers was examined. The study was limited to primary tumors which represented a large homogenous sample not previously subjected to any anticancer treatment. Premenopausal patients were those regularly menstruating; postmenopausal ones had had their last menstrual period at least 1 year before. Out of the 146 patients, 140 belonged to one of these two categories; 46 (32%) were premenopausal, 94 (64%) postmenopausal.

TABLE 3. *Proportion of ER- and PgR-positive primary breast cancers—influence of menopausal status*

	ER		PgR	
Premenopausal	35/46 (76%)	$\chi^2 = 3.29$	26/46 (56%)	$\chi^2 = 0.08$
Postmenopausal	84/94 (89%)	$0.1 < p < 0.05$	57/94 (61%)	ns

As shown in Table 3, ER-positive tumors were slightly less frequent in pre- than in postmenopausal women, but the difference was close to statistical significance ($0.1 < p < 0.05$). In contrast, the rate of PgR-positive tumors was similar in both types of patients. With regard to the mean receptor concentrations, ERs were significantly lower in pre- than in postmenopausal women ($p = 0.001$), whereas the contrary was true for PgR, although for the latter the difference was only close to statistical significance ($0.1 < p < 0.05$) (Fig. 3).

Finally, in both pre- and postmenopausal women ER and PgR concentrations were scattered along an apparently log-normal distribution (Fig. 3).

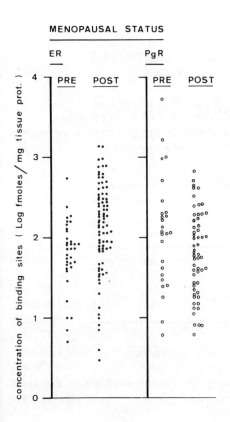

FIG. 3. ER and PgR concentrations in primary breast cancers in relation to menopausal status. Mean ER concentration was lower in pre- than in postmenopausal patients ($p = 0.001$, Mann-Whitney U test). Mean PgR was higher in pre- than in postmenopausal women ($0.1 < p < 0.05$, Mann-Whitney U test).

Qualitative Relationship Between ER and PgR in Individual Malignant and Benign Lesions of the Breast

MALIGNANT MAMMARY LESIONS

Table 4 shows that in primary and metastatic breast cancers, the presence of PgR was essentially restricted to ER-positive lesions, and in this group its relative frequency was higher in primary tumors (83/124) than in metastases (27/52) ($\chi^2 = 2.91$; $0.1 < p < 0.05$). Elimination of metastases from patients previously subjected to endocrine or cytotoxic treatments did not alter this pattern.

TABLE 4. *Relationship between ER and PgR in individual breast cancers (female)*

ER	PgR	Primary	Metastatic
+	+	83 (57%)	27 (46%)
+	−	41 (28%)	25 (42%)
−	+	4 (3%)	1 (2%)
−	−	18 (12%)	6 (10%)
		146	59

INFLUENCE OF MENOPAUSAL STATUS ON PRIMARY CANCERS

Influence of menopausal status was analyzed in primary cancers only for reasons already given (Frequency of Presence and Concentrations . . .). Table 5 shows that this factor did not influence the distribution of tumors into the four classes defined by presence or absence of the two receptors.

TABLE 5. *Relationship between ER and PgR in individual breast cancers—absence of influence of menopausal status*

ER	PgR	Premenopausal	Postmenopausal
+	+	23 (50%)	56 (59%)
+	−	12 (26%)	28 (30%)
−	+	3 (7%)	1 (2%)
−	−	8 (17%)	9 (9%)
		46	94

BENIGN MAMMARY LESIONS

Table 6 indicates that benign mammary lesions containing ER only represented the largest class. The remaining lesions were equally distributed in the three other classes.

TABLE 6. *Relationship between ER and PgR in individual benign mammary lesions*

ER	PgR	
+	+	6 (23%)
+	−	12 (46%)
−	+	3 (12%)
−	−	5 (19%)
		26

Types of lesions (benign dysplasia, fibroadenoma, giant fibroadenoma, gynecomastia): +/+: 2, 3, 0, 1; +/−: 4, 4, 0, 4; −/+: 1, 1, 0, 1; and −/−: 3, 1, 1, 0.

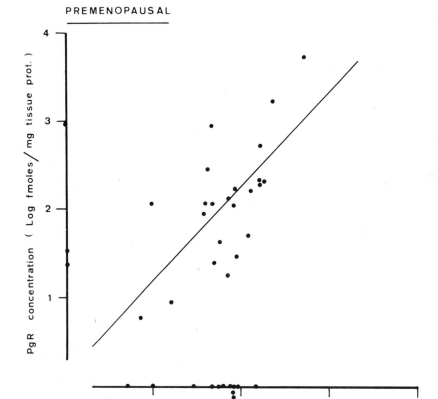

FIG. 4. Correlation study between ER and PgR concentrations in individual primary breast cancers from premenopausal women. In tumors positive for both receptors, ER and PgR concentrations were significantly correlated ($r = 0.67$, $p = 0.001$; equation of the regression line: $Y = 0.14 + 1.05 \times$).

Quantitative Relationship Between ER and PgR in Individual Primary Breast Cancers of Pre- and Postmenopausal Patients

Correlation studies between ER and PgR concentrations in individual primary cancers were carried out separately in the pre- and postmenopausal groups, in view of the quantitative differences associated with menopausal status (Frequency of Presence and Concentrations . . .). Receptor concentrations appeared normally distributed after logarithmic transformation making valid such studies.

In premenopausal patients, ER and PgR concentrations in tumors containing both receptors were positively significantly correlated ($r = 0.67$; $p = 0.001$) (Fig. 4). In postmenopausal patients the correlation was close to statistical significance ($r = 0.25$; $0.1 < p < 0.05$) (Fig. 5).

In both groups of patients, ER-positive, PgR-negative tumors were char-

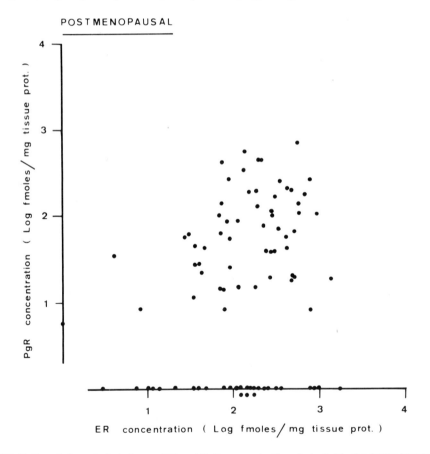

FIG. 5. Correlation study between ER and PgR concentrations in individual primary breast cancers from postmenopausal women. In tumors positive for both receptors, ER and PgR concentrations were significantly correlated ($r = 0.25$; $0.1 < p < 0.05$).

acterized by ER expending along the whole range of concentrations. ER-negative, PgR-positive tumors were too rare to deserve any interpretation.

Clinical Correlation

Seven patients were studied for a possible correlation between ER and PgR concentrations in tumor biopsies and response to endocrine therapy. One patient was premenopausal and was subjected to oöphorectomy. The remaining patients were postmenopausal and subjected to adrenalectomy (two cases) or to additive therapies [1-(p-β-dimethylaminoethoxyphenyl-1,2-diphenyl-1-ene) (Tamoxifen), three cases; 7β,17-dimethyltestosterone (Calusterone), one case]. The biopsies for receptor assay were obtained less than 1 month before initiation of the therapy. Three patients had biopsies positive for both ER and PgR; three others had ER only; the remaining one, PgR only (Table 7).

TABLE 7. ER and PgR in relation to response to endocrine treatments

Patient	Receptor concentration (fmoles/mg tissue protein)		Response	Treatment
	ER	PgR		
1	652	245	+	Tamoxifen
2	146	0	+	Tamoxifen
3	137	306	−	Adrenalectomy
4	19	0	−	Oöphorectomy
5	14	0	−	Calusterone
6	12	35	−	Adrenalectomy
7	0	43	−	Tamoxifen

Objective remissions were recorded in two patients (29%); both had been treated with Tamoxifen. These remissions were associated with the highest levels of ER. One was PgR positive and the other was PgR negative. Three failures were associated with PgR-positive tumors. One of them contained the highest level of PgR of this series.

CONCLUSIONS AND COMMENTS

The present study led to the following conclusions.

1. In accordance with the observations of Horwitz and McGuire (5) and Raynaud et al. (19), ^{3}H-R5020 was a convenient molecule to assay specific PgR in breast tumor tissue.

2. ER and PgR were found in a large proportion of malignant and benign mammary lesions. Rate of positivity and receptor concentrations were much higher in the former.

3. In breast cancers, the presence of PgR was essentially restricted to ER-positive samples confirming earlier studies (6,19). It was apparently more frequent in ER-positive primary than metastatic tumors, suggesting a loss of progesterone receptivity during the metastatic process. In tumors containing both receptors, ER and PgR concentrations were significantly correlated, although the correlation was much stronger in pre- than in postmenopausal women. Absence of PgR in ER-positive tumors was associated with ER concentrations distributed along the whole scale.

4. Menopausal status influenced ER and PgR concentrations. Mean ER concentrations were lower in pre- than in postmenopausal women as previously shown (3,14). The contrary was found for PgR.

5. Correlation with clinical results revealed that objective response to endocrine treatments was remarkably related to the level of ER. This was in agreement with previously reported results (4). In contrast, there was no relationship between response and presence or concentration of PgR. Of the two responses, one occurred in a PgR-positive patient, the other in a PgR-negative one. Three failures were associated with PgR-positive cases. Although this series is small, it suggests that PgR is not an absolute predictor of hormone dependency.

ACKNOWLEDGMENTS

We thank the Service d'Anatomie Pathologique, Institut Jules Bordet, for examining the tumors. We also thank Drs. R. De Jager (Service de Médecine, Institut Jules Bordet), M. Malarme, and P. Wettendorf (Service de Médecine, Institut Médico-Chirurgical d'Ixelles) who participated in reviewing the patients' records. We are indebted to Drs. J. P. Raynaud and D. Philibert (Roussel-Uclaf) who supplied ^{3}H-R5020. This work was supported by a grant from the Fonds Cancérologique de la Caisse Générale d'Epargne et de Retraite de Belgique and performed under Contract of the Ministère de la Politique Scientifique within the framework of the Association Euratom, University of Brussels, and University of Pisa.

REFERENCES

1. E.O.R.T.C. Breast Cancer Cooperative Group (1973): Standards for the assessment of estrogen receptors in human breast cancer. *Eur. J. Cancer,* 9:379–381.
2. Heuson, J. C., Engelsman, E., Blonk-van der Wijst, J., Maas, H., Drochmans, A., Michel, J., Nowakowski, H., and Gorins, A. (1975): Comparative trial of Nafoxidine and ethinyloestradiol in advanced breast cancer: an E.O.R.T.C. study. *Br. Med. J.,* 2:711–713.
3. Heuson, J. C., Leclercq, G., Longeval, E., Deboel, M. C., Mattheiem, W. H., and Heimann, R. (1975): Estrogen receptors: prognostic significance in breast cancer. In: *Estrogen Receptors in Human Breast Cancer,* edited by W. L. McGuire, P. P. Carbone, and E. P. Vollmer, pp. 57–72. Raven Press, New York.
4. Heuson, J. C., Longeval, E., Mattheiem, W. H., Deboel, M. C., Sylvester, R. J., and Leclercq, G. (1976): Significance of quantitative assessment of estrogen receptors for endocrine therapy in advanced breast cancer. *Cancer (in press).*

5. Horwitz, K. B., and McGuire, W. L. (1975): Specific progesterone receptors in human breast cancer. *Steroids,* 25:497–505.
6. Horwitz, K. B., McGuire, W. L., Pearson, O. H., and Segaloff, A. (1975): Predicting response to endocrine therapy in human breast cancer: an hypothesis. *Science,* 189:726–727.
7. Jensen, E. V., Block, G. E., Smith, S., Kyser, K., and DeSombre, E. R. (1971): Estrogen receptors and breast cancer response to adrenalectomy. *Natl. Cancer Inst. Monogr.,* 34:55–79.
8. Leclercq, R., Copinschi, G., and Franckson, J. R. M. (1969): Le dosage par compétition du cortisol plasmatique. Modification de la méthode de Murphy. *Rev. Fr. Et. Clin. Biol.,* 14:815–819.
9. Leclercq, G., Heuson, J. C., Deboel, M. C., and Mattheiem, W. H. (1975): Oestrogen receptors in breast cancer: a changing concept. *Br. Med. J.,* 1:185–189.
10. Leclercq, G., Heuson, J. C., Schoenfeld, R., Mattheiem, W. H., and Tagnon, H. J. (1973): Estrogen receptors in human breast cancer. *Eur. J. Cancer,* 9:665–673.
11. Leclercq, G., Verhest, A., Deboel, M. C., Van Schoubroeck, F., Mattheiem, W. H., and Heuson, J. C. (1976): Oestrogen receptors in male breast cancer. *Biomedicine,* 25:327–330.
12. Lowry, O. H., Rosebrough, N. J., Farr, A. L., and Randall, R. J. (1951): Protein measurement with the Folin phenol reagent. *J. Biol. Chem.,* 193:265–275.
13. Mancini, G., Carbonara, O. A., and Heremans, J. F. (1965): Immunochemical quantitation of antigens by single radial immunodiffusion. *Immunochemistry,* 2:235–254.
14. McGuire, W. L., Carbone, P. P., Sears, M. E., and Escher, G. C. (1975): Estrogen receptors in human breast cancer: an overview. In: *Estrogen Receptors in Human Breast Cancer,* edited by W. L. McGuire, P. P. Carbone, and E. P. Vollmer, pp. 1–8. Raven Press, New York.
15. Mester, J., Robertson, D. M., Feherty, P., and Kellie, A. E. (1970): Determination of high affinity oestrogen receptor sites in uterine supernatant preparations. *Biochem. J.,* 120:831–836.
16. Nemoto, T., and Dao, T. L. (1966): Significance of liver metastasis in women with disseminated breast cancer undergoing endocrine ablative surgery. *Cancer,* 19:421–427.
17. Philibert, D., and Raynaud, J. P. (1973): Progesterone binding in the immature mouse and rat uterus. *Steroids,* 22:89–98.
18. Philibert, D., and Raynaud, J. P. (1974): Progesterone binding in the immature rabbit and guinea pig uterus. *Endocrinology,* 94:627–632.
19. Raynaud, J. P., Bouton, M. M., Philibert, D., Delarue, J. C., Guerinot, F., and Bohuon, C. (1975): Les récepteurs estrogène et progestogène dans le cancer du sein. In: *Hormones and Breast Cancer,* edited by M. Namer and M. Lalanne, pp. 71–81. Inserm, Paris.
20. Scatchard, G. (1949): The attraction of proteins for small molecules and ions. *Ann. N.Y. Acad. Sci.,* 51:660–672.

Progesterone Receptors in Normal and Neoplastic Tissues, edited by W. L. McGuire et al. Raven Press, New York © 1977.

Estrogen and Progestin Receptors in Normal and Cancer Tissue

B. Ramanath Rao and John S. Meyer

Section of Cancer Biology, Division of Radiation Oncology, Mallinckrodt Institute of Radiology, Washington University School of Medicine; and Department of Pathology, Jewish Hospital, St. Louis, Missouri 63110

It is generally accepted that an essential characteristic of steroid hormone target tissues is the capacity to bind the steroid hormones to which they respond specifically and with high affinity. The presence of specific steroid hormone binding protein molecules, called receptors,[1] in various steroid hormone target tissues for each of the steroid hormones has been demonstrated (1,5,12,18,19,23,26). Moreover, this hormone interaction with the receptors precedes alterations in the rates of transcription and translation in target tissues (2,13).

Endocrine therapy is a standard component in the management of breast cancer. The first treatment for recurrent or metastatic breast cancer is usually hormone manipulation. However, two out of three patients will not respond to endocrine therapy. If a distinction could be made at the earliest possible time between those who may respond to hormone therapy or to endocrine organ ablative surgery and those who are nonresponders, valuable time could be saved and appropriate therapy followed from the outset. Presently, there is good evidence to suggest that remission of breast cancer after endocrine therapy occurs more frequently in women whose breast tumors contain estrogen receptors than in those whose cancer tissue did not have measurable estrogen receptors (7). However, there are still a number of "false"-positive and "false"-negative results. Animal studies have shown that some murine mammary cancers may have cytoplasmic estrogen receptors but lack the mechanism for transporting the steroid-receptor complex into the nucleus. Those mammary cancers are not estrogen dependent since they do not regress after ovariectomy (8). A comparable situation may exist in some human breast cancer; therefore, a

---

[1] The experimental demonstration of specific cytoplasmic steroid binding proteins in comparable process to those outlined here has given rise to the use of "receptor" as a convenient short term. Use of the term in this chapter is not meant to imply a definition of "receptor" as currently used in pharmacology.

marker for estrogen action more sensitive than the cytosol estrogen receptor is desirable. This may increase the response rate to endocrine therapy. We have demonstrated that progesterone receptors in the uterus are dependent on estrogen priming (18,26); a similar physiological response may be anticipated in breast tissue. Analysis of breast cancer tissue for progesterone receptors may be a more sensitive marker of hormone dependency of the cancer than analysis for estrogen receptors.

Despite increasing cure rates of endometrial cancer with radiation and surgery, this cancer is increasing in incidence and recurrences remain a significant clinical problem. Administration of high doses of progestin is a standard procedure; remission has been reported in 35 to 45% of cases (21). Selection of patients based on analysis of the cancer tissue for progesterone receptors may increase this response rate.

Identification of progesterone receptors in tissue cytosol preparations has been difficult due to the presence of binders other than the receptors. However, the presence of specific progesterone receptors has been demonstrated in several mammalian species and in chick oviduct (4,12,14–19a, 23, 25,26). Availability of a progestin that can either bind only to specific progesterone receptors or interact with them in a fashion which permits a clear distinction of binding between specific receptor and nonspecific binders will be a great advantage.

Studies of the thymidine labeling index (TLI) of primary human breast carcinomas [i.e., the number of cells per 100 that flash label with tritiated thymidine (^{3}H-TdR) and that constitute the fraction of cells that are synthesizing DNA] indicate a broad range of from near zero to 19 (10). The wide range in TLI is thought to reflect a wide range of growth rates for breast carcinomas. In view of the dependence of the TLI of normal human breast ducts on the phase of the menstrual cycle in which a significantly higher TLI has been detected shortly after the days of high circulating levels of estrogens and progesterone (9), the question of a similar dependence of breast carcinoma TLI on menstrual phase was raised. It seemed reasonable that this dependence should be sought in tumors containing specific receptor proteins for estrogens and for progesterone.

In this chapter, we present data characterizing the binding of R5020 (17,21-dimethyl-19-nor-pregna-4,9-diene-3,20-dione), dydrogesterone (10α-pregna-4,6-diene-3,20-dione), and DU-41164 (6-fluoro-17-hydroxy-1β,2-methylene-9β-10α-pregna-4,6-diene-3,20-dione-17-acetate) to progesterone receptors in cytosols prepared from the uteri of a number of mammalian species as well as human breast cancer tissue. Furthermore, we have attempted to correlate the presence of estrogen and the progesterone receptors in human breast cancer tissue, infiltrated lymph nodes, and tissues from endometrial carcinoma. This report also includes data relating the TLI of breast carcinomas to the tissue content of receptors for estrogen and progesterone.

MATERIALS AND METHODS

1,2,6,7-^{3}H-progesterone (103.7 Ci/mmole) and 2,4,6,7-^{3}H-estradiol-17β (98.5 Ci/mmole) were obtained from New England Nuclear Corporation. Radiopurity of at least 97% was confirmed by paper chromatography. 1,2-^{3}H-R5020 (51.4 Ci/mmole) and R5020 was a gift from Dr. J. P. Raynaud, Roussel-Uclaf, 93230 Romainville, France. Dydrogesterone and ^{3}H-DU-41164 (8 Ci/mmole) were kindly provided by Dr. L. C. Post, Phillips Duphar, B. V., Weesp, The Netherlands. 1,2-^{3}H-dydrogesterone (54 Ci/mmole) was prepared specifically for our studies by Amersham. Radiopurity of this compound checked in our laboratory and by Phillips Duphar, The Netherlands, was found to be over 98%. Other nonradioactive steroids used in this study were obtained from Sigma Chemical Company and recrystallized prior to use. Norit A was purchased from Matheson, Coleman, and Bell. Dextran (grade D, m.w. 20,000) was obtained from Mann Research Laboratories. Carbowax® (polyethylene glycol, m.w. 6,000), monothioglycerol, and glycerol were purchased from Fisher Scientific Company. Ultrapure sucrose (ribonuclease-free) was obtained from Schwarz/Mann.

Six-week-old female rabbits weighing 700 to 900 g were purchased from Eldridge Rabbitry, St. Louis. Rapidly frozen uteri from mature rabbits were purchased from Pel-Freez Biologicals and stored at −80°C prior to cytosol preparation.

Human uteri removed during hysterectomy were immediately chilled in ice cold buffer (0.01 M Tris-HCl at pH 8.0 and containing 10% glycerol and 0.012 M monothioglycerol) and transported to the laboratory in ice. Breast cancer tissue and lymph nodes were frozen as soon as they were excised and stored at −80°C until use.

Preparation of Tissue Fraction

Frozen tissues (breast cancer and lymph node) were weighed and pulverized over dry ice to powder using a Thermovac tissue pulverizer. The powder was suspended in 3 volumes of buffer A (0.01 M Tris-HCl at pH 8.0 and containing 0.001 M disodium EDTA, 0.012 M monothioglycerol, and 10% glycerol). Homogenization was performed in an ice bath with a Polytron homogenizer (Brinkman) using three or four 5- to 10-sec pulses at a speed setting of 5; a minimum of 2 min was allowed for cooling between pulses. The entire homogenate was centrifuged at 40,000 rpm for 90 min in an IEC Model B-60 ultracentrifuge at 4°C. The supernatant fraction (or cytosol) was assayed the same day with unused cytosol being stored at −80°C.

Clotted blood from the human uterine segment was removed by washing repeatedly in cold buffer. Endometrium was scraped from the myometrium with a scalpel. To ensure complete removal of blood, we suspended pieces of endometrium in buffer and centrifuged them at low speed. After the supernatant was discarded, the pellet was more finely minced and homogenized in buffer A. Homogenization and the preparation of cytosol were performed similarly to that described for breast cancer tissue.

Assay Procedure and Evaluation of Steroid Binding Activity

To assay cytosol for the presence of specific progesterone receptors, we first added cortisol (5 µl) dissolved in ethanol to each tube and allowed the ethanol to evaporate under a gentle stream of air. Radiolabeled steroid (5 µl) in ethanol was added to each tube followed by 400 µl of buffer. A 100-µl aliquot of cytosol was added, mixed gently, and incubated at room temperature for 40 min. Tubes containing samples were then transferred to an ice bath, cooled for 15 min, and 0.5 ml of a charcoal suspension (0.5% Norit, 0.025% dextran in 0.01 M Tris-HCl buffer, pH 8.0, plus 0.001 M disodium EDTA) was added to each sample to separate both unbound and weakly bound progesterone as described previously (19). In some assays Carbowax® at a concentration of 0.025% was used instead of dextran.

A procedure similar to that described above for progesterone receptors was followed to assay estrogen receptors in cytosol using ^{3}H-estradiol-17β without the addition of cortisol to the incubation mixture. Quantitative evaluation of the number of binding sites in cytosols and the dissociation constant (K_d) of the steroid-receptor complex was determined in some samples by the method of Scatchard (22) as described previously (18).

Sucrose Gradient Centrifugation

Linear gradients of 5 to 20% sucrose were prepared in buffer A in either polyallomer or polycarbonate centrifuge tubes using an ISCO gradient maker. Appropriate nonradioactive steroids dissolved in ethanol were added to the incubation tube and the ethanol was dried under a gentle stream of air. Radiolabeled steroids dissolved in ethanol (2 µl) were added followed by cytosol (250 µl) to the tube and incubated at 4°C for a minimum of 3 hr before 200 µl of the sample was layered on the gradient. Centrifugation was performed in an IEC Model B-60 centrifuge with SB-405 rotor at 50,000 to 55,000 rpm for 15 to 16.5 hr at 3°C. Fractions were collected directly into scintillation vials from the bottom of the tube for radioactivity counting. The protein content of reference standards (bovine serum albumin and human gamma globulin) was estimated by the method of Lowry et al. (6).

Radioactivity Measurement

Radioactivity was measured with a Packard Tri-Carb Liquid Scintillation Spectrometer (Model 3320) in 10 ml of scintillation fluid containing toluene,

Triton X-100, and 25 X scintillator (Research Products International) in the ratio of 2,640:1,200:160 ml. The cocktail was added, mixed, and cooled before counting. In some samples, 10 ml of Scintiverse (Fisher) was used instead of the first scintillation cocktail. Maximum counting efficiency was 27%, and a sufficiently long counting time was used to achieve a counting error of <2%. Disintegrations per minute were computed using an automatic external standard and tritiated toluene quenched standards.

Ammonium Sulfate Precipitation

Saturated ammonium sulfate maintained at neutral pH by adding ammonium hydroxide was used for the concentration of receptor proteins. Assuming that 53 g of $(NH_4)_2SO_4$ dissolved in 100 ml of buffer at 4°C at a specific gravity of 0.53, we slowly added saturated $(NH_4)_2SO_4$ solution to the cytosol with gentle mixing to a final concentration of 30%. The pH of the mixture was maintained at 7; the precipitated protein was allowed to settle for 1 hr and packed into a pellet by centrifugation at 15,000 rpm for 20 min. The supernatant solution was removed and dialyzed in a cellulose bag overnight in a large volume of ice cold buffer. The dialysate was reduced to one-third of its original volume by vacuum dialysis in a collodion bag. The pellet was suspended in buffer A to one-third of the original volume of cytosol and dialyzed. The supernatant and the solubilized pellet were used in the evaluation of the sedimentation constant of progesterone receptors.

Thymidine Labeling Index

TLIs were measured as previously reported (9,10). In brief, a portion of the tumor was minced into blocks not more than 1 mm³. Approximately 25 such blocks were incubated with $25\mu Ci$ of ^{3}H-TdR (specific activity 6 Ci/mmole) in 4.5 ml Hanks' balanced salt solution under 3 atm oxygen tension. Samples were then incubated in a shaker bath at 36°C for 1 or 2 hr. Some specimens were incubated in the presence of 1.8×10^{-4} M 5-fluorouracil or 5-fluoro-2'-deoxyuridine to block endogenous thymidylate synthesis and thus enhance the uptake of ^{3}H-TdR by neoplastic cells (11). Following incubation, tissues were fixed in 4% aqueous formaldehyde, embedded in paraffin, and sectioned at 3 microns. The sections on glass slides were dipped in Kodak NTB emulsion, developed after 2 weeks of exposure, and stained with hematoxylin and eosin. The TLI was measured as the mean number of labeled nuclei per 100 tumor cell nuclei based on a count of at least 1,000 nuclei. A labeled nucleus was defined as having five or more overlying grains; the background showed fewer than one grain per nucleus.

RESULTS

Characterization of Binding of Progestins

Since cytosol preparations could be expected to contain some serum proteins despite extensive washings, ^{3}H-progesterone could be anticipated to bind to corticosteroid binding globulin (CBG) and albumin, in addition to specific progesterone receptors.

In order to see whether R5020, dydrogesterone, and DU-41164 competed with progesterone for binding sites on serum proteins, we performed sucrose gradient analyses with human serum. It is apparent from Fig. 1 that dydrogesterone did not compete with ^{3}H-progesterone for binding sites on serum proteins, whereas cortisol did compete with ^{3}H-progesterone. R5020 and DU-41164 at concentrations of 1×10^{-8} M showed a displacement of ^{3}H-progesterone similar to the level with cortisol. Results similar to those observed with human serum were also obtained when sera of rabbits, rats, and hamsters were analyzed. At this stage, it was imperative that radiolabeled dydrogesterone be tested for binding to serum proteins and cytosol

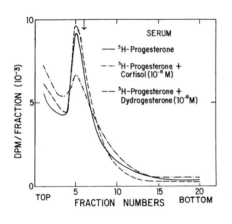

FIG. 1. Sedimentation profiles of human serum incubated with ^{3}H-progesterone, ^{3}H-progesterone plus cortisol (10^{-8} M), and ^{3}H-progesterone plus dydrogesterone for 8 hr at 4°C; 100 µl of the samples was layered on 5 to 20% sucrose gradients. Samples were centrifuged at 50,000 rpm for 16 hr at 3°C, and 30 fractions were collected from each tube. Radioactivity in the first 20 fractions is shown in this figure. The peak fraction containing bovine serum albumin (↓) is shown. A similar representation is followed in the other figures unless otherwise indicated.

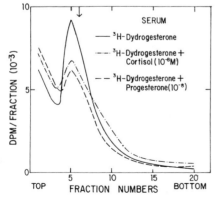

FIG. 2. Sedimentation profiles of human serum as in Fig. 1 except that ^{3}H-dydrogesterone was used in place of ^{3}H-progesterone.

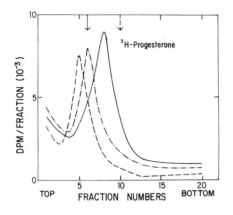

FIG. 3. Sedimentation profiles of uterine cytosols (250 μl = 2.5 mg) prepared from rabbits treated with 100 μg DES daily for 3 (-·-·-) and 10 (———) days and equilibrated with ³H-progesterone plus cortisol (10^{-8} M). Sedimentation of ³H-progesterone serum protein complex (----) was slightly slower than that of bovine serum albumin (↓). The peak fraction containing human gammaglobulin (↓), centrifuged simultaneously, is also shown.

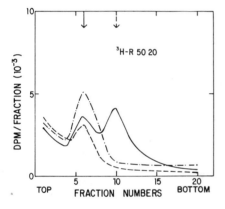

FIG. 4. Sedimentation profiles of uterine cytosols (250 μl = 2.5 mg) prepared from 3- (-·-·-) and 10- (———) day DES-treated rabbits showing a 4.5S and a 7.5S peak with ³H-R5020 plus cortisol (10^{-8} M). Sedimentation profile of binding for ³H-R5020 to serum protein (-----) is shown. Peak fractions containing serum albumin (↓) and gammaglobulin (↓) centrifuged in separate tubes are also shown.

progesterone receptors. ³H-dydrogesterone was found to bind to human serum proteins, and both cortisol and progesterone inhibited the binding (Fig. 2). A similar result was obtained with serum diluted to 1:10 of the original protein concentration and also with serum prepared from other mammalian species.

³H-progesterone and ³H-R5020 binding in the presence of cortisol to uterine cytosol prepared from rabbits administered 100 μg of diethylstilbestrol (DES) daily for 3 and 10 days and to serum are shown in Figs. 3 and 4, respectively. ³H-progesterone binding to cytosol prepared from uteri of 3- and 10-day DES-treated animals sedimented as a single peak in the 4.6S and 6.1S region, respectively. A 7.5S peak in addition to a 4.6S peak was observed with ³H-R5020 binding to the cytosols. The sedimentation profile of binding of the radiolabeled steroids to rabbit serum protein in the absence of cortisol is shown in the same figures. The steroid-protein complex sediments slightly slower than bovine serum albumin.

Figures 5 and 6 show the sedimentation profile of 10-day DES-treated rabbit uterine cytosol which was precipitated with $(NH_4)_2SO_4$ and incu-

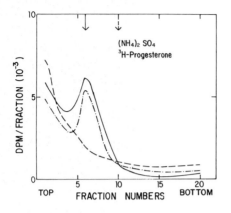

FIG. 5. Sedimentation profiles of an ammonium sulfate precipitate (30%) and supernatant (Materials and Methods) prepared from the uterine cytosol of a 10-day DES-treated rabbit. ³H-progesterone was equilibrated with a 30% resuspended $(NH_4)_2SO_4$ pellet (–·–·–), supernatant (- - - - -), and a 1:1 combination of the pellet and supernatant (———). Arrows as in Fig. 3.

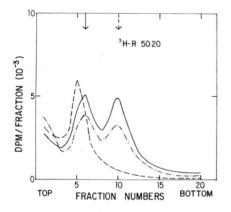

FIG. 6. Sedimentation profiles as in Fig. 5 except that ³H-R5020 was used as the radiolabeled progestin.

bated with either ³H-progesterone plus cortisol or ³H-R5020 plus cortisol, respectively. ³H-progesterone binding was observed with the 30% $(NH_4)_2SO_4$ pellet with little or no binding being detected in the supernatant. A recombination of the 30% $(NH_4)_2SO_4$ pellet and the supernatant resulted in a 4.6S peak similar to that observed with the 30% pellet alone; however, in this case, there was a slight increase in ³H-progesterone binding. In contrast, the analysis with ³H-R5020 gave a different result. Binding of ³H-R5020 was detected as a single 4.6S peak with either the 30% $(NH_4)_2SO_4$ pellet or the supernatant. Their recombination showed two peaks with sedimentation constants of 4.6S and 7.5S.

Human breast cancer tissue cytosol which showed distinct 4.6S and 7.5S peaks with ³H-R5020 was treated with ammonium sulfate (*result not presented*). Only a single 4.6S peak was observed in the 30% pellet as well as in the supernatant and in their recombination. It appears that the ability of ³H-R5020 complex to reform into a 7.5S peak is lost after ammonium sulfate treatment of human breast cancer cytosol.

Steroid Receptors in Breast Cancer

We have analyzed human breast cancer tissues for their cytosol content of estrogen and progesterone receptors. The results are presented in Table 1. The only cytosols scored as positive for estrogen receptors were those in which there was either a specific 8S estrogen binder or a 4S estrogen binder with a dissociation constant for the estrogen receptor complex that was lower than 1×10^{-9} M as calculated from Scatchard plots. Cytosol was considered to exhibit progesterone binding when there was either a 7.5S peak with ^{3}H-R5020 or a 4.6S peak with ^{3}H-progesterone in the presence of a 1,000-fold excess of cortisol. In some cases when there was still some doubt, the dissociation constant of the ^{3}H-progesterone receptor complex was first calculated by Scatchard plot; only those cytosols in which the dissociation constant was lower than 1×10^{-9} were scored as positive for progesterone receptors.

Out of the 57 breast cancer tissues analyzed, 35 (62%) were positive for estrogen receptors and 22 (39%) were positive for progesterone receptors. In three progesterone-positive cases (14%), estrogen receptors were absent. Estrogen content as measured in the cytosols of these three special cases was low, from which we infer that the failure to detect estrogen receptors is probably not due to masking of these receptors by high endogenous levels of estrogen (Fig. 7).

In addition to breast cancer tissue, seven noninfiltrated and five infiltrated lymph nodes were analyzed for steroid receptors (Table 2). Estrogen and progesterone receptors were undetectable in all of the noninfiltrated lymph nodes and in two of the infiltrated nodes. In the remaining three infiltrated nodes, both estrogen and progesterone receptors were detected.

Out of the 22 cases of breast cancer tissues which were positive for progesterone receptors, a distinct 7.5S peak with ^{3}H-R5020 was observed in only eight of the samples. We did not observe a case in which the ^{3}H-progesterone receptor complex migrated in the region of gamma globulin in the sucrose gradient.

Since we observed three cases in which progesterone receptors were present while estrogen receptors were undetectable, we thought that there

TABLE 1. *Breast cancer tissue analysis for steroid receptors*

Estrogen (E) +	35/57	(62%)
Progesterone (P) +	22/57	(39%)
Progesterone + / Estrogen +	19/35	(54%)
Progesterone + / Estrogen −	3/22	(14%)

+, Receptor positive; −, receptor negative.

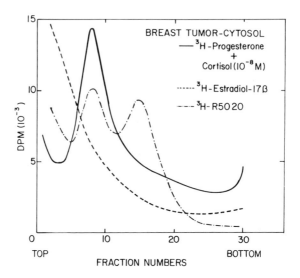

FIG. 7. Human breast tumor cytosol equilibrated with ^{3}H-progesterone plus cortisol (10^{-8} M), ^{3}H-estradiol-17β, and ^{3}H-R5020 analyzed after centrifugation at 55,000 rpm for 16.5 hr at 3°C.

TABLE 2. *Lymph node analysis for steroid receptors*

Steroid	Infiltrated	Noninfiltrated
E +	3	None
E −	2	7
P +	3	None
P −	2	7
R5020 +	2	None
R5020 −	3	7

+, Receptor positive; −, receptor negative.

TABLE 3. *Analysis of small portions of large cancerous tissue*

Case no.	E	P	E	P	E	P	E	P	E	P	E	P
1	+	+	−	+	−	+	+	+	+	+	+	+
2	−	−	−	−	−	−	−	−	−	−	−	−
3	+	−	+	−	+	−	+	−	−	−	+	−
4	+	+	+	+	+	+	−	+	−	+		
5	−	+	+	+	+	+	+	+	+	+		

+, Receptor positive; −, receptor negative.

may be differences in the distribution of these two receptors, particularly in large cancerous tissues. Therefore, portions of large tissues were analyzed for steroid receptors and results are presented in Table 3. Of the five samples analyzed, three had progesterone receptors and four had estrogen receptors. What was striking was the uniform distribution of progesterone receptors with estrogen receptors being absent in some portions. These particular tissues contained neither necrotic nor normal breast tissue.

Steroid Receptors in Human Endometrium

The presence of estrogen and progesterone receptors was detectable in the cytosol of seven normal endometrial samples. Cytosol progesterone binding was not detected in the four endometrial cancer tissue samples examined. None of the patients had undergone irradiation; however, the presence of both normal and malignant cells in the tissues used for cytosol preparation was established by histological examination. Only estradiol-17β binding was observed in the latter tissues as shown in Table 4. Analysis of endometrial cytosol for progesterone receptors was performed only with ^{3}H-progesterone.

TABLE 4. *Steroid receptors in endometrium*

Cytosol	Estrogen	Progesterone
Normal ($N = 7$)	+	+
Cancer ($N = 4$)	+	−

+, Receptor positive; −, receptor negative.

Steroid Receptors and TLI in Breast Cancer Tissue

As shown in Table 5, the absence or presence of estrogen and progesterone receptors could not be correlated with the means of the TLI of these tissues. Differences in TLI of all four groups were not statistically significant.

TABLE 5. *Steroid receptors and TLI in breast cancer tissue*

TLI	E+	E−	P+	P−
Geo. mean	3.59	3.95	6.55	4.21
Range	0.1–16.1	0.4–16.7	2.3–18.2	0.4–10.9
p values	<0.8		<0.2	

+, Receptor positive; −, receptor negative.

DISCUSSION

The three areas of investigation pursued in this study pertain to (a) the characterization of progesterone receptors, (b) the possible correlation of the presence of estrogen with progesterone receptors in cancer tissues, and (c) the possible correlation of estrogen and progesterone receptors with TLI in breast cancer tissue.

Progestins used in the characterization of progesterone receptors in the tissue cytosol preparations in this study were selected on the basis of each progestin's known characteristics. Reeves et al. (20) indicated that dydrogesterone did not compete successfully with ^{3}H-corticosterone for binding sites on CBG, which prompted us to test this particular progestin. It appears from the results presented in this chapter that nonlabeled dydrogesterone does not compete with ^{3}H-progesterone for serum protein binding, but that progesterone and cortisol effectively inhibit the binding of ^{3}H-dydrogesterone to serum proteins. These data indicate that there may be differences in binding sites for dydrogesterone from those specific for progesterone and cortisol, and that the latter two may have a higher affinity for serum proteins than dydrogesterone.

Terenius (26) showed that DU-41164 had the highest affinity for cytosol progesterone receptors. Although we have confirmed that DU-41164 has a higher affinity for cytosol progesterone receptors than any other progestin (*results not shown*), the DU-41164 receptor complex behaves in a manner similar to dydrogesterone. Thus, no advantage is provided in the identification of progesterone receptors using dydrogesterone or DU-41164 over that obtained using progesterone.

Compound R5020, on the other hand, has been shown to bind to cytosol progesterone receptors in several species (3,4,14–16). The advantage of using R5020 in the identification of progesterone receptors is that the radiolabeled progestin receptor complex sediments in the region of gamma globulin (7.5S). This sedimentation profile has never been observed for the binding of ^{3}H-R5020 to serum proteins.

We have previously reported the presence of an 8S progesterone binding component in the uterine cytosol of mature spayed rabbits which were estrogen primed and injected with labeled progesterone *in vivo* (18,26). We have also observed that daily DES priming of immature rabbits increases the concentration of uterine cytosol progesterone receptors to over 100-fold by day 10 (19a). We have never observed the 8S component, however, in *in vitro* incubations of labeled progesterone with uterine cytosols of either mature rabbits (either spayed or primed with estrogen) or immature, estrogen-primed rabbits. However, it is clear from our present report that prolonged treatment with estrogen not only resulted in an increased number of progesterone receptors but also changed the qualitative characteristics of the ^{3}H-progesterone cytosol receptor complex when analyzed by sucrose gradient centrifugation. Since our analysis was performed with identical protein concentrations in the cytosol samples (following either 3 or 10 days of DES treatment), the change in the sedimentation constant from 4.6S to 6.1S cannot be due to differences in protein concentrations (24). Analysis of uterine and breast cancer tissue cytosol following ammonium sulfate precipitation showed the 6.1S receptor to be very labile and incapable of being reformed once disrupted. It appears that there is a factor in the uterine cytosol to which progesterone does not bind. This factor is present only in the estrogen-primed uterus and is probably responsible for the higher progesterone receptor sedimentation constant. It is not clear from our studies whether the 7.5S component observed using ^{3}H-R5020 is due to a similar interaction or even to the same factor. In this regard, it is important to note that the recombination of the 30% ammonium sulfate pellet with the supernatant of breast tumor cytosol (*result not presented*) does not result in the formation of a 7.5S component.

Results of our analysis of breast cancer tissue for steroid receptors showed that 62 and 39% of cases had either estrogen or progesterone receptors, respectively. This result closely corresponds with other reports (3,4). Included in our results on progesterone receptors in breast tumor cytosol are those positive cases obtained using ^{3}H-progesterone as well as those obtained using ^{3}H-R5020. We have not made a correlation between the presence of progesterone receptors in the breast tumor cytosol and the resultant response to endocrine therapy. We plan to follow those patients in whom endocrine therapy was the therapy of choice.

The striking results in our present study of steroid receptors in breast cancer cytosol are: firstly, that progesterone receptors are more uniformly distributed than estrogen receptors throughout the tissue analyzed; and, secondly, that in some cases progesterone receptors could be present in the absence of detectable levels of estrogen receptors. At present we cannot offer any definitive explanation for these observations.

Although a relationship exists between the presence of progesterone receptors and the response to progestin therapy in endometrial cancer, experimental evidence of its applicability is totally lacking. The presence of progesterone receptors would certainly support progestin therapy in these cases.

In the present chapter, we have included the correlation between steroid receptors and TLI of breast carcinomas in order to point out that the detection of estrogen and progesterone cytosol receptors may not depend on the growth rate of the breast carcinoma, and that receptors may be found in both rapidly and slowly growing tumors.

ACKNOWLEDGMENTS

This investigation was supported by grant no. 1POCA13053–05, awarded by the National Cancer Institute, DHEW, and USPHS, HD-05680. The technical assistance of Mary Ann Rafle, Susan Aycock, Dan S. Reiner, and Glenn Fry is gratefully acknowledged. We wish to thank Ms. Geraldine Coleman for help extended in the preparation of this manuscript and to Dr. Alexander Nakeff for his valuable critical review. Sincere thanks are extended to Dr. J. P. Raynaud, Roussel-Uclaf, France, and to Dr. L. C. Post, Phillips Duphar, The Netherlands, for providing the synthetic progestins used in this investigation.

REFERENCES

1. Bruchovsky, N., and Wilson, J. D. (1968): The intranuclear binding of testosterone and 5 α-androstan-17β-ol-3-one by rat prostate. *J. Biol. Chem.,* 243:5953–5960.
2. Hamilton, T. H. (1968): Control by estrogen of genetic transcription and translation. *Science,* 161:649–661.
3. Horwitz, K. B., and McGuire, W. L. (1975): Specific progesterone receptors in human breast cancer. *Steroids,* 25:497–505.
4. Horwitz, K. B., McGuire, W. L., Pearson, O. H., and Segaloff, A. (1975): Predicting response to endocrine therapy in human breast cancer: A hypothesis. *Science,* 189:726–727.
5. Jensen, E. V., and DeSombre, E. R. (1972): Mechanism of action of the female sex hormones. *Annu. Rev. Biochem.,* 41:203–230.
6. Lowry, O. H., Rosebrough, N. J., Farr, A. L., and Randall, R. J. (1951): Protein measurement with the Folin phenol reagent. *J. Biol. Chem.,* 193:265–275.
7. McGuire, W. L., Carbone, P. P., and Vollmer, E. P. (Eds.) (1975): In: *Estrogen Receptors in Human Breast Cancer.* Raven Press, New York.
8. McGuire, W. L., and Chamness, G. C. (1973): Studies on the estrogen receptor in breast cancer. In: *Receptors for Reproductive Hormones,* edited by B. W. O'Mally and A. R. Means, pp. 113–136. Plenum Press, New York.
9. Meyer, J. S. (1973): Cell proliferation in ducts of normal human breast and fibroadenomas. *Lab. Invest.,* 34:327–330.
10. Meyer, J. S., and Bauer, W. C. (1975): In vitro determination of tritiated thymidine labeling index (LI). Evaluation of a method utilizing hyperbaric oxygen and observations of the LI of human mammary carcinoma. *Cancer,* 36:1374–1380.
11. Meyer, J. S., and Facher, R.: Thymidine labeling index of human breast carcinoma. Enhancement of in vitro labeling by 5-fluorouracil and 5-fluoro-2'-deoxyuridine. *Cancer* (*in press*).
12. Milgrom, E., and Baulieu, E. E.: Progesterone in uterus and plasma. I. Binding in rat uterus 105,000 g supernatant. *Endocrinology,* 87:276–287.
13. O'Mally, B. W., Rosenfield, G. C., Comstock, J. P., and Means, A. R. (1972): Induction of specific translatable messenger RNA's by oestrogen and progesterone. In: *Karolinska Symposia No. 5 Gene Transcription in Reproductive Tissue,* edited by E. Diczfulusy. Bogtrykkcriet Forum, Copenhagen.
14. Philibert, D., and Raynaud, J. P. (1973): Progesterone binding in the immature mouse and rat uterus. *Steroids,* 22:89–98.
15. Philibert, D., and Raynaud, J. P. (1974a): Binding of progesterone and R5020, a highly potent progestin, to human endometrium and myometrium. *Contraception,* 10:457–466.
16. Philibert, D., and Raynaud, J. P. (1974b): Progesterone binding in the immature rabbit and guinea-pig uterus. *Endocrinology,* 94:627–632.
17. Rao, B. R., and Wiest, W. G. (1974): Receptors for progesterone. *Gynecol. Oncol.,* 2:239–248.
18. Rao, B. R., Wiest, W. G., and Allen, W. M. (1973): Progesterone "receptor" in rabbit uterus. I. Characterization and estradiol-17β augmentation. *Endocrinology,* 92:1229–1240.
19. Rao, B. R., Wiest, W. G., and Allen, W. M. (1974): Progesterone "receptor" in human endometrium. *Endocrinology,* 95:1275–1281.
19a. Rao, B. R., and Katz, R. M. (1977): Progesterone receptors in rabbit uterus. II. Characterization and estrogen augmentation. *J. Steroid Biochem.* (*in press*).

20. Reeves, B. D., deSauza, M. L. A., Thompson, I. E., and Diczfalusy, E. (1970): An improved method for the assay of progesterone by competitive protein binding. *Acta Endocrinol.* (Kbh.), 63:225–241.
21. Reifenstein, E. C. (1974): The treatment of advance endometrial cancer with hydroxyprogesterone caproate. *Gynecol. Oncol.*, 2:377–414.
22. Scatchard, G. (1949): The attraction of proteins for small molecules and ions. *Ann. N.Y. Acad. Sci.*, 51:660–672.
23. Sherman, M. R., Corvol, P. E., and O'Mally, B. W. (1970): Progesterone-binding components of chick oviduct. I. Preliminary characterization of cytoplasmic components. *J. Biol. Chem.*, 245:6085–6096.
24. Stancel, G. M., Leung, K. M. T., and Gorski, J. (1973): Estrogen receptors in the rat uterus. Multiple forms produced by concentration-dependent aggregation. *Biochemistry*, 12:2130–2136.
25. Terenius, L. (1974): Affinities of progestogen and estrogen receptors in rabbit uterus for synthetic progestogens. *Steroids*, 23:909–919.
26. Wiest, W. G., and Rao, B. R. (1971): Progesterone binding proteins in rabbit uterus and human endometrium. In: *Advances in the Biosciences 7, Schering Workshop on Steroid Hormone "Receptors,"* edited by G. Raspe, pp. 251–266. Pergamon Press, New York.

Progesterone Receptors in Normal and Neoplastic Tissues, edited by W. L. McGuire et al. Raven Press, New York © 1977.

Estrogen and Progestin Receptors in Human Breast Cancer

J. P. Raynaud, T. Ojasoo, *J. C. Delarue, **H. Magdelenat, †P. Martin, and D. Philibert

Centre de Recherches Roussel-Uclaf, 93230 Romainville, France;
* Institut Gustave Roussy, 94330 Villejuif, France;
** Fondation Pierre-Marie Curie, 75005 Paris, France; and
† Centre Hospitalier Régional, 13005 Marseille, France

An explanation for the hormone dependency of about 30% of cases of human breast cancer has been sought in differences in endogenous steroid metabolism (1,2) and more recently, since the partial elucidation of hormone mechanism of action in target organs, in the presence and concentration of hormone receptors in malignant tissue (3). A vast number of studies has shown that the presence of estrogen receptor in human breast tumors has some, albeit insufficient, prognostic value for endocrine ablation or hormonal therapy (4,5). More recent studies (6,7) have postulated that the presence of progesterone receptor might be an additional sensitive marker for predicting clinical response since estradiol induces the synthesis of this receptor in animal tissues (8,9; Philibert and Raynaud, *this volume*). The detection and measurement of the progesterone receptor are, however, beset with difficulties arising mainly from the instability of the progesterone-receptor complex and from interference by plasma binding proteins. This interference is especially marked when the receptor concentrations are very low as in the cytoplasm of many human mammary tumors.

The problems in the detection of the progestin receptor have been partly solved by the use of synthetic hormones (10), in particular R5020 (17,21-dimethyl-19-nor-4,9-pregnadiene-3,20-dione), which does not bind specifically to animal and human corticosteroid binding globulin (CBG) (11) and which stabilizes the progestin receptor by forming a slowly dissociating complex (Philibert and Raynaud, *this volume*). R5020 is being used by various teams to assay progestin binding sites in tumors, and a few preliminary results have already been published (7,12–16).

The sensitivity and general suitability of current assay techniques remain, however, questionable, and as yet it is not certain that an accurate quantitative assessment of receptor levels is necessary as a prognostic for endocrine therapy. It may be the *quality,* and not the *quantity,* of the receptor, that is all important in predicting clinical response. Even if the hormone-receptor complex is present in the cytoplasm, a defect may exist in the sequence of events leading to the cell response. As suggested by Horwitz et al. (7), the functionality of the estrogen receptor could be gauged by the presence of progestin receptor, but this hypothesis raises two questions: first, it is necessary to quantify accurately the progestin receptor to estimate the functionality of the estrogen receptor; and, second, is it necessary to have some idea of the functionality of the progestin receptor?

An indication of the quality of a cytoplasmic hormone-receptor complex is given by its ability to enter the nucleus, to interact with chromatin, and to induce a response. Available models for such studies are, however, still highly experimental (17–19), and at the moment only one routine technique, namely, sucrose density gradient analysis, yields information on the nature of the receptor complex. Problems arise, however, in the interpretation of specific complexes sedimenting in different regions (7S and 4S) of the gradient, and it is likely that the drastic conditions of the method lead to dissociation of the 7S progestin-specific complex during migration. Although technical improvements limiting such dissociation are underway, sucrose gradient analysis cannot become a method of choice for routine studies. It requires large amounts of tissue, which are not always available, and expensive equipment; is time consuming; and, moreover, cannot be used to quantify the receptor. It should be remembered that if receptor determinations are to be a means of diagnosis, the method finally selected will have to be adaptable for use in hospital laboratories and therefore will have to be simple, reliable, rapid, and inexpensive.

If the accurate measurement of the tissue receptor turns out to be the ultimate aim, the most promising technique is probably the assay of specific binding sites by a dextran-coated charcoal (DCC) adsorption technique since this method is fast and simple. Traditionally, in the absence of high levels of nonspecific binding, such results are analyzed by a Scatchard plot. However, since binding site levels vary markedly from tumor to tumor, it is necessary when using this method to have some foreknowledge of the approximate receptor level in a particular tumor in order to define a suitable range of radioligand concentrations to trace a precise binding curve.

The DCC adsorption technique can be much simplified by applying the principle of the exchange assay as already described for the measurement of estrogen binding sites in rat uterus cytoplasm (20). In this case, a single saturating radioligand concentration and highly controlled incubation conditions are chosen, but they remain applicable to the determination of a wide range of binding site concentrations. The exchange assay not only gives an accurate notion of quantity of cytoplasmic receptor since it determines the total (free and occupied) number of binding sites, whatever the endogenous hormone concentration, but, in view of the rapid progress being made in nuclear binding assays, it may soon also give some indication of the amount of cytoplasmic receptor which enters the nucleus to induce a response.

An exchange assay technique has been developed for the measurement of progestin binding sites in cytosol from rabbit and mouse uterus (13, Philibert and Raynaud, *this volume*), and, in the present chapter, it has been extended to human mammary tumor cytosol. A similar assay is also being perfected for the measurement of estrogen binding sites using a synthetic ligand, R2858 (11β-methoxy-17-ethynyl-estra-1,3,5(10)-triene-3,17β-diol), which, unlike estradiol, does not bind to human sex steroid binding protein. As observed by several authors (21,22), this plasma protein is a major interference in tissue receptor assays.

MATERIALS

Labeled Steroids

6,7 [^{3}H]estradiol (specific activity 51 Ci/mmole), 6,7 [^{3}H]R5020 (17,21-dimethyl-19-nor-pregna-4,9-diene-3,20-dione) (specific activity 51 Ci/mmole), 6,7[^{3}H]R2858 (11β-methoxy-17-ethynyl-estra-1,3,5(10)-triene-3,17β-diol) (specific activity 52 Ci/mmole), and [1-^{3}H]dihydrotestosterone (24 Ci/mmole) were synthesized by the Centre de Recherches Roussel-Uclaf. 1[^{3}H]Progesterone (27 Ci/mmole) and 1,2[^{3}H]cortisol (56 Ci/mmole) were purchased from the C.E.A. (France). Radiochemical purity was checked by thin-layer chromatography (TLC).

Radioinert Steroids

The following radioinert steroids were used: *estrogens:* estradiol, ethynylestradiol, estriol, diethylstilbestrol, R2858, and RU16117 (11α-methoxy-17-ethynyl-1,3,5(10)-estratriene-3,17β-diol); *progestins:* R5020, norgestrel (13-ethyl-17-hydroxy-18,19-dinor-17α-pregn-4-en-20-yn-3-one), and chlormadinone acetate (6-chloro-17α-hydroxy-pregna-4,6-diene-3,20-dione acetate); *androgens:* dihydrotestosterone (DHT) and R1881 (17β-hydroxy-17α-methyl-estra-4,9,11-trien-3-one); *glucocorticoids:* cortisol and dexamethasone (9-fluoro-11β,17,21-trihydroxy-16α-methyl-pregna-1,4-diene-3,20-dione); and *mineralocorticoids:* aldosterone.

ROUTINE ASSAYS

Method

Primary breast cancer specimens were obtained from 281 women and 8 men patients at the Institut Gustave Roussy (Villejuif) who had not undergone any hormonal treatment or radiotherapy prior to surgery. Malignancy was verified by histological examination. Within 15 min of excision, the samples were crushed at 0°C in 5 volumes of 0.01 M Tris-HCl (pH 7.4), 0.012 M dithiothreitol buffer containing 10% of glycerol. They were crushed with an ultraturrax (three to four 5-sec bursts at 15-sec cooling intervals) and homogenized in a Teflon-glass homogenizer. Cytosol (supernatant) was obtained by centrifugation of homogenate for 1 hr at 105,000 × g in a 50 Ti rotor of a Beckman $L_2$65B centrifuge.

Binding sites were measured by a DCC adsorption technique as follows. Cytosol (250 µl) was incubated for 20 hr at 0°C with 5 nM [^{3}H]estradiol or [^{3}H]R5020 in the presence or absence of 2,500 nM radioinert hormone. Total radioactivity was measured on a 100-µl aliquot. Another 100-µl aliquot was shaken for 10 min at 4°C with 100 µl of DCC suspension (0.625 to 1.25%) in a microtiter plate, then centrifuged for 10 min at 800 g. The radioactivity of the supernatant was counted. Since the labeled hormone bound in the presence of a large excess of radioinert hormone is nonspecifically bound, the difference in cytosol radioactivity in the presence and absence of radioinert hormone corresponds to the concentration of specifically bound hormone.

Plasma progesterone levels were determined by radioimmunoassay using an antibody produced by immunization of rabbits with progesterone-3-carboxymethyloxime bovine serum albumin (Roussel-Uclaf). On account of the high degree of specificity of this antibody, a chromatography step was considered unnecessary.

Aliquots (50 µl) of plasma were added to 100 µl of an aqueous solution containing 4,000 cpm [^{3}H]progesterone. The volume was made up to 0.5 ml and the hormone was extracted with 2 ml redistilled ether while shaking for 5 min at room temperature. The aqueous phase was frozen in methanol-dry ice; the organic phase was removed and evaporated under a stream of nitrogen at 40°C. The residue was taken up in 0.5 ml of 0.1 M phosphate buffer (pH 6.9) containing 0.2% gelatin, 0.1% sodium azide, and a suspension of progesterone antibody (dilution 1/25,000), sheep antiserum to rabbit anti-γ-globulin (dilution 1/50), and normal rabbit serum (dilution 1/250). The mixture was stored overnight at 4°C and then centrifuged at 3,000 × g for 20 min. The radioactivity in the precipitate was measured after addition of 2 ml of scintillation cocktail (8 g Omnifluor, 100 ml Soluene 350 in 2 liters of toluene). The sensitivity of the assay was 20 pg and the coefficient of variation less than 10%.

Results

In Figs. 1 and 2, estrogen binding site levels have been plotted against progestin binding site levels for pre- and postmenopausal patients. Women without menses for less than 1 year were considered to be still premeno-

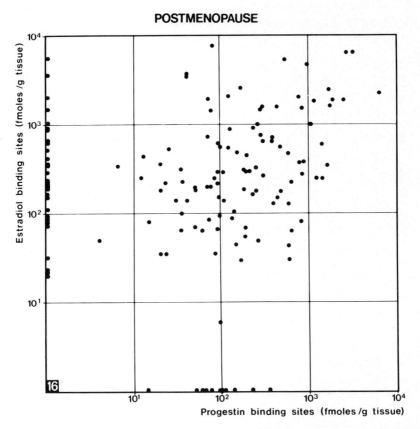

FIG. 1. Correlation between estradiol and progestin (R5020) binding sites in cytosol from primary mammary tumors from 164 postmenopausal women.

pausal. There was no evident correlation between the concentrations of the two receptors in either category of patient, although in premenopausal women a tendency toward high progestin and low estrogen receptor levels was noted, and in postmenopausal patients the tendency was toward high estrogen and low progestin receptor levels. This observation was confirmed by the ratio of the number of post- to premenopausal patients, which increased with high estradiol receptor concentrations (Table 1), and by the fact that an estrogen/progestin binding ratio greater than 1 was obtained in 69% of postmenopausal patients and in only 34% of premenopausal patients.

Plasma progesterone levels were determined in 200 patients, but, as shown in Fig. 2, there was no clear relationship between luteal phase progesterone levels (>2 ng/ml) and progestin binding site levels.

Binding site concentrations were also determined in eight cases of primary breast cancer in men patients and six cases of gynecomastia (Table 2).

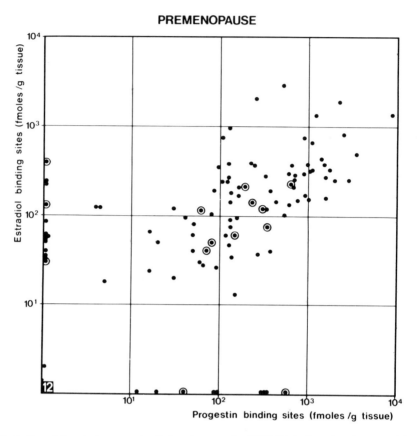

FIG. 2. Correlation between estradiol and progestin (R5020) binding sites in cytosol from primary mammary tumors from 117 premenopausal women. The encircled tumors were from women with plasma progesterone levels exceeding 2 ng/ml plasma at mastectomy.

TABLE 1. *Distribution of 281 primary human mammary tumors in pre- and postmenopausal patients according to concentration of estradiol and progestin specific binding sites*

Binding sites (fmoles/g tissue)	Estradiol No. of tumors			Progestin No. of tumors		
	Pre-menopause	Post-menopause	Post/pre	Pre-menopause	Post-menopause	Post/pre
0–100	58	60	1	52	96	1.8
100–500	49	56	1.1	35	40	1.1
500–1,000	5	18	3.6	15	14	0.9
1,000–5,000	5	25	5.0	14	12	0.8
>5,000	—	5		1	2	
Total	117	164		117	164	

TABLE 2. *Hormone-specific binding sites in male patients*

	Estradiol	Progestin
Primary mammary cancer		
Ler	135	1,170
Leg	500	190
Mau	2,490	550
Jou	500	790
Per	1,940	6,530
Rap	200	0
Cle	2,049	5,170
Nad	705	3,515
Gynecomastia		
Pio	0	0
Gom	0	0
Dro	0	0
Bug	20	15
Cou	59	0
Zec	29	0

Values are in fmoles/g tissue.

TABLE 3. *Hormone-specific binding in benign mammary tumors*

Fibroadenomas		Phyllode tumors		Sclerocystic mastosis	
Estradiol	Progestin	Estradiol	Progestin	Estradiol	Progestin
Cas 48	103	Bur..... 15	62	Gue... 0	0
Mai..... 0	0	L'Hu ... 20	32	Nic 60	14
Jus..... 0	95	Saa 23	0	Dan ... 183	24
Var..... 18	0	Cai 39	91	Pas.... 144	1,490
Bou 40	33	Mah....1,046	0	Das ... 0	0
Gui 71	62	Gau 10	10	Tyr 0	0
Dra..... 28	0	Gau.... 32	236	Mar ... 0	0
Bou....112	134	Mar 0	0	Mag... 16	112
Fou 51	0	Can 13	268	Bru.... 0	51
Mar 0	117	Gra..... 66	59	Duc ... 0	0
Rau..... 19	30			Sai 0	0
Gou.... 29	508			Iaf 21	0
Lou 5	368			Cha ... 14	0
Gui..... 48	0			Bug... 0	5
Dal 213	157			Mey... 0	0
Par..... 51	27			Ode...3,776	793
Gat..... 37	0			Ben ... 25	58
Bro..... 0	0				
Mey.... 0	0				
Pal 71	101				
Pom ... 50	26				
Bru..... 22	58				
Rey 30	153				
Hun..... 33	334				
Gre..... 22	37				

Values are in fmoles/g tissue.

Both receptors were present in relatively high levels in seven of the cancer patients; the remaining patient had estradiol receptor only. Neither receptor was present in the gynecomastia patients.

Finally, binding sites were determined in various types of benign tumors: 25 fibroadenomas, 10 phyllode-type tumors, and 17 cases of sclerocystic mastosis (Table 3). Except for a few rare exceptions (three cases of estradiol binding sites greater than 200 fmoles/g tissue and seven cases of progestin binding sites greater than this value), the values recorded were in general well below the limits of the sensitivity of the technique.

FURTHER CHARACTERIZATION OF THE TUMOR CYTOSOL RECEPTOR TO IMPROVE BINDING SITE ASSAY

Method

Primary breast cancer specimens were obtained from women patients at the Curie Foundation, approximately 30% of whom had undergone radiotherapy prior to surgery. Excised tumors were placed in liquid nitrogen and delivered to the Centre de Recherches Roussel-Uclaf.

For analysis, the tumors were thawed and homogenized at 0°C in 0.01 M Tris-HCl (pH 7.4), 0.25 M sucrose buffer. They were crushed with an ultra-turrax (four to five 5-sec bursts at 30-sec cooling intervals) and homogenized in a glass-glass homogenizer. Cytosol (final dilution = 1/7 wt/vol) was prepared by centrifuging for 45 min at 105,000 g in the 50 Ti rotor of a Beckman $L_2$65B centrifuge. Binding sites were assayed by the DCC adsorption technique described above. Further experimental details are given in the figure legends.

Results

ASSAY OF PLASMA SEX STEROID BINDING PROTEIN

Plasma sex steroid binding protein (SBP) levels were measured in plasma from 14 women with primary breast cancer (Fig. 3). Levels ranged from 30 to 100 pmoles/ml plasma with a mean value of 48 ± 22 pmoles/ml plasma and were well within the limits reported in the literature for healthy women (23). Assuming that contamination of cytosol from an average mammary tumor (300 fmoles of estrogen binding sites per gram tissue) is as low as 1%, this nevertheless means that in a given cytosol sample there is a high chance that the concentration of specific plasma binder exceeds that of tissue binder. For this reason, synthetic ligands, which are not specifically bound by plasma proteins, were used in tumor binding site assays.

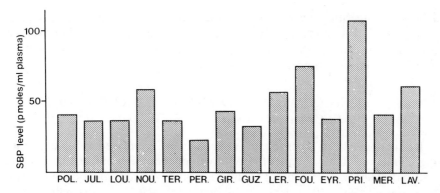

FIG. 3. Plasma sex steroid binding protein levels in 14 women with breast cancer. Plasma was incubated with 20 nM [^{3}H]DHT for 20 hr at 0°C. Radioinert DHT (5,000 nM) was then added and bound radioactivity was measured by DCC adsorption.

LACK OF SPECIFIC PLASMA BINDING OF SYNTHETIC TAGS

The absence of specific binding of R5020 to the corticosteroid binding globulin present in human plasma has been established previously (14). In

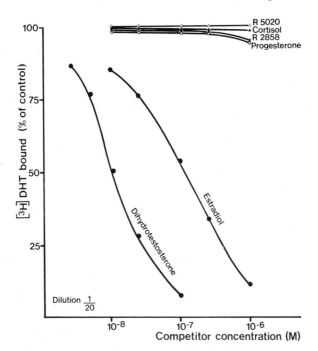

FIG. 4. Competition for labeled dihydrotestosterone binding in human plasma. Aliquots of 250 μl of human plasma, diluted to 1/20 with 10 mM Tris-HCl (pH 7.4), 0.25 M sucrose buffer, were incubated for 2 hr at 0°C with 2.5 nM [^{3}H]DHT in the presence of increasing concentrations of radioinert competitor. Bound DHT was measured by DCC adsorption.

Fig. 4, competition by various steroids for labeled dihydrotestosterone binding to SBP in human plasma is represented. Whereas estradiol exerts a marked competitive effect, about 10 times less than that of DHT, neither R2858 nor R5020 competes for this binding.

SPECIFICITY OF BINDING TO ESTROGEN AND PROGESTIN RECEPTORS IN TUMOR CYTOSOL

Results have been recorded in Table 4 and are expressed as relative binding affinities (RBAs), i.e., as the relative concentrations of test compound and radioinert ligand required to displace 50% of bound radioligand, [^{3}H]-estradiol or [^{3}H]R5020, from its binding sites. RBAs for the natural hormones are taken as 100.

Neither progestins, androgens, gluco- nor mineralocorticoids have any effect on [^{3}H]estradiol binding. The antiestrogens RU16117 (24) and ICI 46474; trans-1-(p-β-dimethylaminoethoxyphenyl)-1,2-diphenylbut-1-ene (Tamoxifen) exert a weak competitive effect; estrogens such as diethylstilbestrol (DES) and estriol have a definite action but not as marked as

TABLE 4. Competition for [^{3}H]estradiol and [^{3}H]R5020 binding in cytosol from human mammary tumors

	[^{3}H]estradiol 0°C	[^{3}H]estradiol 25°C	[^{3}H]R5020 0°C
Estradiol	100	100	~4
Ethynylestradiol	50	110	
Moxestrol® (R2858)	40	100	~5
DES	26	33	
Estriol	14	6	
RU16117	12	4	
Tamoxifen	0.35	0.15	
Progesterone	<0.1	<0.1	100
R5020	<0.1	<0.1	760
Norgestrel			760
Chlormadinone acetate			275
17α-Hydroxyprogesterone			<0.1
R1881	<0.1	<0.1	660
DHT	<0.1	<0.1	<0.1
Cortisol	<0.1	<0.1	<0.1
Dexamethasone	<0.1	<0.1	1.5
Aldosterone	<0.1	<0.1	

Standard displacement curves were obtained by incubating replicate aliquots of 100 μl of cytosol with 2.5 nM [^{3}H]estradiol or [^{3}H]R5020 and increasing concentrations of unlabeled competitor for 24 hr at 0°C or 5 hr at 25°C. Bound radioactivity was measured by DCC adsorption. The ratio of the concentration of ligand over the concentration of competitor required to displace radioligand binding by 50% (relative binding affinity) was determined.

that of more potent estrogens such as ethynylestradiol and R2858. The discrepancies in the values recorded at 0° and 25°C for R2858 are explained by the different dissociation rates recorded at these temperatures, compared to those of estradiol.

In competition for [^{3}H]R5020 binding, R5020 and norgestrel compete to similar extents and much more markedly than chlormadinone acetate. The androgen R1881 also exerts a very pronounced effect; its competition for binding to the progestin receptor has already been noted in animal uterine cytosol. Dexamethasone, on the other hand, has a negligible effect.

From these results, it would appear that the specificity of binding to the estrogen and progestin receptors in cytosol from mammary tumors is similar to that previously recorded in animal uterine cytosol, although relative binding affinities are not entirely comparable since in the present experiments cytosol was incubated for 20 rather than 2 hr. Under these conditions, ligands which dissociate slowly from the receptor have a higher relative binding affinity than at 2 hr. Moreover, tumor tissue differs from castrated animal uterus in at least two ways: the receptors corresponding to the different classes of hormones are present in different proportions, and a fair number of tumor binding sites are occupied by endogenous hormones.

CHOICE OF A SATURATING RADIOLIGAND CONCENTRATION FOR AN EXCHANGE ASSAY

To saturate 0.8 pmoles of estrogen binding sites per milliliter cytosol, a concentration of 3 nM [^{3}H]estradiol and 7 nM [^{3}H]R2858 was required (Fig. 5). Above these concentrations, the binding curves obtained in the

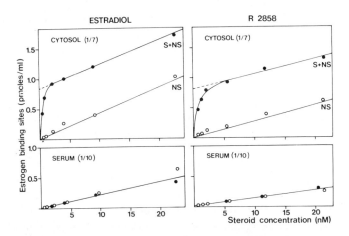

FIG. 5. Determination of a saturating radioligand concentration (estrogen). Cytosol or serum was first incubated with 100 nM DHT, then for 2 hr at 0°C with different concentrations of [^{3}H]estradiol or [^{3}H]R2858 in the presence or absence of 5,000 nM radioinert estradiol or R2858, respectively. Bound radioactivity was measured by DCC adsorption. S, specific binding; NS, nonspecific binding.

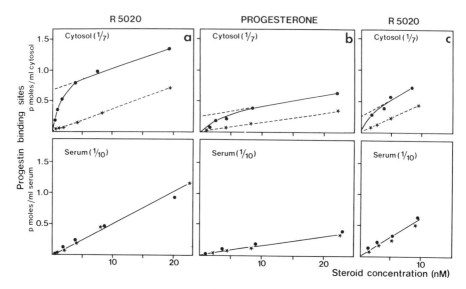

FIG. 6. Determination of a saturating radioligand concentration (progestin). Cytosol from two different tumors or serum was first incubated with 100 nM cortisol, then as above with progesterone or [^{3}H]R5020. Diagrams **b** and **c** refer to the same tumor.

absence and presence of radioinert ligand were parallel; in other words, specific binding was no longer dependent on radioligand concentration. No specific binding was detected in serum since R2858 does not bind to SBP and since, in the estradiol experiments, 100 nM DHT had been added. Nonspecific binding was lower with R2858 than with estradiol.

To saturate 0.7 pmoles of progestin binding sites per milliliter cytosol, a concentration of 5 nM [^{3}H]R5020 was required (Fig. 6a). In a tumor containing considerably fewer binding sites (Fig. 6b and c), it can be seen that approximately twice as much progesterone as R5020 is required to saturate the same number of sites. Nonspecific binding was higher with R5020 than with progesterone.

DISSOCIATION RATES OF THE STEROID-RECEPTOR COMPLEXES IN TUMOR CYTOSOL

As shown in Fig. 7, both estradiol and R2858 dissociate very slowly from the estrogen receptor at 0°C ($k_{-1} = 1.0 \times 10^{-3}$ min^{-1} for estradiol and 1.8×10^{-3} min^{-1} for R2858). On increasing the temperature to 25°C, the dissociation of estradiol increases markedly ($k_{-1} = 7.9 \times 10^{-3}$ min^{-1}), whereas that of R2858 remains fairly slow ($k_{-1} = 3.5 \times 10^{-3}$ min^{-1}). In the case of the progestin receptor (Fig. 8), progesterone dissociates very rapidly, even at 0°C ($k_{-1} = 45 \times 10^{-3}$ min^{-1}), whereas R5020 has a dissociation rate which is five times slower ($k_{-1} = 9 \times 10^{-3}$ min^{-1}). These results

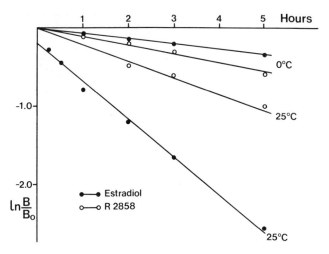

FIG. 7. Dissociation rates of estrogen-receptor complexes. Cytosol was incubated with 2.5 nM [³H]estradiol or [³H]R2858 for 2 hr at 0° or 25°C. Radioinert steroid (5,000 nM) was then added and bound radioactivity was measured at different times by DCC adsorption.

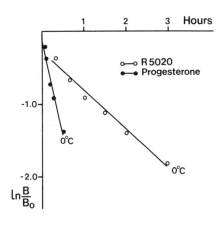

FIG. 8. Dissociation rates of the progestin-receptor complexes. The experimental conditions are given in the legend of Fig. 7.

would seem to imply that for an exchange assay, where it is necessary that the natural hormone should dissociate rapidly in order to be replaced by a slowly dissociating synthetic ligand, it would be advisable to choose conditions of incubation at 25°C with [³H]R2858 for estrogen receptor assay and 0°C with [³H]R5020 for progestin receptor assay.

CORRELATION BETWEEN ESTRADIOL AND R2858 BINDING SITES AND PROGESTERONE AND R5020 BINDING SITES IN TUMOR CYTOSOL

Binding sites were measured as described in Methods using either the synthetic ligand alone or the natural hormone in the presence of a 100-nM

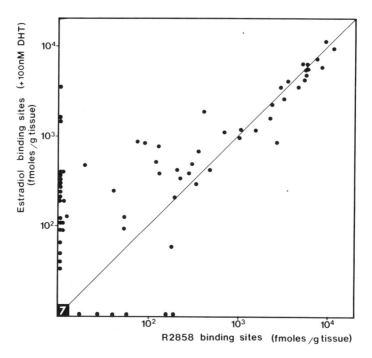

FIG. 9. Correlation between estrogen binding sites assayed with estradiol (in the presence of 100 nM DHT) and with R2858 alone in cytosol from 81 primary mammary tumors.

concentration of DHT or cortisol to minimize interference by plasma binding. Values for estradiol binding were plotted against those for R2858 binding (Fig. 9) and values for progesterone binding against those for R5020 binding (Fig. 10). For high binding site concentrations, a good correlation seemed to exist between the two measurements. For low binding site concentrations, the points diverged away from the correlation curve toward the estradiol axis, suggesting that in spite of the presence of 100 nM DHT in the assay medium, part of estradiol binding to SBP might have been measured by exchange. Analysis of a couple of tumor samples with high estradiol binding sites but no R2858 binding sites by sucrose density gradient analysis revealed the presence of a 4S radioactivity peak only. A similar divergence toward the ordinate at low binding site concentrations was observed when comparing measurements with progesterone and R5020. It was less marked than in the case of the estrogen binding sites partly because of the relatively high nonspecific binding of R5020. These results suggest that borderline tumors which are considered receptor positive when using the natural hormone could in fact be merely contaminated by plasma.

No attempt was made to correlate R2858 binding sites with R5020 binding sites in view of the small number of cases available in this pilot study,

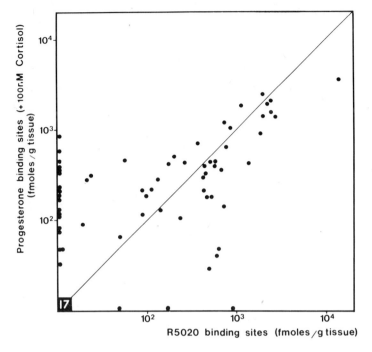

FIG. 10. Correlation between progestin binding sites assayed with progesterone (in the presence of 100 nM cortisol) and with R5020 alone in cytosol from 81 primary mammary tumors.

the heterogeneity of the group of patients under study (30% had undergone radiotherapy), and their undefined menopausal status.

TISSUE HANDLING

Method

Tumor tissues from the Centre Hospitalier Regional de Marseille were excised, trimmed of fat, cut into 500-mg pieces of which 100 mg was sent to the pathologist for histological examination, frozen in liquid nitrogen, and stored until assay. They were powdered in the frozen state with a Thermovac tissue pulverizer and homogenized in 5 ml of buffer per gram tissue with a polytron PT10 homogenizer (setting 4; five 2-sec bursts at 5-sec cooling intervals). Cytosol was prepared by centrifuging homogenate at 105,000 g for 50 min at 4°C in a Spinco L65B.

For the determination of binding site concentration, 200 µl of cytosol was incubated with 30 nM of [³H]estradiol and 100 nM DHT for 6 hr at 20°C or with 20 nM [³H]R5020 for 18 hr at 0°C in the presence or absence of

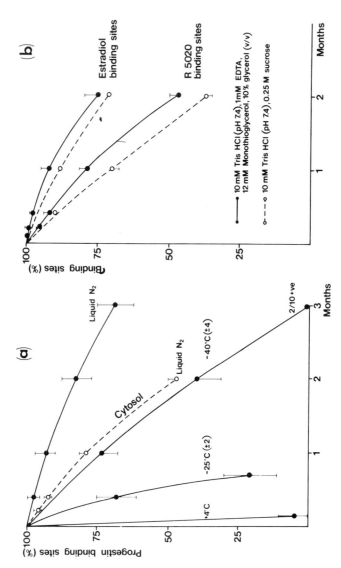

FIG. 11. Stability of progestin binding sites (± confidence interval) on storage of tumor specimens (a) and of estrogen and progestin binding sites on storage of cytosol prepared in two different buffers (b).

a 1,000-fold excess of radioinert ligand. Dextran-coated charcoal (0.1 to 1%) (500 µl) was then added. After it was shaken for 30 min, the mixture was centrifuged for 10 min at 1,000 g. Supernatant radioactivity was measured on 400-µl aliquots.

Results

Samples (500 mg) from large tumors, delivered in liquid nitrogen, were stored under different conditions (in liquid nitrogen, at $-40°C, -25°C$) for up to 3 months. On arrival of the tumors and at different times during storage, a sample was powdered for the determination of specific binding capacity (R5020) as described above. Initial binding capacity ranged between 150 and 400 fmoles/mg protein. Approximately 6 samples were analyzed for each point (Fig. 11a).

Whereas about 80% of binding capacity was retained after 2 months when the samples were stored in liquid nitrogen, only 40% was recovered at $-40°C$. At higher temperatures $(-25°C)$, virtually all binding sites were degraded within 1 month, and at $+4°C$ sites were detected after 48 hr only in tumors with high progesterone receptor concentrations.

If the cytosol prepared immediately on arrival of the tumor was stored in liquid nitrogen, a far greater loss in binding capacity was observed with time, and only 50% of the sites were recovered at 2 months (mean of 30 tumors). It was interesting to note that, in general, tumors with a high density of malignant cells appeared to resist storage better than tumors with inflammatory cells. Since binding site determinations were always performed under identical conditions, it would seem that this difference might reflect a higher proteolytic activity in low-density samples after excision and prior to storage in liquid nitrogen.

Cytosol from homogenate (same tumor) prepared in two different buffers [10 mM Tris-HCl (pH 7.4), 1 nM EDTA, 12 mM monothioglycerol, 10% glycerol (vol/vol), and 10 mM Tris-HCl (pH 7.4), 0.25 M sucrose] was stored for 2 months. The cytosol was prepared on arrival of the frozen tissue (in liquid N_2). The process of freezing and thawing tissue led to an approximately 10% loss in estrogen receptor and 16 to 20% loss in progestin receptor compared to unfrozen fresh tissue. After various time intervals, estrogen and R5020 binding sites were determined as described in Methods. According to Fig. 11b, it would appear that the degradation of progestin binding sites is faster than that of estrogen binding sites, and that conservation in the (10%) glycerol-containing buffer is slightly better.

DISCUSSION

The development of an exchange assay for the measurement of estrogen and progestin binding sites in cytosol from human breast tumors is beset with many difficulties, but the outcome would be rewarding insofar as this would be a simple technique most appropriate for routine use in hospital laboratories.

The present studies have established that synthetic ligands such as R5020 and R2858, which do not bind tightly to plasma proteins and yet are specific to a single class of hormone receptor, are particularly useful in the assay of tissue binding sites since they eliminate all problems of interference, however marked, of plasma contamination of cytosol. A distinction between tissue and plasma binding is essential since tissue receptors are known to have some predictive value in determining response to endocrine therapy and since several authors have suggested that plasma binding may also have some pathological significance in breast cancer (25,26).

To perform an exchange assay, it is necessary first to define the conditions for the rapid dissociation of the endogenous hormone from its binding sites in order that it may be replaced by a slowly dissociating radioligand. As has already been demonstrated in the uterine cytosol of several species (8; Philibert and Raynaud, *this volume*), progesterone dissociates very rapidly from its binding sites even at 0°C, whereas the dissociation rate (k_{-1}) of R5020 is about 5 to 10 times slower according to species. In human mammary tumor cytosol, the k_{-1} is 45×10^{-3} min^{-1} for progesterone and 9×10^{-3} min^{-1} for R5020 at 0°C. On the other hand, the dissociation of estradiol, and also of R2858, from estrogen binding sites in tumor cytosol at 0°C is relatively slow. At 25°C, however, estradiol dissociates twice as fast as R2858 ($k_{-1} = 8 \times 10^{-3}$ min^{-1} and 3.5×10^{-3}, respectively). Conditions of incubation at 0°C with R5020 and 25°C with R2858 therefore seem the most appropriate for the measurement of total (free and occupied) binding sites. Experiments on animal uterine cytosol have indeed shown that total exchange of bound hormone is obtained in 4 hr at 0°C (27; Philibert and Raynaud, *this volume*) when using a saturating [^{3}H]R5020 concentration and in 5 hr at 25°C when using a saturating [^{3}H]R2858 concentration (*unpublished*).

In part of the above studies, estrogen- and progestin-specific binding sites in cytosol from primary human breast tumors were measured by a routine technique which, in the light of present results, needs improving. No changes in the experimental conditions were made, however, during the study in order that assays on all tumors be comparable. Incubation was carried out with [^{3}H]R5020 at 0°C, but with [^{3}H]estradiol (in the absence of DHT) at 0°C instead of with [^{3}H]R2858 at 25°C. The conclusions confirm those of the literature. Considerable variations were recorded from patient to patient. There was a tendency toward high estrogen binding site values in postmenopausal patients and toward high progestin binding site values in premenopausal patients. In seven out of eight male patients, high binding site levels were recorded, which could possibly be related to low circulating estrogen and progesterone levels since, as suggested by animal studies, receptor levels are high when plasma progesterone levels are low.

However, upon measurement of plasma progesterone levels in premenopausal women at mastectomy, no relationship was found between high plasma progesterone and low progestin receptor levels. Neither receptor was present in male gynecomastia patients. These may have been cases not of hormonal disequilibrium but merely of adipomastia.

No attempt was made to draw a definite cutoff point to classify patients into receptor-positive and receptor-negative categories since in the absence of data on responsiveness to endocrine therapy there was no criterion for such a choice. Moreover, our routine technique was not considered sensitive enough to discriminate among patients with low receptor levels, and for greater sensitivity estrogen binding sites will have to be assayed with R2858 at 25°C. On the basis of a cutoff level of 100 fmoles/g tissue, which is at the limit of the sensitivity of the routine technique, the patients would be classified as follows: 32% of patients have no receptor. Among receptor-positive patients, 54% have both receptors, 31% estrogen receptor only, and 15% progestin receptor only.

Even though one can increase the sensitivity of the technique and take care to use the best conditions of storage, buffer, etc. to prevent degradation of binding sites during handling and storage, the problem of the heterogeneity of the experimental material remains unsolved and it would be wise to introduce a correction factor for the presence of nonmalignant inflammatory cells (28,29). Preliminary experiments comparing binding sites in heterogeneous tumors and homogeneous lymph nodes from the same patient would seem to confirm the usefulness of histological grading for a definition of the "malignancy" of a tumor sample.

In conclusion, the introduction of the synthetic ligand R2858 will enable the clear-cut discrimination between tissue and plasma estrogen receptors as R5020 has already done in the case of the progestin receptor. The estrogen exchange assay will, however, have to be performed at 25°C instead of 0°C in order to measure a number of binding sites as close to the total number as possible. Although in the light of present results the significance of establishing the hormonal status of the patient by plasma radioimmunoassay remains unknown, it is certain that a more accurate definition of the nature of the tissue sample and greater care in its handling are called for. The next step will then be the development of a nuclear assay.

ACKNOWLEDGMENT

The competent technical assistance of J. Humbert is gratefully acknowledged.

REFERENCES

1. Dao, T. L. (1975): Pharmacology and clinical utility of hormones in hormone related neoplasms. *Handbk. Exp. Pharmakol.*, 38:170–192.
2. Laumas, K. R., and Verma, U. (1976): Binding and metabolism of 1,2 [^{3}H]-progesterone

in human breast cancerous tissue. Abstracts, Vth International Congress of Endocrinology, Hamburg, July 1976, p. 54, No. 134.
3. Anonymous (1976): Hormone receptors and breast cancer. *Br. Med. J.,* 6027:67–68.
4. Jensen, E. V., Block, G. E., Smith, S., Kyser, K., and DeSombre, E. R. (1971): Estrogen receptors and breast cancer response to adrenalectomy. Prediction of response to cancer therapies. *Nat. Cancer Inst. Monogr.,* 34:55–79.
5. McGuire, W. L., Carbone, P. P., and Vollmer, E. P. (Eds.) (1975): *Estrogen Receptors in Human Breast Cancer.* Raven Press, New York.
6. Terenius, L. (1973): Estrogen and progestogen binders in human and rat mammary carcinoma. *Eur. J. Cancer,* 9:291–294.
7. Horwitz, K. B., McGuire, W. L., Pearson, O. H., and Segaloff, A. (1975): Predicting response to endocrine therapy in human breast cancer: an hypothesis. *Science,* 189:726–727.
8. Feil, P. D., Glasser, S. R., Toft, D. O., and O'Malley, B. W. (1972): Progesterone binding in the mouse and rat uterus. *Endocrinology,* 91:738–746.
9. Rao, B. R., Wiest, W. G., and Allen, W. M. (1973): Progesterone "receptor" in rabbit uterus. I. Characterization and estradiol-17β augmentation. *Endocrinology,* 92:1229–1240.
10. Murugesan, K., and Laumas, K. R. (1973): Binding of 6,7-^{3}H-norethynodrel to the rat uterine cytosol and to human endometrium and myometrium. *Steroids,* 8:451–470.
11. Philibert, D., and Raynaud, J. P. (1975): Binding of progesterone and R 5020, a highly potent progestin, to human endometrium and myometrium. *Contraception,* 10:457–466.
12. Horwitz, K. B., and McGuire, W. L. (1975): Specific progesterone receptors in human breast cancer. *Steroids,* 25:497–505.
13. Raynaud, J. P., Bouton, M. M., Philibert, D., Delarue, J. C., Guerinot, F., and Bohuon, C. (1975): Les récepteurs estrogène et progestogène dans le cancer du sein. Proceedings, International Symposium on Hormones and Breast Cancer, Nice, May 1975. *Inserm,* 55:71–82.
14. Raynaud, J. P., Bouton, M. M., Philibert, D., Delarue, J. C., Guerinot, F., and Bohuon, C. (1976): Progesterone and estradiol binding sites in human breast carcinoma. *Res. Steroids* (in press).
15. Huber, P. R., Geyer, E., Wyss, H. I., and Eppenberger, U. (1976): Progesterone and estradiol receptors in human breast cancer. *Experientia,* 32:798.
16. Trams, T., and Henning, H. (1976): Specific progesterone receptors in hormone-dependent tumors. Abstracts, Vth International Congress of Endocrinology, Hamburg, July 1976, p. 55, No. 135.
17. Arbogast, L. Y., and DeSombre, E. R. (1975): Estrogen dependent in vitro stimulation of RNA synthesis in hormone-dependent mammary tumors of the rat. *J. Natl. Cancer Inst.,* 54:483–485.
18. Daehnfeldt, J. L., and Schülein, M. (1975): High affinity oestradiol receptors and the activity of glucose-6-phosphate dehydrogenase and lactose synthetase in mammary carcinoma of postmenopausal women. *Br. J. Cancer,* 31:424–428.
19. Martucci, C., Fishman, K., and Hellman, L. (1976): A chromatin binding assay for estrogen receptors in human breast tumor cytosols. Abstracts, Vth International Congress of Endocrinology, Hamburg, July 1976, p. 54, No. 132.
20. Katzenellenbogen, J. A., Johnson, H. J., Jr., and Carlson, K. E. (1973): Studies on the uterine cytoplasmic estrogen binding protein. Thermal stability and ligand dissociation rate. An assay of empty and filled sites by exchange. *Biochemistry,* 12:4092–4099.
21. Hähnel, R., Ratajczak, T., and Twaddle, E. (1976): Problems in quantitation and isolation of estrogen receptors. Abstracts, Vth International Congress of Endocrinology, Hamburg, July 1976, p. 54, No. 133.
22. Sherman, M. R., and Miller, L. K. (1976): Resolution of human breast tumor estrogen receptors from sex steroid binding globulin. *Fed. Proc.,* 35:71, 1560.
23. Vermeulen, A., and Verdonck, L. (1968): Studies on the binding of testosterone to human plasma. *Steroids,* 11:609–635.
24. Bouton, M. M., and Raynaud, J. P. (1976): Impaired nuclear translocation and regulation: A possible explanation of anti-estrogenic activity. *Res. Steroids* (in press).
25. Tisman, G., and Wu, S. J. G. (1976): Oestrogen-binding protein in blood to predict response of breast cancer to hormone manipulation. *Lancet,* 7977:145–146.

26. Amaral, L., and Werthamer, S. (1976): Identification of breast cancer transcortin and its inhibitory role in cell-mediated immunity. *Nature,* 262:589–590.
27. Raynaud, J. P., Bouton, M. M., Philibert, D., and Vannier, B. (1976): Steroid binding in the hypothalamus and pituitary. In: *Hypothalamus and Endocrine Functions,* edited by F. Labrie, J. Meites, and G. Pelletier, pp. 171–189. Plenum Press, New York.
28. Rosen, P. P., Menendez-Botet, C. J., Nisselbaum, J. S., Urban, J. A., Mike, V., Fracchia, A., and Schwartz, M. K. (1975): Pathological review of breast lesions analyzed for estrogen receptor protein. *Cancer Res.,* 35:3187–3194.
29. Longo, S., Lerner, H., and Fisher, B. (1976): Correlation of histopathologic grading and degree of estrogen binding in carcinoma of the breast. *Cancer Res. Proc.,* 17:198.

Interactions of R5020 with Progesterone and Glucocorticoid Receptors in Human Breast Cancer and Peripheral Blood Lymphocytes *In Vitro*

*Marc Lippman, *Karen Huff, *Gail Bolan, and **James P. Neifeld†

*Medical Breast Cancer Section, Medicine Branch, and **Surgery Branch, National Cancer Institute, National Institutes of Health, Bethesda, Maryland 20014*

Work performed in many laboratories in the past decade has established the primary role of specific receptors for steroid hormones as mediators for most if not all steroid-mediated effects. Thus, the presence of a specific receptor has generally been correlated with hormonal responsiveness. For example, quantification of estrogen receptors in human breast cancer has significantly improved the clinician's ability to select patients suitable for hormonal therapy (1). More recently, it has been suggested that determination of progesterone receptor levels in breast cancer may provide even greater accuracy in predicting which patients with metastatic breast cancer are likely to respond to hormonal therapy (2,3). Data supporting this hypothesis are reviewed extensively elsewhere in this volume. Obviously, if determinations of progesterone receptor values are to be of any value, specific methodologies for their accurate measurement must be available. Unfortunately, specific progesterone receptor binding may be extremely difficult to quantitate. Many progestational agents bind not only to progesterone receptor but to plasma binding components (4) and glucocorticoid receptor (5) as well. Recently a synthetic progestational agent, 17,21-dimethyl-19-nor-pregna-4,9-diene-3,20 dione (R5020), has been reported to show high binding affinity for progesterone receptors but not significant binding to plasma binding components (6,7). Thus, it appeared appropriate to evaluate this agent under controlled conditions which might allow detailed analysis of its binding properties and biologic effects. We chose to examine human breast cancer cell lines *in vitro* as well as purified popula-

---

† Current address: Department of Surgery, Medical College of Virginia, Richmond, Virginia 23225.

tions of human peripheral blood lymphocytes. These two systems were chosen for a variety of reasons. Aside from the evidence mentioned above (2,3), there are many reasons to believe that progestational agents and glucocorticoids interact with mammary gland. Glucocorticoids are required for normal rodent mammary gland morphogenesis (8). Human mammary gland also appears to require glucocorticoids for differentiation (9). Glucocorticoids induce an increase in mouse mammary tumor virus in tissue culture (10). Specific glucocorticoid receptors have been detected in some rodent mammary carcinomas (11). Recently Teulings and co-workers (12) have reported that some human breast cancers contain specific glucocorticoid receptors. In as yet unpublished work we have confirmed this finding. We have reported that some human breast cancer cell lines[1] in long-term tissue culture contain estrogen (13,14), androgen (15,16), and glucocorticoid and progesterone (17) receptors. The human and mammary nature of these cell lines has recently been reviewed (18). By a variety of genetic, morphologic, and biochemical criteria, these cell lines appear to be suitable models for the study of the nature of the interactions between various progestational agents and glucocorticoids and specific receptors. In addition, these cell lines have the advantage of greatly simplifying the interpretations of the effects of various hormonal agents. By using defined media, free of serum components, in which single hormones may be added to well-characterized, uniformly viable populations of cells, investigators can avoid many difficulties of more complex systems. Thus, a given hormone's effects, which are mediated by alterations in either the level or activity of another, can be avoided, as can indirect effects of a given hormone. For example, effects on supporting stromal elements or on the immune system may thus be eliminated from consideration.

Purified populations of human peripheral blood lymphocytes provide a second attractive system for the evaluation of the interaction of progestational agents and glucocorticoids with specific receptors. Corticosteroids have pronounced inhibitory effects on lymphoid cells (19). Glucocorticoid receptors have been extensively characterized in rodent thymocytes (20) and leukemia (21). In these and other systems progesterone has pronounced antiglucocorticoid activity and little intrinsic activity of its own (21,22). Recently we have reported that normal human peripheral blood lymphocytes contain glucocorticoid but not progesterone receptor sites (24,25). Glucocorticoids dramatically inhibit mitogen-stimulated nucleoside incorporation in these cells (25). Thus, they appear to provide an ideal complementary human system for the evaluation of the binding properties and biologic

---

[1] The ZR-75-1 cell line was initially established by Young and Engel at the National Cancer Institute (*manuscript in preparation*). The MCF-7 cell line (28) was provided by Marvin Rich of the Michigan Cancer Foundation. The Evsa T cell line was established in our own laboratory as previously described (14). MDA-231, HT-39, BT-20, G11 and 496 were provided by Dr. Ronald Herberman of the National Cancer Institute.

effects of various progestational agents. Although many of the following points will be expanded considerably, a few important aspects of these cells should be mentioned here. Both systems are human. The breast cells contain both glucocorticoid and progesterone receptors, whereas lymphocytes have only glucocorticoid receptor. Breast cancer cells are inhibited by glucocorticoids but essentially unaffected by progesterone. Lymphocytes are also inhibited by glucocorticoids; progesterone is also inhibitory but about 100-fold less potent. Therefore, these cell populations should provide an interesting vehicle for the evaluation of R5020.

Some of the work to be discussed in this chapter has been previously published elsewhere (17,18,25–27).

RESULTS

Breast Cancer Cell Lines

The inhibitory effects of glucocorticoids on the MCF-7 cell line are documented in Fig. 1. As shown, following addition of 10^{-7} M dexamethasone to replicately plated cells, there is a marked inhibition of cell division. By phase microscopic inspection, cells growing in the presence of glucocorticoid do not appear grossly different from control cells. Particularly, cell vacuola-

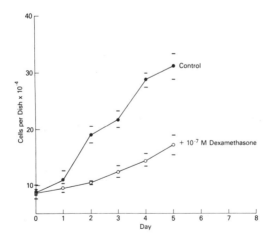

FIG. 1. The effects of 10^{-7} M dexamethasone on cell division in MCF-7 cells. Cells growing in Eagle's minimal essential medium (MEM) supplemented with 2X glutamine and antibiotics plus 5% calf serum (North America Biological) were harvested in trypsin-EDTA and replicately plated into multiwell plastic dishes. Then 24 hr later the medium was changed to fresh MEM with or without the addition of 10^{-7} M dexamethasone added as a 1,000X concentrate in ethanol (final volume ethanol <0.1%). This volume of ethanol was added to control dishes. Dishes were harvested daily in EDTA solution and cells counted in a hemocytometer. Results are means of triplicate determinations ± 1 SD.

TABLE 1. *Effects of dexamethasone on thymidine and leucine incorporation in MCF-7 human breast cancer*

Dexamethasone concentration (M)	Leucine incorporation (dpm $\times$ 10^{-2}/mg protein/hr) + 1 SD	Thymidine incorporation (dpm $\times$ 10^{-2}/μg/hr) + 1 SD
0	39.6 ± 0.68	52.3 ± 5.1
10^{-9}	39.1 ± 2.3	49.7 ± 3.2
10^{-8}	46.6 ± 6.8	40.4 ± 3.6
10^{-7}	37.3 ± 2.9	30.8 ± 2.4
10^{-6}	37.3 ± 2.9	32.1 ± 3.1

Cells growing as described in Fig. 1 were replicately plated into multiwell Linbro dishes. After 24 hr the medium was exchanged for serum-free MEM; 24 hr later hormone was added as a 1,000X concentrate in ethanol. Cells were harvested 24 hr later, and incorporation of precursor into acid-insoluble material was assessed as previously described (17).

tion, detachment, and death are not apparent. [This is in contrast to treatment with antiestrogens which are lethal to cells (13,14).]

These effects of cell growth are reflected in an inhibition of thymidine incorporation into DNA as shown in Table 1. Interestingly, the incorporation of leucine into amino acids is not inhibited at 24 hr following glucocorticoid addition, a time at which thymidine incorporation is already inhibited.

Furthermore, as shown in Table 2 there is good agreement between presence of specific glucocorticoid receptors in various cell lines and their inhibition by glucocorticoids. Neither cell line lacking glucocorticoid receptor is inhibited by dexamethasone, whereas all four cell lines with receptor show a fall in thymidine incorporation ranging from 23 to 76%.

A binding curve of [^{3}H]dexamethasone to cytoplasmic extracts from MCF-7 cells is shown in Fig. 2. A high-affinity receptor is seen. The data are replotted in the inset of this figure using the Scatchard technique (31). The straight line obtained suggests that within the concentration range of ligand employed, the [^{3}H]dexamethasone is binding to a single class of receptor sites of uniform affinity.

TABLE 2. *Comparison of glucocorticoid receptor levels and inhibition of thymidine incorporation in human breast cancer cells in tissue culture*

Cell line	Glucocorticoid receptor activity (fmoles dexamethasone bound/mg cytoplasmic protein)	% Inhibition of thymidine incorporation by 10^{-7} M dexamethasone ± 1 SD
G-11	146	76 ± 4
MCF-7	199	46 ± 7
ZR-75-1	102	43 ± 2
MDA-231	77	23 ± 6
496	5.3	4 ± 8
EVSA-T	0	0 ± 6

Thymidine incorporation was measured in the presence or absence of 10^{-7} M dexamethasone as previously described (17). Glucocorticoid receptor activity was determined using minor modifications (29) of a previously published method (30).

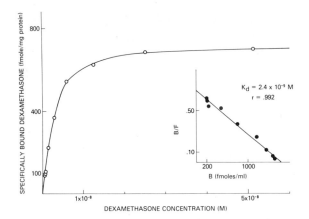

FIG. 2. The binding of [^{3}H]dexamethasone to cytoplasmic extracts from MCF-7 human breast cancer cells. The binding data shown are plotted in the inset using the Scatchard technique (31). Methods for preparation of cytosols and binding assays are supplied elsewhere (29,30). The straight line drawn in the Scatchard plot (correlation coefficient $r = 0.992$) is derived using computer-assisted methods (32).

However, one cannot exclude that binding to cytoplasmic extracts at 4°C gives values for binding affinity different from those which obtain using intact cells under more physiologic conditions. This point was examined directly by performing competitive binding studies on intact cells at 37°C incubated in tissue culture medium identical to that used for incorporation studies. Results are shown in Fig. 3. Once again, a high-affinity receptor for [^{3}H]dexamethasone is found. The K_d of this receptor (6.8×10^{-9} M) is at the upper limit of values obtained in eight experiments on cytoplasmic

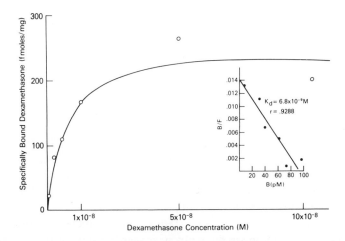

FIG. 3. The binding of [^{3}H]dexamethasone to intact MCF-7 human breast cancer cells at 37°C. The binding data shown are plotted in the inset using the Scatchard technique (31). Methods for this whole-cell assay technique are supplied elsewhere (16,17).

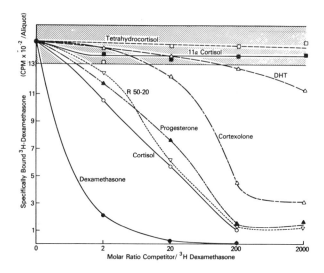

FIG. 4. Ability of various steroids to compete with [^{3}H]dexamethasone for specific binding sites in MCF-7 human breast cancer cells. In these experiments, cytoplasmic extracts were prepared and aliquoted into separate tubes. Unlabeled steroids were added, followed immediately by sufficient concentrated [^{3}H]dexamethasone to yield a final concentration of 5×10^{-9} M. Following overnight incubation at 4°C, specifically bound steroid was assessed as previously described (29,30) (DHT, 5α-dihydrotestosterone).

receptor. We conclude that for these cells there is little difference between binding affinities obtained on cytoplasmic extracts at 4°C and those obtained in intact cells.

The binding specificities of this receptor were next examined. Figure 4 illustrates the ability of various concentrations of unlabeled steroids to compete with [^{3}H]dexamethasone for specific binding. As anticipated, dexamethasone and cortisol both compete well. Progestational agents such as R5020 and progesterone also both compete, although less potently than dexamethasone. The antiglucocorticoid cortexolone (33) competes less effectively, but, at a 200-fold molar excess, has reduced specific binding to approximately one-third of that seen in the absence of competitor. The reduced metabolite tetrahydrocortisol and the biologically inactive stereo-isomer 11α-OH-cortisol do not compete even when present at 2,000-fold molar excess. Note that 5α-dihydrotestosterone competes for less than 25% of the specific binding sites even when present in 2,000-fold molar excess. In these competition studies, 5×10^{-9} M [^{3}H]dexamethasone was used. At this concentration greater than 75% of all protein-bound steroid is associated with specific binding sites (competable).

However, the fact that an unlabeled steroid reduces binding of [^{3}H]-dexamethasone to receptors does not necessarily indicate true competition for a common binding site. One might, for example, imagine that an un-

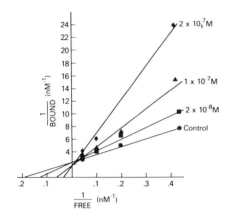

FIG. 5. Double reciprocal analysis of specific binding of [^{3}H]dexamethasone to receptor from MCF-7 human breast cancer cells in the presence of various concentrations of unlabeled cortisol. The concentration of unlabeled cortisol used is indicated adjacent to each line. Specifically bound and free [^{3}H]dexamethasone are in units of fmoles/ml. Specific binding was measured and analyzed as previously described (17,32).

labeled steroid increased the nonspecific binding of the labeled ligand (dexamethasone). Under these circumstances an apparent decrease in specific binding would be observed. Similarly, noncompetitive inactivation of receptor by the unlabeled steroid would induce a similar effect. For these reasons, specific binding of increasing concentrations of [^{3}H]dexamethasone was measured in the presence of various concentrations of unlabeled competitors. Results are shown in Fig. 5 through 8. In Fig. 5 is shown a double reciprocal plot of [^{3}H]dexamethasone binding in the presence of various concentrations of cortisol. Competitive inhibition is the expected and observed result. Note that the extrapolation of the lines to infinite dexamethasone concentration yields a constant number of binding sites. In Fig. 6, the same data are represented using Scatchard analysis. The extrapolation of the lines to constant "Bound" provides an alternative graphic representation of the competitive binding phenomenon. In Fig. 7 are shown the results of an analogous experiment except that [^{3}H]dexamethasone binding is assessed in the presence of increasing concentrations of unlabeled progesterone. A double reciprocal plot is used to illustrate the data, and competitive binding kinetics are seen. Finally, in Fig. 8 the experiment is repeated except that R5020 is the unlabeled ligand. The competitive kinetics

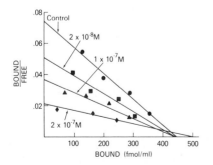

FIG. 6. Scatchard analysis of specific binding of [^{3}H]dexamethasone to receptor from MCF-7 human breast cancer cells in the presence of various concentrations of unlabeled cortisol. Concentrations of cortisol used are indicated adjacent to each line. Methods are given in the legend to Fig. 5.

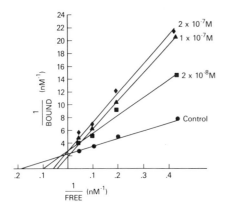

FIG. 7. Double reciprocal analysis of specific binding of [³H]dexamethasone to receptor from MCF-7 human breast cancer cells in the presence of various concentrations of progesterone (shown adjacent to each line). Methods are given in the legend to Fig. 5.

obtained suggest that both progestational agents compete with dexamethasone for a common binding site. We conclude that the glucocorticoid receptor in these cell lines has significant affinity for progestational agents which is 10- to 100-fold lower than for glucocorticoids.

The biologic effects of various glucocorticoids and progestational agents on thymidine incorporation are shown in Fig. 9. Active glucocorticoids such as cortisol and dexamethasone inhibit thymidine incorporation. Note that cortisol appears to be somewhat less potent than dexamethasone. This may occur in part because these experiments are performed in 1% calf serum "stripped" of endogenous steroid by greater than 99% by treatment with dextran-coated charcoal (14). 11α-Cortisol and the reduced metabolite tetrahydrocortisol have essentially no inhibitory effects at concentrations tested. The antiglucocorticoid cortexolone shows inhibitory effects but with only 1/100 the potency of dexamethasone. Thus, for these compounds there is good agreement between biologic effect and ability to bind to glucocorticoid receptor.

R5020 also has inhibitory activity approximately equivalent to that of cortisol. Thus, R5020 is able to compete for glucocorticoid receptor and

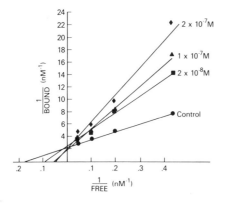

FIG. 8. Double reciprocal analysis of specific binding of [³H]dexamethasone to receptor from MCF-7 human breast cancer cells in the presence of various concentrations of R5020 (shown adjacent to each line). Methods are given in the legend to Fig. 5.

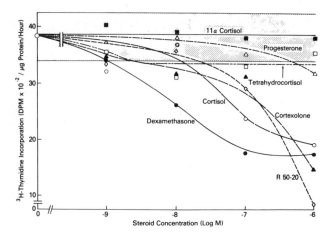

FIG. 9. Ability of various glucocorticoid and progestational agents to inhibit thymidine incorporation in MCF-7 human breast cancer cells. Methods used in these experiments are given in ref. 17. Results given are means of triplicate determinations. The shaded area represents control values ± 1 SD.

is able to induce glucocorticoid effects in MCF-7 human breast cancer cells. A strikingly different result is obtained with progesterone. Although this compound competes for glucocorticoid receptor, it showed essentially no glucocorticoid activity. This result with progesterone is entirely analogous to that reported for progesterone interaction with rat hepatoma tissue culture cells (5). In that system also, progesterone binds to glucocorticoid receptor but does not induce glucocorticoid effects.

On the basis of these competition and binding studies it might be concluded that these cells contain a single class of receptors which by binding affinity and specificity studies appears most compatible with a glucocorticoid receptor. Two kinds of experiments suggested that this might not be the case. First, if cytoplasmic extracts from MCF-7 cells were incubated with relatively high concentrations (10^{-7} M) of either [^{3}H]dexamethasone or [^{3}H]progesterone and the homologous unlabeled steroid at a concentration of 10^{-5} M added to parallel tubes incubated with labeled steroid as well, and specific binding was assessed, then [^{3}H]progesterone binding significantly exceeded that seen with [^{3}H]dexamethasone. The second suggestion that [^{3}H]progesterone might be binding to an additional site was indicated by results shown in Fig. 10. In this experiment, cytoplasmic extracts from MCF-7 cells were incubated with a relatively low concentration of [^{3}H]-progesterone (10^{-8} M) and various concentrations of unlabeled steroids. As anticipated, both R5020 and progesterone compete well with a somewhat higher apparent affinity for the R5020. Some competition is also seen with 5α-dihydrotestosterone (DHT). This is consistent with the known antiprogestational action of androgens. Most interesting, however, is the fact that glucocorticoids compete poorly. Significant competition is not seen

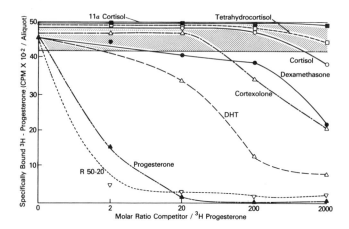

FIG. 10. Ability of various steroids to compete with [³H]progesterone for specific binding sites. Cytoplasmic extracts from MCF-7 human breast cancer cells were incubated with [³H]progesterone (10^{-8} M) with or without unlabeled steroids at the concentrations shown. Following overnight incubation at 0°C, specific binding was assessed using dextran-coated charcoal to separate protein-associated and free steroid (29,30). Results are means of triplicate determinations; the shaded area represents control (uncompeted) binding ± 1 SD.

with dexamethasone unless 2×10^{-5} M (2,000-fold molar excess) is used, and cortisol barely competes even at that concentration. It is difficult to suggest that progesterone is binding only to the glucocorticoid receptor under these circumstances since progesterone competes less well than dexamethasone for glucocorticoid receptor, and yet glucocorticoids essentially fail to compete with [³H]progesterone for binding.

This question was examined in the competition studies summarized in Figs. 11 and 12, in which the ability of either cortisol or R5020 to compete with [³H]progesterone was studied. In Fig. 11 the ability of various con-

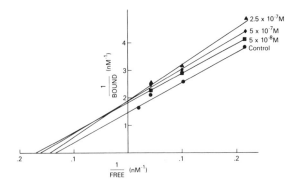

FIG. 11. Double reciprocal analysis of [³H]progesterone binding to specific receptor sites from MCF-7 human breast cancer cells in the presence of various concentrations of unlabeled cortisol (concentration shown adjacent to each line). Specific binding was measured and analyzed as previously described (17,32).

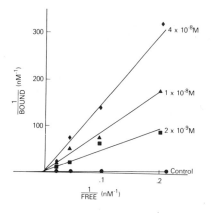

FIG. 12. Double reciprocal analysis of [^{3}H]progesterone binding to specific sites from MCF-7 human breast cancer cells in the presence of various concentrations of unlabeled R5020 (concentration shown adjacent to each line). Specific binding was measured and analyzed as previously described (17,32).

centrations of unlabeled cortisol to compete with [^{3}H]progesterone was examined using a double reciprocal method of kinetic analysis. Unlabeled cortisol essentially shows noncompetitive kinetics. The upward displacement of the three competed lines probably represents a reduction in progesterone binding to glucocorticoid receptor (*vide infra*). The results with R5020 shown in Fig. 12 are in sharp contrast. As anticipated, R5020 shows competitive binding kinetics with [^{3}H]progesterone. These results would appear to strongly suggest that these cells contain two receptors. First, a glucocorticoid receptor to which progestational agents can bind as well. In addition, there is a progesterone receptor having binding specificities limited to progestational agents and possibly androgens. Thus, one would predict that, if done over a sufficiently broad concentration range, it should be possible to demonstrate two receptors for progesterone. The results shown in Fig. 13 show that this prediction was indeed borne out. When the binding

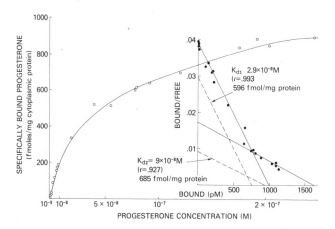

FIG. 13. Binding of [^{3}H]progesterone to receptor sites in cytoplasmic extracts from MCF-7 human breast cancer cells. Binding techniques are outlined elsewhere (17,32). Scatchard analysis (31) of the binding data is plotted in the inset. Quantification of the separate binding components was performed as suggested by Buller, Schrader, and O'Malley (34).

TABLE 3. *Progesterone binding components in MCF-7 human breast cancer cells*

Experiment	Component A			Component B[c]		
	K_d (M)	r^a	No. of sites[b]	K_d (M)	r^a	No. of sites[b]
I	2.9×10^{-8}	0.993	596	9.0×10^{-8}	0.927	685
II	2.1×10^{-8}	0.895	257	9.0×10^{-8}	0.862	704
III	3.1×10^{-8}	0.974	656	7.5×10^{-8}	0.900	219

Summary of binding data of [^{3}H]progesterone to binding components in MCF-7 human breast cancer cells. Cytoplasmic extracts incubated at 0°C overnight were quantified and analyzed as previously described (17,32).
[a] Least squares correlation coefficient.
[b] fmoles/mg cytoplasmic protein.
[c] Component B is quantified by subtracting component A from the total bound (34).

data are plotted using the Scatchard technique (31), a two-component curve is obtained. Note the very low B/F values at which the second site is expressed. This observation indicates the ease with which such lower-affinity binding components can be missed. Furthermore, as shown in Fig. 13, their omission from consideration will significantly compromise the quantification of the higher-affinity binding component. Results of this and other similar experiments are summarized in Table 3 and emphasize the reproducibility of this finding.

The previous competition studies suggest that R5020 is also binding to both glucocorticoid and progesterone receptors. Thus, one would anticipate a two-component Scatchard plot for R5020 binding. Figure 14 reveals the expected result. The binding data are best fitted by two straight lines suggesting that R5020 is binding to two receptor sites of differing affinities. Actual quantification of each binding component requires subtraction of the contribution to total binding of one component from the other (34). Results of three separate experiments are shown in Table 4. Note that R5020 binds

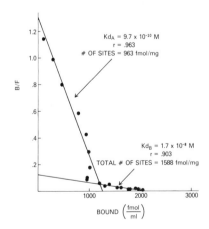

FIG. 14. Binding of [^{3}H]R5020 to receptor sites in cytoplasmic extracts from MCF-7 human breast cancer cells. Binding techniques are outlined elsewhere (17,32). Only Scatchard analysis of the binding sites is shown. The X intercepts shown represent extrapolations of the best straight line (least squares regression) through each set of points. Actual quantification of each binding component using the method of Buller, Schrader, and O'Malley (34) is shown in Table 4.

TABLE 4. [³H]R5020 binding components in MCF-7 human breast cancer cells

Experiment	Component A			Component B[c]		
	K_d (M)	r^a	No. of sites[b]	K_d (M)	r^a	No. of sites[b]
I	9.7×10^{-10}	0.963	842	1.7×10^{-8}	0.903	746
II	2.9×10^{-9}	0.980	738	1.3×10^{-8}	0.941	460
III	4.2×10^{-9}	0.938	888	3.3×10^{-8}	0.497	712

Summary of binding data of [³H]R5020 binding components in MCF-7 human breast cancer cells. Cytoplasmic extracts incubated at 0°C overnight were quantified and analyzed as previously described (17,32).
[a] Least squares correlation coefficient.
[b] fmoles/mg cytoplasmic protein.
[c] Component B is quantified by subtracting component A from the total bound (34).

to component A (the presumed progesterone receptor) about 10 times more avidly than [³H]progesterone. However, [³H]R5020 binds about five times more tightly to component B (the presumed glucocorticoid receptor) than [³H]progesterone. Thus, R5020 would appear to be about twice as "selective" for progesterone receptor in the presence of glucocorticoid. It must be borne in mind that if two binding components are present simultaneously, there will be a very significant overestimation of the quantity of the high-affinity binding site unless the second component is either also quantified and subtracted out or eliminated from consideration. This latter alternative may be achieved by adding an excess of a ligand to both competed and uncompeted tubes which will bind only to the lower binding affinity component. This can be realized by using unlabeled cortisol which will not compete significantly for progesterone receptor sites while competing for glucocorticoid sites. (An additional benefit of this methodology is that binding to transcortin by progesterone can be removed from consideration.) This consideration of interference by a second binding component is relevant because it has been reported that more than 30% of human breast cancer samples have significant glucocorticoid binding activity (which quantitatively may greatly exceed progesterone binding activity). We conclude from these studies that it may not be sufficient to use R5020 for the quantification of progesterone receptors if other binding components are present. Further evidence of the "glucocorticoid" activity of R5020 is obtained in the human lymphocyte experiments described next.

Human Peripheral Blood Lymphocytes

As a complementary system we examined the binding and biologic effects of glucocorticoids and progestational agents on human peripheral blood lymphocytes. Figure 15 shows a binding curve of [³H]dexamethasone to intact human peripheral blood lymphocytes. A high-affinity dexamethasone

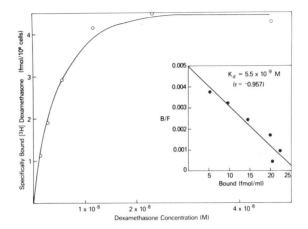

FIG. 15. Binding of [³H]dexamethasone to human peripheral blood lymphocytes. A Scatchard analysis (31) of the binding data is shown in the inset. Methods are given elsewhere in detail (25,32). Essentially, purified lymphocytes are incubated in serum-free buffer with radioactive steroid with or without added competitor. At the end of the incubation (2 hr at 22°C) bound and free steroid are separated by washing of the cells.

binding site is seen. Scatchard analysis suggests that the dexamethasone is binding to a single class of receptors of uniform affinity.

Binding studies performed using [³H]progesterone failed on multiple occasions to demonstrate specific progesterone binding in either purified peripheral blood lymphocytes or lymphocytes treated with phytohemagglutinin (PHA). PHA treatment induces a 2- to 3-fold increase in glucocorticoid receptor activity (24,25).

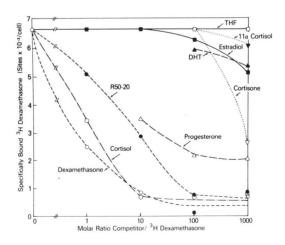

FIG. 16. Competition of various steroids with [³H]dexamethasone for specific binding sites in human peripheral blood lymphocytes. Methods are given elsewhere (25,32). DHT, 5α-dihydrotestosterone; THF, tetrahydrocortisol.

Further characterization of this receptor is provided by the specificity studies shown in Fig. 16. In these experiments, the ability of various steroids to compete with [^{3}H]dexamethasone for specific binding sites is examined. As expected, glucocorticoids such as dexamethasone and cortisol compete well for specific binding. The inactive stereoisomer 11α-OH-cortisol and the reduced metabolite tetrahydrocortisol do not compete significantly. Sex steroids also show little competitive ability. Note that cortisone is more than 100 times less potent than cortisol as a competitor. This is consistent with our previous results in human acute leukemic lymphoblasts (23). Of particular interest are the results with progestational agents. Both R5020 and progesterone appear to compete with [^{3}H]dexamethasone for receptor sites, although R5020 has an apparently higher affinity for the receptor. This is similar to the results shown for glucocorticoid receptor in human breast cancer cells (see Fig. 4).

There is a good correlation between concentrations of dexamethasone which inhibit nucleoside incorporation in PHA-treated lymphocytes and concentrations of dexamethasone which bind to receptor. Results are illustrated in Fig. 17. It should be noted that both binding and incorporation studies are performed on intact cells incubated in an identical buffered medium. This close dose-response relationship suggests a direct role for these receptor sites in mediating glucocorticoid action in these cells.

Effects of R5020 and progesterone on [^{3}H]thymidine incorporation in PHA-stimulated human peripheral blood lymphocytes are given in Fig. 18. Both R5020 and progesterone have glucocorticoid effects in lymphocytes.

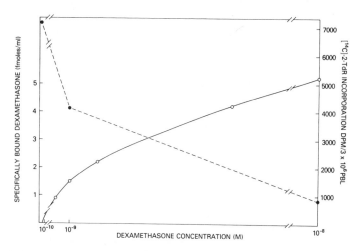

FIG. 17. Comparison of glucocorticoid binding to receptor in human peripheral blood and inhibitory effects on [^{3}H]thymidine incorporation. Methods for binding and incorporation studies are provided elsewhere (25). Results of binding studies are means of quadruplicate determinations. Results of incorporation studies are means of triplicate determinations.

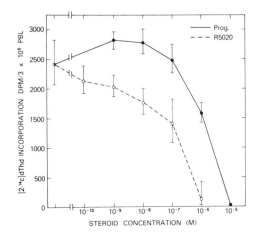

FIG. 18. Effects of progesterone and R5020 on inhibiting nucleoside incorporation in PHA-stimulated human peripheral blood lymphocytes. Results are means of quadruplicate cultures ± 1 SD. Methods are given in ref. 25.

R5020 is clearly about 10-fold more potent. This result closely correlates with the concentrations of these steroids which compete with [^{3}H]dexamethasone for binding to receptor.

Thus, in human lymphocytes which lack progesterone receptor but do contain glucocorticoid receptor, R5020 has considerable glucocorticoid activity. These results tend to confirm our findings in breast cancer cell lines and suggest that care must be employed in interpretation of either binding or biologic effects in experiments performed with R5020.

REFERENCES

1. McGuire, W. L., Carbone, P. P., Sears, M. E., and Escher, G. C. (1975): Estrogen receptors in human breast cancer: An overview. In: *Estrogen Receptors in Human Breast Cancer,* edited by W. L. McGuire, P. P. Carbone, and E. P. Vollmer, pp. 1–8. Raven Press, New York.
2. Horwitz, K. B., and McGuire, W. L. (1975): Specific progesterone receptors in human breast cancer. *Steroids,* 25:497–505.
3. Horwitz, K. B., McGuire, W. L., Pearson, O. H., and Segaloff, A. (1975): Predicting response to endocrine therapy in human breast cancer: A hypothesis. *Science,* 189:726–727.
4. Kontula, K., Janne, O., Vihko, R., Jager, E., Visser, J., and Zeelen, F. (1975): Progesterone-binding proteins: in vitro binding and biological activity of different steroidal ligands. *Acta Endocrinol. (Kbh.),* 78:574–592.
5. Rousseau, G. G., Baxter, J. D., and Tomkins, G. M. (1972): Glucocorticoid receptors: Relations between steroid binding and biological effects. *J. Mol. Biol.,* 67:99–116.
6. Philibert, D., and Raynaud, J. P. (1973): Progesterone binding in the immature mouse and rat uterus. *Steroids,* 22:89–98.
7. Philibert, D., and Raynaud, J. P. (1974): Progesterone binding in the immature rabbit and guinea pig uterus. *Endocrinology,* 94:627–634.
8. Vonderhaar, B. K., and Topper, Y. J. (1973): Critical cell proliferation as a prerequisite for differentiation of mammary epithelial cells. *Enzyme,* 15:340–350.

9. Flaxman, B. A. (1973): In vitro studies of the human mammary gland: Effects of hormones on proliferation on primary cell culture. *J. Invest. Dermatol.,* 61:67–71.
10. Parks, W., Scolnick, E., and Kuzihowski, I. (1974): Dexamethasone stimulation of murine mammary tumor virus: A tissue culture source of virus. *Science,* 184:158–160.
11. Goral, J. E., and Wittliff, J. L. (1975): Comparison of glucocorticoid binding proteins in normal and neoplastic mammary tissue of the rat. *Biochemistry,* 14:2944–2952.
12. Teulings, F. A. G., Treurniet, R. E., and VanGilse, H. A. (1976): Glucocorticoid receptors in human breast carcinomas. Proceedings of the V International Congress of Endocrinology, Abstract #136:55, Hamburg.
13. Lippman, M. E., Bolan, G., and Huff, K. (1976): The effects of estrogens and antiestrogens on hormone responsive human breast cancer in long term tissue culture. *Cancer Res.,* 36:4595–4601.
14. Lippman, M. E., and Bolan, G. (1975): Estrogen responsive human breast cancer in continuous tissue culture. *Nature,* 256:592–593.
15. Lippman, M. E., Bolan, G., and Huff, K. (1975): Androgen responsive human breast cancer in long term tissue culture. *Nature,* 258:339–341.
16. Lippman, M. E., Bolan, G., and Huff, K. (1976): The effects of androgens and antiandrogens on hormone responsive human breast cancer in long term culture. *Cancer Res.,* 36:4610–4618.
17. Lippman, M. E., Bolan, G., and Huff, K. (1976): The effects of glucocorticoids and progesterone on hormone responsive human breast cancer in long term tissue culture. *Cancer Res.,* 36:4602–4609.
18. Lippman, M. E., Osborne, C. K., Knazek, R., and Young, N. (1977): In vitro model systems for the study of hormone dependent human breast cancer. *N. Engl. J. Med.,* 296:154–159.
19. Claman, H. N. (1972): Corticosteroids and lymphoid cells. *N. Engl. J. Med.,* 287:388–397.
20. Munck, A., and Wira, C. (1971): Glucocorticoid receptors in rat thymus cells. In: *Advances in the Biosciences, No. 7: Schering Workshop on Steroid Hormone Receptors,* edited by G. Raspe, pp. 301–324. Pergamon Press, Oxford.
21. Baxter, J. D., Harris, A. W., Tomkins, G. M., and Cohn, M. (1971): Glucocorticoid receptors in lymphoma cells in culture: relationship to glucocorticoid killing activity. *Science,* 171:189–191.
22. Samuels, H. H., and Tomkins, G. M. (1970): Relation of steroid structure to enzyme induction in hepatoma tissue culture cells. *J. Mol. Biol.,* 52:57–65.
23. Lippman, M. E., Halterman, R. M., Leventhal, B. G., Perry, S., and Thompson, E. B. (1973): Glucocorticoid binding proteins in human acute lymphoblastic leukemic blast cells. *J. Clin. Invest.,* 52:1715–1725.
24. Neifeld, J. P., Lippman, M. E., and Tormey, D. C. (1976): Induction of glucocorticoid receptor activity in human peripheral blood lymphocytes by PHA stimulation. Proceedings of the 16th Annual Meeting of the American Society of Clinical Investigation, Atlantic City.
25. Neifeld, J. P., Lippman, M. E., and Tormey, D. C. (1976): Steroid hormone receptors in normal human lymphocytes: Induction of glucocorticoid receptor activity by PHA stimulation. *J. Biol. Chem. (in press).*
26. Lippman, M. E., Huff, K., and Bolan, G. (1976): Progesterone and glucocorticoid interactions with receptor in breast cancer cells in long term tissue culture. *Proc. N.Y. Acad. Sci. (in press).*
27. Lippman, M. E., Bolan, G., and Huff, K. (1976): Hormone responsive human breast cancer in long term tissue culture. In: *Breast Cancer: Trends in Research and Treatment,* edited by J. C. Heuson, W. H. Mattheiem, and M. Rozenzweig, pp. 111–139. Raven Press, New York.
28. Soule, H. D., Vazquez, J., Long, A., Abert, S., and Brennan, M. (1973): A human cell line from a pleural effusion derived from a breast carcinoma. *J. Natl. Cancer Inst.,* 51:1409–1413.
29. Lippman, M. E., and Huff, K. (1976): A demonstration of androgen and estrogen receptors in a human breast cancer using a new protamine sulfate assay. *Cancer,* 38:868–874.
30. McGuire, W. L., and DeLaGarza, M. (1973): Improved sensitivity in the measurement of estrogen receptor in human breast cancer. *J. Clin. Endocrinol. Metab.,* 36:548–552.

31. Scatchard, G. (1949): The attraction of proteins for small molecules. *Ann. N.Y. Acad. Sci.,* 51:660–672.
32. Aitken, S. C., and Lippman, M. E. (1976): A simple computer program for quantification and Scatchard analysis of steroid receptor proteins. *J. Steroid Biochem. (in press).*
33. Munck, A., and Brinck-Johnsen, T. (1965): Specific and nonspecific physiochemical interactions of glucocorticoids and related steroids with rat thymus cells in vitro. *J. Biol. Chem.,* 243:5556–5565.
34. Buller, R. E., Schrader, W. T., and O'Malley, B. W. (1976): Steroids and the practical aspects of performing binding studies. *J. Steroid Biochem.,* 7:321–326.

Estrogen and Progesterone Receptors and Glucose Oxidation in Mammary Tissue

Joseph Levy and Seymour M. Glick

Soroka Medical Center and University, Center for the Health Sciences, Ben Gurion University of the Negev, Beersheva, Israel

Estimation of estrogen receptor (ER) in the cytosol of biopsy tissue from breast cancer is currently being used for the prediction of the results of endocrine therapy (17).

Recently, McGuire and his group (11) have suggested that the presence of progesterone receptor (PgR) in the cytosol of mammary tumor tissue may be an additional marker of hormonal responsiveness of the tissue because in estrogen target tissue the synthesis of PgR depends on the action of estrogen. The measurement of both PgR and ER thus increases significantly the accuracy of prediction of the results of endocrine therapy for breast cancer.

The presence of a specific receptor for hormones in any target tissue is a prerequisite for the responsiveness to the hormone, but the receptor-hormone interaction is only an early step in hormone action. It may very well be possible that the failure of response to endocrine therapy by some tumors that are not endocrine dependent but contain ER may be due to a defect at a later step in hormone action. An ideal marker of endocrine responsiveness would therefore be a measurable product of hormone action rather than the initial binding step.

The aim of this study was to look for new methods to study the biochemical expression of hormonal action in isolated breast cancer tissue. Because energy requirements of the cell may reflect changes in its overall metabolism, glucose oxidation was examined as a potential indicator possibly affected by hormones in responsive tissue.

MATERIALS AND METHODS

Materials

17β-Estradiol-[2,4,6,7(n)]-^{3}H, specific activity 85 Ci/mmole, and D[u-^{14}C]glucose, specific activity 3.0 to 3.8 mCi/mmole, were obtained from the Radiochemical Centre, Amersham, England. R5020 (12,21-dimethyl-19-nor-4,9-pregnadiene-3,20-dione) and R5020-6,7^{3}H, specific

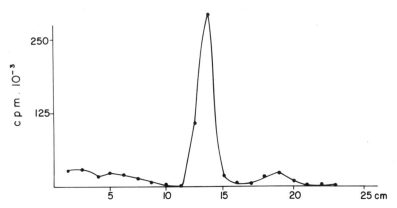

FIG. 1. Determinations of steroid purity by thin-layer chromatography on Silica Gel in benzene ethyl acetate (7:3, vol/vol).

activity 51.4 Ci/mmole, were a gift from Roussel-Uclaf (Romainville, France). The purity of R5020 was checked by thin-layer chromatography on Silica Gel and developed in benzene ethyl acetate 7:3. As can be noted in Fig. 1, some impurities exist in the labeled material. Hydroxide of Hyamine® 10X (methylbenzethonium chloride), 2,5-diphenyloxazolol, and 1,4 bis [2(4-methyl-5-phenyloxazol)] benzene were obtained from Packard (Downey Grove, Illinois). Radioinert steroids were obtained from Ikapharm (Ramat Gan, Israel). All other products were analytical grade.

Tissue Handling

For steroid receptor assays human tumor samples were obtained as biopsy or mastectomy specimens within 30 min after removal and were quick frozen in liquid nitrogen. Rat mammary tissue was dissected immediately after death by cervical dislocation. Both human and rat tissues were stored at −70°C until assay.

For the metabolic studies human tumor tissue and rat mammary tissue were transferred to Eagle's minimum essential medium (6), which was kept in an ice bath during all manipulations.

Receptor Assay

The frozen tissue was pulverized and then homogenized in a Heidolph homogenizer set at medium speed, in a buffer containing 0.01 M Tris-HCl, 15 mM EDTA, 0.5 mM dithiothreitol, pH 7.4. In some of the assays for 17β-estradiol receptors and in the assay for progesterone receptors, β-lipoprotein isolated from human plasma was added to improve the solubility of the steroids. The homogenate was centrifuged at 105,000 g in a Spinco SW 41 Ti rotor for 60 min at 0 to 4°C to obtain the supernatant cytosolic fraction.

To analyze saturation kinetics, we incubated 200 μl cytosol (2 to 10 mg soluble protein per milliliter) assayed by the method of Lowry et al. (14) in duplicate with increasing quantities of ^{3}H-hormone added in 50 μl homogenization buffer (0.1 to 3 nM final concentration). Parallel incubations contained 100-fold excess nonradioactive hormone. After 16 hr of incubation at 4°C, 0.5 ml of a charcoal suspension was added (250 mg Norit A, 2.5 mg dextran in 100 ml buffer). After vigorous shaking the mixture was kept at 4°C for 30 min and was then centrifuged for 15 min at 2,000 g. Then 0.5 ml of the supernatant was added to 4 ml of a counting mixture of 667 ml toluene, 333 ml Triton X-100, and 5.5 g 2,5-diphenyloxazolol + 70 mg 1,4 bis [2(4-methyl-5-phenyloxazol)]-benzene, counted in a Tri Carb liquid scintillation counter (Packard).

The data were analyzed according to Scatchard (20) after subtraction of nonspecific binding calculated from the incubation containing excess unlabeled or labeled hormone, with extrapolation of the nonspecific binding line in the original binding curve.

Glucose Oxidation

Fresh tissue in Eagle's minimum essential medium at 4°C was cut into small pieces which were transferred to a preincubation medium (Eagle's) containing steroids dissolved in ethanol. Control preincubation mixtures contained the ethanol solvent only. Approximately 50 pieces of tissue were preincubated, with continuous shaking at 37°C for 1 hr in the case of human breast tumors and for 2 hr in the case of rat mammary tissue. The flasks were occasionally gassed with O_2-CO_2 mixture (95:5). After the preincubation period, the tissue was cooled in ice and washed four times with ice cold bicarbonate buffer (15) to remove the glucose from the preincubation medium. One to two pieces of about 3 to 7 mg each of tumor tissue, or about three pieces of 2 mg each of normal mammary tissue, were blotted carefully and transferred to 5 × 50 mm test tubes containing 150 μl bicarbonate buffer, supplemented with 0.75 μCi of ^{14}C-glucose (U) to a final concentration of 1 mM (unless otherwise specified) and with various steroids. The tubes were inserted into normal counting vials and stoppered with rubber stoppers. The vials were aerated occasionally with O_2-CO_2 mixture and shaken during incubation at 37°C for 120 min in the case of tumor tissue, and for 20 min in the case of rat mammary tissue.

Immediately after incubation the reaction was stopped by injecting through the rubber stopper 200 μl of HCl 0.2 N into the inner tube and 0.5 ml of hydroxide of methylbenzethonium chloride into the outer incubation flask. After the tubes were shaken at room temperature for 60 min, the tubes containing the tissue were taken out and 10 ml of toluene-based scintillation mixture was added.

The $^{14}CO_2$ recovered in the methylbenzethonium chloride was calculated as picomoles glucose/incubate/incubation time, after subtracting for blanks. The results were expressed as a mean ± SEM. Significance was checked according to the Student t-test.

RESULTS

R5020 Binding by Human Breast Cancers

Progesterone and estradiol receptors were assayed routinely in every cytosol obtained from human breast cancer biopsy. Scatchard plot analyses of two such cytosols containing progesterone receptors are illustrated in Fig. 2.

Twenty-one cytosols of human mammary tumors were fully evaluated for the presence of estradiol and progesterone receptors. Ten cases (~47%) contained significant concentrations of receptors to progesterone and estradiol. Six cases (~28%) were positive for estradiol but contained no detectable progesterone receptors in their cytosols. Five cases (~25%) contained neither receptor (Fig. 3).

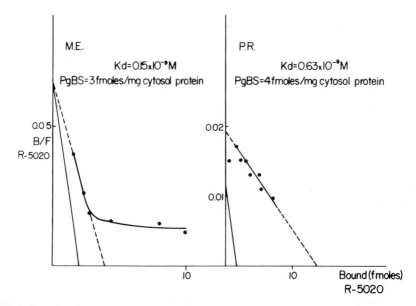

FIG. 2. Scatchard analyses of Pg binding sites (PgBS) and apparent dissociation constant in two cytosols obtained from biopsies of human breast cancer. Cytosol was incubated with 0.4 to 3 nM ^{3}H-hormone. Nonspecific binding was corrected for as indicated in Methods.

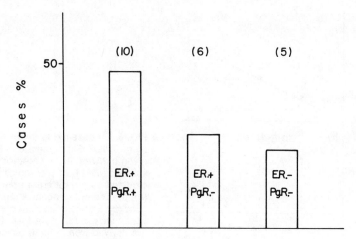

FIG. 3. Comparison of ER and PgR in 21 breast tumors. An assay was considered positive if there were more than 3 fmoles/mg of cytosol protein. The number of cases in each group is given in parentheses.

Table 1 summarizes the results in ten cases which contained significant estradiol cytosolic receptors. Progesterone binding sites varied from 3 to 73 fmoles/mg cytosolic protein. Two cases (Table 1—M.R., B.L.) were evaluated twice, and the variation of the results was in an acceptable range. The average K_d calculated from these results, $0.63 \pm 0.16 \times 10^{-9}$ M, is a little lower than the results previously obtained (10).

TABLE 1. *Progesterone receptors in human breast tumors*

	ER	PgR $K_d \cdot 10^{-9}$ M	fmoles/mg	PBS cyto. protein
L.D.	+	0.53	27	
M.R.	+	0.38	9	
		0.53	5	
B.R.	+	0.35	10	
P.R.	+?	0.63	4	
Z.F.	+	0.21	4	$K_d = 0.63 \pm 0.16 \times 10^{-9}$ M
A.B.	+	0.09	22	$N = 12$
K.J.	+	0.63	12	
M.E.	+	0.15	3	
B.L.	+?	0.90	10	
		0.93	17	
I.M.	+	2.2	73	

Summary of K_d and Pg binding sites in 10 cases described in Fig. 3. Some of the assays were performed more than once. (?) depicts results in specimens in which no estradiol receptors were detected—on one occasion out of 3 or 4 tested.

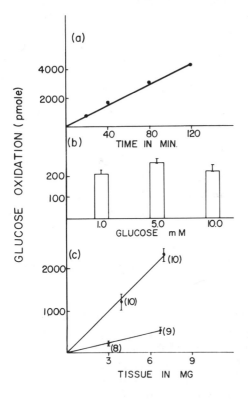

FIG. 4. Glucose oxidation in rat mammary tissue. Experimental details are given in Materials and Methods. Mean values (± SEM) of picomoles glucose oxidized during different time periods **(a)**; in different concentrations of glucose **(b)**; and as a function of different tissue weight **(c)**. Data in **a**, **b**, and **c** are derived from 7 to 10 observations each.

Glucose Oxidation by Rat Mammary Tissue

As a system parallel to human breast tumor tissue, we used normal mammary tissue from pregnant and lactating rats. Figure 4 illustrates some experimental features of the system. The oxidation of glucose is not dependent on the glucose concentration in the medium between 1 and 10 mM (Fig. 4b). Identical results were obtained with human tumorous tissue (*not shown*). We used a 1 mM concentration of glucose in all our metabolic experiments. The glucose oxidation in rat mammary tissue was linear with time (Fig. 4a) up to 120 min. In the experiments with rats a 20-min incubation period was sufficient, but in human breast tissue studies a 120-min period was needed because of the relatively low rate of oxidation. The oxidation of glucose was dependent on the weight of tissue slices present and was linear at least up to 7 mg of tissue in the incubate. Our results are expressed per incubate rather than per milligram of tissue, but routine checks of tissue weights were made to ensure uniformity of quantity of tissue in each incubate.

In the rat mammary tissue system, we measured glucose oxidation to CO_2 in tissue isolated on various days of gestation (Fig. 5). As can be seen, glucose oxidation is very low in tissue slices isolated from rats in their first

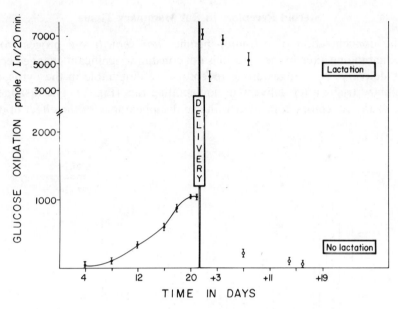

FIG. 5. Glucose oxidation in rat mammary tissue. Mean values (± SEM) of picomoles glucose oxidized per incubate per 20 min during and after pregnancy. In the no lactation group, the litters were removed from the mother approximately 10 hr after birth.

days of pregnancy, but it increases in an exponential manner until 1 day before delivery. Immediately after delivery, glucose oxidation rises 4- to 7-fold over predelivery values and then stabilizes at this level at least for the first 8 days of lactation. In rats separated from their litters immediately after birth, glucose oxidation decreased abruptly to low values. These results suggest that glucose oxidation may reflect closely the physiological changes in mammary tissue and may serve as a reliable index for overall metabolism. It may well be worthwhile to check hormonal effects on this index in view of the known effect of hormones on mammary development and lactation.

No effect of estradiol, progesterone, or R5020 on glucose oxidation was detected during the first 18 days of pregnancy (Fig. 6). In all these experiments, the tissue was preincubated in Eagle's minimum essential medium, supplemented with the above-mentioned hormones (2×10^{-8} to 2×10^{-9} M). Usually the hormones were present also in the incubation medium. The results were expressed as percent of control. The levels of oxidation, expressed as picomoles per incubate, varied from day to day. Modest but significant effects of progesterone and R5020 on glucose oxidation can be detected on the 20th and 21st days of gestation, and on the first day after delivery. The inhibition by these two steroids was in contrast to the lack of effect produced by 17β-estradiol at the same concentration. R5020 had no effect on glucose oxidation after delivery in nonlactating rats.

Steroid Receptors in Rat Mammary Tissue

The concentration of estradiol-specific (*not shown*) and progesterone-specific binding sites in the cytosol is not constant. Significant change occurs after delivery. No progesterone receptors are detectable in the cytosol of mammary tissue after delivery in nonlactating rats (Fig. 7). The disappearance of the receptors coincided with the disappearance of the effect of pro-

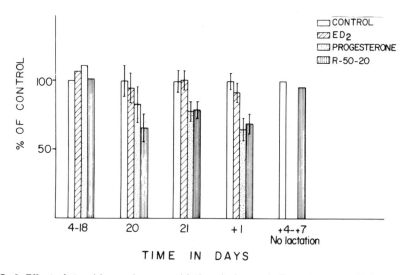

FIG. 6. Effect of steroids on glucose oxidation during and after pregnancy. Rat mammary tissue was preincubated with the indicated steroids (2×10^{-8} M) for 2 hr, and, after extensive wash, was incubated for 20 min with labeled glucose and the same steroids. The results were expressed as percent of control (preincubated and incubated with ethanol). Mean values ($\pm$ SEM) for more than 15 observations are given. In the no lactation group, the litters were removed from the mother approximately 10 hr after birth.

gesterone and R5020 on *in vitro* oxidation of glucose in this tissue, as described in Fig. 6.

These results support the previously well-established concept that hormonal effects in target tissues are mediated through the specific receptors in the tissue, and that changes in the concentration of the receptor in a given tissue regulate the responsiveness of this tissue to the hormone.

Glucose Oxidation in Human Breast Cancer

The observation of *in vitro* responsiveness of target tissue—rat mammary gland—to various steroids suggests the possibility that biochemical manifestations of hormone action can also be detected *in vitro*, and that the response is related to the presence of specific receptors and can supply valuable information for the responsiveness of the particular tissue to the hormones checked.

Thus far we have checked this system in 16 breast tumors which were analyzed for estradiol (Ed) and Pg receptors, and for the effects of Ed, Pg, and R5020 on glucose oxidation. In some instances the tissue responded, whereas in others glucose oxidation was unaffected by these hormones.

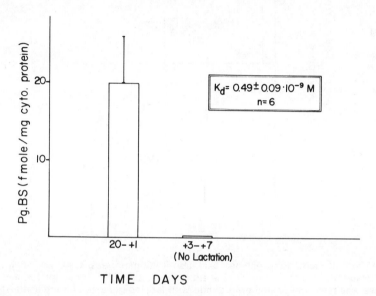

FIG. 7. Progesterone binding sites in rat mammary tissue. The data were derived from Scatchard analysis as described in Materials and Methods. The mean values (± SEM) of progesterone binding sites (PgBS) are derived from 5 to 6 rats in each group.

Figure 8 depicts two cases which responded with a change of glucose oxidation. The lower part of the figure shows glucose oxidation stimulated in tumor slices by Ed, Pg, and R5020. The tissue was preincubated for 1 hr in Eagle's medium enriched with various steroids in concentrations of 2×10^{-8} M. After the preincubation, the pieces were washed and transferred to a bicarbonate buffer medium, which contained 1 mM labeled glucose and a steroid in the same concentration as in the preincubation, and incubated for 2 hr. Occasionally hormone was present only in the preincubation phase (as can be seen on the bar on the far right in Fig. 8). The results were corrected for blanks and were expressed as picomoles glucose oxidized per incubate per 120 min of incubation.

The upper part of Fig. 8 depicts another responsive tissue in which progesterone and R5020 inhibited glucose oxidation.

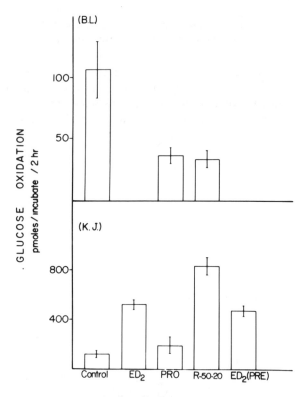

FIG. 8. Effect of steroids on glucose oxidation in human breast cancer. Human breast cancer tissue was preincubated with the indicated steroids (2×10^{-8} M) for 1 hr and, after extensive wash, was incubated for 120 min with labeled glucose and the same steroid. In another experimental group steroid was present in the preincubation period only: (PRE)−lower panel, far right. Control:ethanol only. Mean values (± SEM) for 7 to 9 observations are given.

Comparison of ER and PgR Content and Responsiveness of Glucose Oxidation to Steroids

Tables 2 and 3 summarize the results obtained in 16 different tumors with respect to ER, PgR, and the response of glucose oxidation *in vitro* to estradiol, progesterone, and R5020. Table 2 demonstrates that this *in vitro* response cannot be detected in tumors lacking progesterone receptors in their cytosol, even in the presence of estradiol receptors. Table 3 illustrates the positive cases. A clear correlation can be seen between the presence of specific receptors to progesterone and estradiol and the effects on glucose oxidation by these hormones. In five of the tumors glucose oxidation was stimulated by the steroids examined; in three of the tumors glucose oxidation was inhibited. But whether the hormones stimulated or suppressed glucose oxidation, the presence of a significant effect was invariably associated with the presence of steroid binding receptors.

TABLE 2. ER, PgR content, and responsiveness of glucose oxidation to steroids (nonresponsive tumors)

	ER	PgR	Effect on glucose oxidation		
			Ed_2	Pr	R5020
Z.F.	+	±	−		
M.A.	+	−	−	−	−
B.S.	+	Not done	−	−	−
H.S.	+	−	−		
L.L.	−	−?	−		
A.K.	−	−	−	−	−
T.R.	−	Not done	−	−	↑
G.N.	−	−	−	−	−

As in Table 3. (±), assay revealed PgR, but at lower limits of detectability. (?), assay revealed high K_d and high binding capacity.

TABLE 3. ER, PgR content, and responsiveness of glucose oxidation to steroids (responsive tumors)

	ER	PgR	Effect on glucose oxidation		
			Ed_2	Pr	R5020
M.R.	+	+	↑		
U.R.	+	Not done	↑	↑	
A.B.	+	+	↑	↑	
K.J.	+	+	↑	↑ (ns)	↑
M.E.	+	+	↑	↑	↑
I.M.	+	+	↓	↓	↓
B.R.	+	+	↓		
B.L.	+?	+		↓	↓

↑, increment in glucose oxidation caused by the steroids control. ↓, inhibition in glucose oxidation caused by the steroids control. (?), results in specimens in which on one occasion (out of 3 or 4 tested) no estradiol receptors were detected.

DISCUSSION

R5020 in Detection of Cytosolic Progesterone Receptors

The binding of progesterone to its specific receptors in rat and human tissue is masked by nonspecific binding, such as binding to cytoplasmic glucocorticoid receptors and to corticosteroid binding globulin, which is present in plasma (7,22).

R5020 is a synthetic progestin whose binding specificity in the cytosol is restricted to the progesterone receptor (18). R5020 was used recently to detect specific progesterone receptors in human breast cancer tissue (10) and has similarly been applied by us.

The dissociation constant for R5020 obtained by McGuire et al. (10) using Scatchard analysis of results obtained with a dextran-coated charcoal assay was approximately 2×10^{-9} M, a little higher than that found in our studies, which used mammary tumorous tissue (Table 1) and rat mammary tissue isolated from the pregnant rat (Fig. 7).

Steroid Receptors and Metabolism of Rat Mammary Gland

The dynamic changes of glucose oxidation by rat mammary tissue during pregnancy and lactation (Fig. 5) may be inferred from various histological and metabolic changes which have previously been reported (3) during these stages of the reproductive cycle.

Glucose oxidation is not the only process in this tissue which supplies the changing energy requirements. For example, Davis and Ben Meptan (4) observed catabolism of certain essential amino acids in the isolated perfused lactating guinea pig mammary gland. However, in the lactating goat the oxidation of glucose and acetate accounts for more than two thirds of the CO_2 production of the mammary gland in fed animals (1).

Various hormones including insulin and prolactin affect the proliferating mammary tissue during pregnancy and the active gland during lactation.

Specific receptors for 17β-estradiol (8) were detected in the 105,000 g supernatants of the lactating mammary gland of rats; much lower levels of ER were observed in glands obtained from pregnant and virgin rats.

In our studies (*unpublished results*) with the pregnant rat, we confirm these data.

The detection of progesterone receptors in this study in mammary tissue isolated from pregnant rats is not surprising, since progesterone is known to affect mammary development.

Our findings that progesterone and R5020 inhibit glucose oxidation in mammary gland isolated from the rat during the last 2 days of gestation and immediately after delivery, periods in which progesterone receptors are clearly demonstrable in this tissue (Fig. 7), are in accord with previous results which show that progesterone has an inhibiting effect on various aspects of mammary function in the pregnant rat.

Progesterone injected daily at different stages of pregnancy in the rabbit reduces or temporarily blocks mammary secretion measured either by the incorporation of ^{14}C-glucose into lactose or by the lactose content of mammary tissue (5), since in rat peripheral blood progesterone levels drop sharply between days 19 and 21 of gestation (23). This period may be most suitable for demonstrating the *in vitro* effect of progesterone, as endogenous progesterone is not available to mask the effect.

It is generally accepted that the disappearance of the inhibitory effect of progesterone has a decisive effect on lactogenesis in the pregnant rat. Glucose oxidation in the nonlactating rat is no longer affected by R5020 and progesterone (Fig. 6). Simultaneously, progesterone-specific receptors can no longer be found in this tissue (Fig. 7). These results support the concept that the effects of steroid hormones are mediated by the presence of specific cytosolic receptors in the target tissue.

Recently it was suggested by Liu and Greengard (13) that steroid hormones may exert some of their effects by affecting the phosphorylation of a special protein which they name SCARP—steroid and cyclic adenosine-3':5'-monophosphate regulated phosphoprotein. They studied the effect of some steroid hormones in their respective target tissue on the phosphorylation of this specific protein. The effect of the steroid hormones on decreasing the phosphorylation of SCARP was specific for the steroid's respective target tissue. This raises the possibility that this biochemical action may be a component of the mechanism by which these steroids achieve some of their biological effects. These effects can be linked directly to changes in the energy requirements of the cell. The inhibition of glucose oxidation by progesterone and R5020 as described in this chapter is compatible with the SCARP hypothesis, but further work is needed to confirm this relationship.

Similar Effects of Steroids on Glucose Metabolism in Nonmammary Tissue

Other studies suggest that glucose metabolism may be a marker for hormonal steroid action on target tissues other than the mammary gland.

Smith and Gorski (21) demonstrated that uteri from rats treated for 1 hr with 5 g of 17β-estradiol showed a 135% increase of phosphorylation of 2-deoxyglucose above control levels. Cycloheximide injected 30 min prior to estradiol abolished the effect of the hormone, implying that the development of the hormone's metabolic response was dependent on protein synthesis.

Lan and Katzenellenbogen (12) demonstrated a correlation between the metabolic effects, such as 2-deoxyglucose phosphorylation of various derivatives of estriol, and the long-term maintenance of hormone-receptor complexes in the uterine nucleus.

In *in vivo* studies in the rat uterus it was shown that estradiol and estriol are equipotent as stimulators of early uterotropic responses such as the increment of uterine wet weight and CO_2 production from glucose (2). With respect to inhibition of glucose oxidation by progestins, Hetlendy et al. (9) reported inhibition of glucose oxidation to CO_2 in a rat fat cell preparation obtained from animals treated *in vivo* with medroxyprogesterone acetate.

Metabolic Markers for Hormone Action in the Prediction of Responsiveness of Human Breast Cancer to Endocrine Therapy

Since endocrine-responsive tumors constitute only 20 to 40% of cases, the success rate with endocrine therapy can be improved to 55 to 60% by selecting for endocrine therapy only those patients whose tumors contain estrogen receptors (17).

If, as McGuire et al. (11) suggested, the presence of cytosolic progesterone receptor is dependent in turn on the responsiveness of target tissue to estradiol, one would expect that tissue lacking estradiol receptor would rarely contain progesterone receptors, but that tissue containing estradiol receptors would be divided between those which contain and those which do not contain progesterone receptors. Our results (Fig. 3) are in accordance with this theory and confirm the published data by these authors.

In our search for additional markers for endocrine responsiveness in tumorous tissues, the metabolic marker examined was glucose oxidation. Previous work (19) suggested that *in vitro* histochemical assessment of pentose shunt activity of human breast tumors maintained in the presence and in the absence of hormones can provide a reliable index for response to endocrine therapy, but this study was not confirmed by another group of investigators (16). Our results suggest that a correlation exists between the presence of ER and PgR in cytosol of human mammary tissue and the effect of those hormones on glucose oxidation by isolated tissue (Table 3). These data suggest that glucose oxidation may serve as a metabolic marker for hormonal expression in isolated breast tumor tissue.

ACKNOWLEDGMENTS

This work was supported in part by a grant from the Israel Cancer Association and the Jewish Agency. The skillful technical assistance of Sholom Barami, Oswaldo Pudhadzer, and Miriam Marbach is gratefully acknowledged.

REFERENCES

1. Annison, E. F., and Linzel, J. L. (1964): The oxidation and utilization of glucose and acetate by the mammary gland of the goat in relation to their overall metabolism and to milk formation. *J. Physiol. (Lond.)*, 175:372–385.
2. Clark, J. H., and Peck, E. J. (1976): Estrogen-receptors binding: Relationship to estrogen-induced responses. *J. Toxicol. Environ. Health*, 1:561–586.
3. Cowie, A. T. (1974): Hormonal factors in mammary development and lactation. In: *Mammary Cancer and Endocrine Therapy*, edited by B. A. Stoll, pp. 3–24. Butterworths, London.
4. Davis, S. R., and Ben Meptan, T. (1976): Metabolism of L-(U-^{14}C) valine, L-(U-^{14}C) leucine, L-(U-^{14}C) histidine, and L-(U-^{14}C) phenylalanine by isolated perfused lactating guinea-pig mammary gland. *Biochem. J.*, 156:553–560.
5. Denamur, R., and Delouis, C. (1972): Effects of progesterone and prolactin on the secretory activity and the nucleic acid content of the mammary gland of pregnant rabbits. *Acta Endocrinol. (Kbh.)*, 70:603–618.
6. Eagle, H. (1959): Amino acid metabolism in mammalian cell culture. *Science*, 130:432–437.
7. Feil, P. D., Glasser, S. R., Toft, D. O., and O'Malley, B. W. (1972): Progesterone binding in the mouse and rat uterus. *Endocrinology*, 91:738.

8. Gardner, D. G., and Wittliff, J. L. (1973): Specific estrogen receptors in the lactating mammary gland of the rat. *Biochemistry*, 12:3090.
9. Hetlendy, F., Dahm, C. H., and Mueller, E. J. (1976): Effect of medroxyprogesterone acetate on rat fat cell metabolism. *Horm. Metab. Res.*, 8:82–83.
10. Horwitz, K. B., and McGuire, W. L. (1975): Specific progesterone receptors in human breast cancer. *Steroids*, 25(4):497.
11. Horwitz, K. B., McGuire, W. L., Pearson, O. H., and Segaloff, A. (1975): Prediction response to endocrine therapy in human breast cancer. A hypothesis. *Science*, 189:726–727.
12. Lan, N. C., and Katzenellenbogen, B. S. (1976): Temporal relationships between hormone receptor binding and biological responses in the uterus. Studies with short and long acting derivatives of estriol. *Endocrinology*, 98:1.
13. Liu, A. Y. C., and Greengard, P. (1976): Regulation by steroids hormones of phosphorylation of specific protein common to several target organs. *Proc. Natl. Acad. Sci. U.S.A.*, 73(2):568–572.
14. Lowry, O. H., Rosebrough, N. J., Farr, A. L., and Randall, R. J. (1951): Protein measurement with the Folin phenol reagent. *J. Biol. Chem.*, 193:265.
15. Malaisse, W., Malaisse-Lagae, F., and Wright, P. H. (1967): A new method for the measurement in vitro of pancreatic insulin secretion. *Endocrinology*, 80:99.
16. Masters, J. R. W., Sangster, K., Smith, I. I., and Forrest, A. P. M. (1976): Human breast carcinomata in organ culture. The effect of hormones. *Br. J. Cancer*, 33:564–566.
17. McGuire, W. L., Pearson, O. H., and Segaloff, A. (1975): Predicting hormone responsiveness in human breast cancer. In: *Estrogen Receptors in Human Breast Cancer*, edited by W. L. McGuire, P. P. Carbone, and E. P. Volmer, p. 17. Raven Press, New York.
18. Philibert, D., and Raynaud, J. P. (1973): Progesterone binding in the immature mouse and rat uterus. *Steroids*, 22:89–98.
19. Salia, H., Flax, H., and Hobbs, J. R. (1972): In vitro oestrogen sensitivity of breast cancer tissue as a possible screening method for hormone treatment. *Lancet*, 1:1198.
20. Scatchard, G. (1949): The attraction of proteins for small molecules and ions. *Ann. N.Y. Acad. Sci.*, 51:660–672.
21. Smith, D. E., and Gorski, J. (1968): Estrogen control of uterine glucose metabolism. *J. Biol. Chem.*, 243(6):4169–4174.
22. Verma, U., and Laumas, K. R. (1973): In vitro binding of progesterone to receptors in the human endometrium and the myometrium. *Biochem. Biophys. Acta*, 317:403.
23. Wiest, W. G., Kidwell, W. R., and Balogh, K. (1968): Progesterone catabolism in the rat ovary. A regulatory mechanism for progestational potency during pregnancy. *Endocrinology*, 82:844–859.

*Progesterone Receptors in Normal and
Neoplastic Tissues,* edited by W. L. McGuire
et al. Raven Press, New York © 1977.

Cytoplasmic Progestin Receptors in Mouse Uterus

D. Philibert and J. P. Raynaud

Centre de Recherches Roussel-Uclaf, 93230 Romainville, France

A progestin-specific receptor was first detected in the uterine cytoplasm of the adult mouse by Feil et al. (1) by stabilization of the rapidly dissociating progesterone-receptor complex with glycerol. A similar receptor was later detected in the uterine cytoplasm of the immature animal by the use of the highly potent synthetic progestin R5020 (17,21-dimethyl-19-nor-4,9-pregnadiene-3,20-dione) (2), which, unlike progesterone, is not specifically bound in plasma (3,4) and which, depending on species, has an affinity for the progestin receptor 3 to 10 times that of progesterone. Competition studies for R5020 binding using a large series of progestins have revealed a similarity in the specificity of the cytoplasmic uterine receptor in the mouse and rabbit, and a close correlation has been established between relative binding affinity and biological activity (5).

In the present chapter, the use of R5020 has been extended from progestin receptor detection and specificity establishment to the assay of the total number of binding sites in uterine cytoplasm by exchange. The exchange assay enables the determination of free plus occupied sites whatever the endogenous hormone concentration and has already been described by Katzenellenbogen et al. (6) for the measurement of estrogen binding sites in rat uterus cytoplasm. It has thus become possible to confirm, in mice of different endocrine statuses, several observations already made in other species and, more important still, to quantify these observations. We have measured the decrease in progestin receptor after castration, the induction of progestin receptor following estrogen priming, and the variations in receptor levels during the cycle and pregnancy.

MATERIALS AND METHODS

Labeled Steroids

[1,2-^{3}H]cortisol (56 Ci/mmole) was supplied by the C.E.A., Saclay, France. [1-^{3}H]progesterone (27 Ci/mmole) and [6,7-^{3}H]17,21-dimethyl-19-nor-pregna-4,9-diene-3,20-dione (R5020) (56 Ci/mmole) were synthesized by Roussel-Uclaf and tested for purity (>98% pure) by thin-layer chromatography.

Radioinert Steroids

Radioinert steroids included cortisol, progesterone, R5020, norgestrel (13-ethyl-17-hydroxy-18,19-dinor-17α-pregn-4-en-20-yn-3-one), estradiol, R2858 (11β-methoxy-17-ethynylestra-1,3,5(10)-triene-3,17β-diol), RU-16117 (11α-methoxy-17-ethynyl-estra-1,3,5(10)-triene-3,17β-diol), testosterone, dihydrotestosterone, and aldosterone.

Buffer

Unless otherwise specified, all tissues were homogenized in 10 mM Tris-HCl (pH 7.4), 0.25 M sucrose buffer, henceforth denoted "standard buffer."

Animals

Female Swiss mice from Iffa-Credo (France) were used. The 5- to 6-week-old mice were castrated and, unless otherwise indicated, received 3 weeks later a subcutaneous injection of 3 µg of estradiol in 0.2 ml of sesame oil. They were killed 40 hr after the injection. Vaginal smears were obtained from 3- to 4-month-old mice prior to sacrifice and were stained in eosin and hematoxylin. The smears were classified as follows: diestrus 1, cornified epithelial cells plus leukocytes; diestrus 2, large numbers of leukocytes; proestrus, large numbers of round epithelial cells; estrus, large numbers of cornified epithelial cells. Pregnant mice were obtained by placing three females in a cage with a fertile male at 5 p.m. If a vaginal plug was present the following morning (9 a.m.), the mice were considered pregnant (day 1).

Blood was collected on Na-heparinate and centrifuged at 800 g for 10 min at 0°C. Plasma was diluted in standard buffer. In some experiments serum was used.

Preparation of the 105,000 g Supernatant (Cytosol)

Following excision, the uteri were weighed, minced, pooled, and homogenized in standard buffer in an ice-cooled Teflon-glass homogenizer. In the case of pregnant mice, the conceptus was removed and the uterus washed prior to treatment. Cytosol (supernatant) was prepared by centrifuging homogenate at 105,000 g for 90 min at 4°C and was diluted to 1/20 (wt/vol). Protein concentration was measured by the method of Lowry et al. (7).

Measurement of Binding by Dextran-Coated Charcoal Adsorption

Details of incubation of cytosol are given in table and figure legends. Total radioactivity was measured on a 100-µl aliquot. Bound radioactivity was measured on a further 100-µl aliquot which was stirred in the presence of 100 µl dextran-coated charcoal (DCC; 0.625 to 1.25%) for 10 min at 0°C in a microtiter plate (Cooke Engineering; Scientific Products), then centrifuged for 10 min at 800 g. The radioactivity of a 100-µl supernatant sample was measured in an Intertechnique Liquid Scintillation Spectrometer (model SL_{30}) by the channel ratio method.

Sucrose Density Gradient Ultracentrifugation

Uteri were homogenized and gradients prepared in 10 mM Tris-HCl (pH 7.4), 12 mM thioglycerol, 1.5 mM EDTA buffer containing 10% glycerol. Aliquots of plasma or cytosol (300 µl) were incubated as described in figure legends and then layered on a 5 to 20% sucrose gradient which was centrifuged at 48,000 rpm for 17 hr at 4°C in an SW 50.1 rotor. The radioactivity of two-drop fractions collected from the bottom of the tubes was counted.

Plasma Radioimmunoassays

Serum progesterone levels were determined by radioimmunoassay using an antibody produced by immunization of rabbits with progesterone-3-carboxymethyloxime bovine serum albumin (Roussel-Uclaf). On account of the high degree of specificity of this antibody, the chromatography step was omitted.

Aliquots (50 µl) of serum were added to 100 µl of an aqueous solution containing 4,000 cpm [^{3}H]progesterone. The volume was made up to 0.5 ml and the hormone was extracted with 2 ml redistilled ether while shaking for 5 min at room temperature. The aqueous phase was frozen in methanol–dry ice; the organic phase was removed and evaporated under a stream of nitrogen at 40°C. The residue was taken up in 0.5 ml of 0.1 M phosphate buffer (pH 6.9) containing 0.2% gelatin, 0.1% sodium azide, and a suspension of progesterone antibody (dilution 1/25,000), sheep antiserum to rabbit anti-γ-globulin (dilution 1/50), and normal rabbit serum (dilution 1/250). The mixture was stored overnight at 4°C and then centrifuged at 3,000 × g for 20 min. The radioactivity in the precipitate was measured after addition of 2 ml of scintillation cocktail (8 g Omnifluor, 100 ml Soluene 350 in 2 liters of toluene). The sensitivity of the assay was 20 pg and the coefficient of variation less than 10%.

RESULTS

Sucrose Gradient Analysis of the Steroid-Receptor Complex in Uterine Cytosol from Estradiol-Primed Castrated Mice

Following layering on a sucrose gradient of either cytosol incubated *in vitro* with [^{3}H]R5020 or obtained after *in vivo* injection of [^{3}H]R5020, a

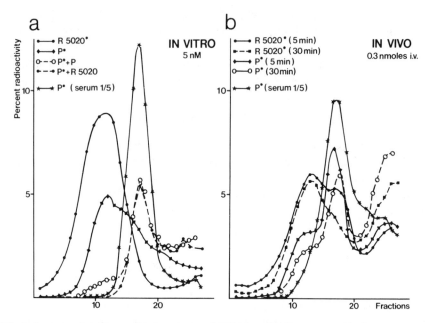

FIG. 1. Sucrose density gradient patterns of R5020 and progesterone binding to uterine cytosol receptor from estradiol-primed castrated mice following *"in vitro"* incubation of hormone (a) or *"in vivo"* injection (b). **(a):** Uterine cytosol was obtained from mice primed by percutaneous application of 10 μg estradiol 5 days prior to sacrifice and was incubated with 5 nM [^{3}H]R5020 or [^{3}H]progesterone for 90 min at 0°C in the presence or absence of 100 nM radioinert hormone. **(b):** Uterine cytosol was obtained from mice similarly primed and which had received an intravenous injection of 0.3 nmoles [^{3}H]R5020 (~15 μCi) or [^{3}H]progesterone (~8 μCi) in 0.1 ml saline 5 or 30 min prior to sacrifice.

marked radioactivity peak was observed in the 7S region. This peak could be detected even with cytosol obtained 30 min after injection (Fig. 1).

A similar radioactivity peak was obtained with cytosol incubated with [^{3}H]progesterone *in vitro,* but only a small shoulder was observed with cytosol obtained 5 min after injection of [^{3}H]progesterone. These results suggest that progesterone dissociates faster from the receptor than R5020 during migration through a gradient.

Determination of the Physicochemical Parameters of Progesterone and R5020 Binding to the Uterine Cytosol Receptor from Estradiol-Primed Castrated Mice

In order to measure these parameters, it was first of all necessary to determine the radioligand concentration which saturates all binding sites and to establish the stability of these sites during incubation.

CHOICE OF RADIOLIGAND CONCENTRATION

According to Fig. 2, a 20 nM concentration of [^{3}H]R5020 and a somewhat higher concentration (>50 nM) of [^{3}H]progesterone were required to saturate approximately 3 pmoles of binding sites per milliliter cytosol, since above these concentrations the number of specific binding sites measured no longer depended on ligand concentration.

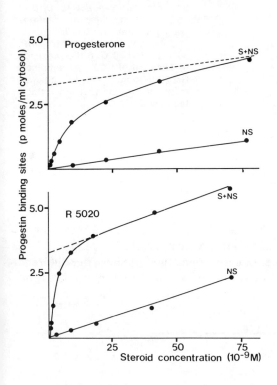

FIG. 2. Determination of a saturating radioligand concentration. Uterine cytosol from estradiol-primed castrated mice was incubated for 4 hr at 0°C with various concentrations of [^{3}H]progesterone or [^{3}H]R5020 in the presence or absence of 5,000 nM radioinert progesterone or R5020, respectively. Bound radioactivity was measured by the DCC adsorption technique. S, specific binding; NS, nonspecific binding.

In Fig. 3, binding sites in cytosol from estradiol-primed castrated mice have been measured with a single concentration (50 nM) of labeled [^{3}H]-R5020 over a range of cytosol dilutions. A linear relationship was found to exist between number of binding sites and dilution, indicating that dilution has no effect on binding site measurement.

STABILITY OF FILLED AND EMPTY BINDING SITES

The stability of empty and progestin-filled binding sites was determined at 0° and 25°C (Fig. 4). Since castrated animals were used, it was assumed that in the absence of exogenous hormone all binding sites present were unoccupied. The saturating R5020 concentration of 25 nM was used to fill the sites.

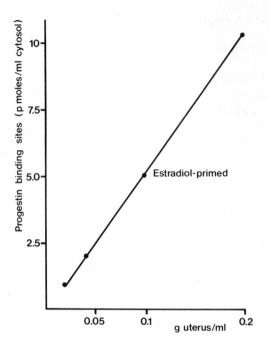

FIG. 3. Linearity of binding measurement with respect to cytosol concentration. Cytosol dilutions were prepared in standard buffer and incubated for 4 hr at 0°C with 50 nM [^{3}H]R5020 in the presence or absence of 5,000 nM radioinert R5020. Bound radioactivity was measured by the DCC adsorption technique.

In the presence of R5020 at 0°C, the binding sites remained stable for at least 24 hr. At 25°C, only 15% of the sites were recovered at 24 hr.

In order to measure the number of undegraded binding sites after different incubation times in the absence of hormone, we incubated the samples for a further 3 hr at 0°C with labeled ligand in the presence or absence of excess

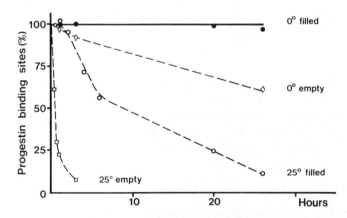

FIG. 4. Receptor stability. Uterine cytosol from estradiol-primed castrated mice was divided into two portions. One portion was incubated for different lengths of time with 25 nM [^{3}H]R5020 at 0° or 25°C to measure the stability of filled sites. The remainder was left for the same time intervals without addition of labeled steroid, then incubated for 3 hr at 0°C with 25 nM [^{3}H]R5020, in the presence or absence of 5,000 nM radioinert R5020, to measure undegraded binding sites. Bound radioactivity was measured by the DCC adsorption technique.

radioinert ligand. In the absence of hormone at 0°C, approximately 70% of the sites were recovered at 24 hr; at 25°C, only 10% of the sites were recovered at 4 hr.

ASSOCIATION AND DISSOCIATION RATES

Figure 5 illustrates the association and dissociation rates between steroid (progesterone and R5020) and receptor in uterine cytosol from estradiol-

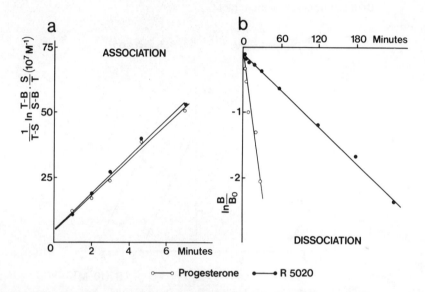

FIG. 5. Rates of association and dissociation of the hormone receptor complex. **(a):** Association rate. Cytosol (100 µl) from estradiol-primed castrated mice was added to microtiter plates containing [^{3}H]R5020 or [^{3}H]progesterone (final concentration = 10 nM). After incubation for different lengths of time, bound radioactivity was measured by the DCC adsorption technique. Nonspecific binding was determined in parallel by addition of 2,500 nM radioinert steroid. The association rate (k_{+1}) is given by the expression $k_{+1}(t) = \frac{1}{T-S} \ln \frac{T-B}{S-B} \times \frac{S}{T}$, where T = initial concentration of radioligand, S = maximum concentration of bound radioligand, and B = concentration of bound radioligand at time t. **(b):** Dissociation rate. Cytosol was incubated with 10 nM radioligand for 2 hr at 0°C. Radioinert hormone (5,000 nM) was added and bound radioactivity was measured at different times by the DCC adsorption technique.

primed castrated mice. Both compounds have similar association rates but progesterone dissociated eight times faster from the receptor than R5020 (Table 1).

PHYSICOCHEMICAL BINDING PARAMETERS

According to a Scatchard analysis of binding as measured by a DCC adsorption technique (Fig. 6), R5020 binds to the uterine cytosol progestin

TABLE 1. *Association rate (k_{+1}), dissociation rate (k_{-1}) and intrinsic association constant (K_a) of binding to the uterine cytosol progestin receptor from estradiol-primed castrated mice*

	k_{+1} 10^7 M^{-1} min^{-1}	k_{-1} 10^{-3} min^{-1}	K_a^a 10^9 M^{-1}	K_a^b 10^9 M^{-1}
Progesterone	6.7	70	0.9	0.15 ± 0.02
R5020	6.4	9	7	1.0 ± 0.3

^{a} Calculated from the ratio k_{+1}/k_{-1}.
^{b} Deduced from a Scatchard plot.

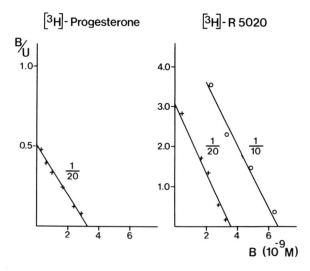

FIG. 6. Scatchard plot of R5020 and progesterone binding to progestin receptor in uterine cytosol. Uterine cytosol from estradiol-primed castrated mice was incubated for 4 hr at 0°C with various concentrations of [^{3}H]progesterone or [^{3}H]R5020. Bound radioactivity was measured by the DCC adsorption technique.

receptor with an intrinsic association constant of 1.0 ± 0.3 nM, which is independent of cytosol dilution, and which is six times higher than the constant for progesterone (0.15 ± 0.02 nM). The same affinity ratio is recorded when calculating the association constants by dividing the association rate by the dissociation rate (third column of Table 1). As often noted, however, the constants deduced from kinetic rates are higher (seven times) than those obtained by direct measurement.

EFFECT OF DISSOCIATION RATE ON RELATIVE BINDING AFFINITY

Competition experiments were performed using either [^{3}H]R5020 or [^{3}H]progesterone as the radioactive ligand (Fig. 7). Measurements with

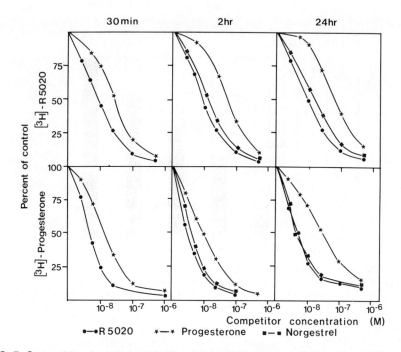

FIG. 7. Competition for labeled R5020 and progesterone binding in uterine cytosol after three different incubation times. Standard displacement curves were obtained by incubating replicate aliquots of 250 µl of cytosol with [^{3}H]R5020 or [^{3}H]progesterone and increasing concentrations (1 to 250 nM) of unlabeled competitor for 30 min, 2 hr, and 24 hr at 0°C. Bound radioactivity was measured by the DCC adsorption technique. The ratio of the concentration of progesterone over the concentration of competitor required to displace radioligand binding by 50% (relative binding affinity) was determined.

[^{3}H]R5020 were more sensitive since in the absence of radioinert substance 30% of the steroid was bound, whereas in the case of [^{3}H]progesterone, only 15% was bound. Whatever the radioligand, the competitive effect exerted by the potent progestin norgestrel was the same.

Relative binding affinities (RBAs) between progesterone and R5020 were measured after three different incubation times: 30 min, 2 hr, and 24 hr. The RBA increased between 30 min and 2 hr, an observation which can be explained by the difference in dissociation rates of the two compounds, but remained constant between 2 and 24 hr.

Progestin Binding Site Concentrations in Mice of Different Endocrine Statuses

Free binding sites in uterine cytosol from castrated mice were measured and then saturated with a large excess of exogenous progesterone. Excess free progesterone was removed by a concentration of dextran-coated char-

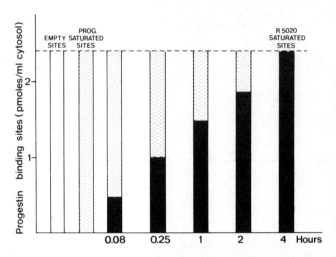

FIG. 8. Determination of percent exchange as a function of time. A 250-μl aliquot of uterine cytosol was incubated with 25 nM [³H]R5020 for different times at 0°C in the presence or absence of 2,500 nM radioinert R5020. The remaining cytosol was preincubated with a saturating concentration of radioinert progesterone (50 nM), treated with DCC (10% vol/vol) to remove free hormone, divided into 100-μl aliquots, and then incubated as above with R5020, but for different time intervals. In each case bound radioactivity was measured by the DCC adsorption technique. Open bar, total number of progestin binding sites in the uterus of the castrated mouse; shadowed bars, number of binding sites still occupied by exogenous progesterone after various incubation times; black bars, number of binding sites previously occupied by exogenous progesterone, now replaced by [³H]R5020.

coal which had no effect on filled sites, and the time required to exchange the bound radioinert progesterone by 25 nM [³H]R5020 was determined (Fig. 8). After 4 hr at 0°C, all binding sites initially saturated with progesterone were occupied by [³H]R5020.

Conditions of incubation with a 25 nM saturating [³H]R5020 concentration for 20 hr at 0°C were used to measure progestin binding site concentration in the uterine cytosol of mice of different endocrine statuses.

ESTRADIOL-PRIMED CASTRATED MICE

Two parameters were investigated: the time-course of progestin-binding site induction and the specificity of induction.

A maximum progestin binding site concentration of 1.3 pmoles/mg protein was recorded 40 hr after a single injection of 3 μg of estradiol (Fig. 9). This represented an approximately threefold increase with respect to the control value of 0.4 pmoles/mg protein. This maximum induction at 40 hr could be obtained with a dose of only 1 μg of estradiol and remained unchanged when increasing the dose to 10 μg of estradiol (Fig. 10). On sucrose density gradient analysis (Fig. 11), a marked 7S peak was detected in the cytosol of estradiol-primed animals incubated *in vitro* with [³H]R5020,

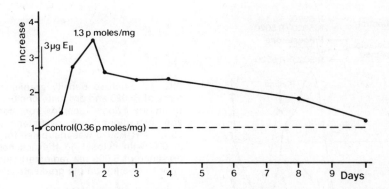

FIG. 9. Relative increase in the number of progestin binding sites in uterine cytosol from castrated mice following priming with 3 µg of estradiol.

but only a small shoulder was seen on incubation with [^{3}H]progesterone. In control unprimed animals, after incubation of [^{3}H]R5020 no definite 7S peak was observed. Radioactivity was spread over several fractions; this radioactivity could be displaced by addition of excess radioinert compound.

In one series of experiments, the mice were primed with compounds other than estradiol and the progestin binding site concentration at 40 hr was measured (Table 2). A dose of 3 µg of R2858 induced the maximum response observed with 3 µg of estradiol; on the other hand, 100 µg of the antiestrogen RU16117 gave rise to an effect which was high but submaximal. The remaining compounds (testosterone, dihydrotestosterone, and aldos-

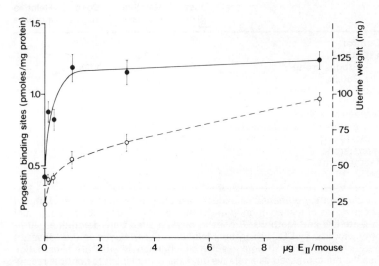

FIG. 10. Progestin receptor induction as a function of the estradiol priming dose in uterine cytosol from castrated mice.

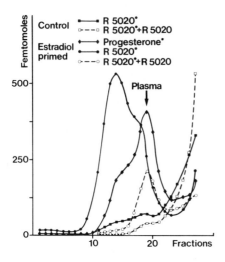

FIG. 11. Sucrose density gradient patterns of R5020 and progesterone binding to uterine cytosol receptor from control and estradiol-primed castrated mice. 25 nM [^{3}H]R5020 was incubated for 2 hr at 0°C with cytosol, in the absence or presence of 5,000 nM radioinert steroid, prior to layering on the gradient.

terone) were virtually devoid of any activity as progestin receptor inducers. When progesterone or R5020 was injected simultaneously with the 3 μg estradiol, the priming effect of estradiol was counteracted in a dose-dependent fashion.

TABLE 2. *Effect of various steroids on the progestin receptor level in castrated mouse uterine cytosol*

	Induction		Inhibition	
	Dose (μg)	Relative increase	Dose (μg)	Relative increase
Control	—	1		
Estradiol	3	3.5		
+ Progesterone			1,000	1.4
+ R5020			10	2.7
			100	1.1
			1,000	0.7
Moxestrol (R2858)	3	3.6		
11α-methoxy ethynylestradiol (RU16117)	3	1.3		
	100	2.9		
Aldosterone	100	1.0		
Testosterone	100	1.1		
DHT	100	1.5		

Induction experiments: immature castrated mice were primed with the above doses of test compound under the conditions described for estradiol in Materials and Methods. Progestin binding sites were determined 40 hr after sacrifice by DCC adsorption. *Inhibition experiments:* various doses of progesterone or R5020 were injected simultaneously with 3 μg of estradiol. Results are expressed as a relative increase with respect to controls receiving solvent only (0.2 ml sesame oil).

ADULT MICE OF DIFFERENT ENDOCRINE STATUSES

Progestin binding site concentration was determined at different times during the cycle. It was found to be highest at proestrus and, at this time, was close to the maximum response induced 40 hr after estradiol injection to castrated mice. Values at diestrus and estrus did not differ significantly among themselves (Fig. 12).

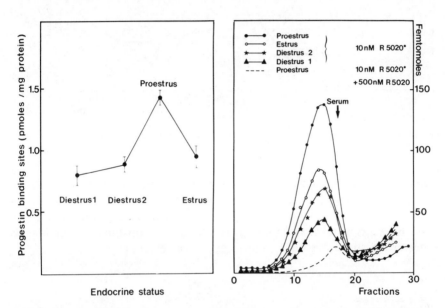

FIG. 12. Progestin binding site concentration and sucrose density gradient pattern of R5020 binding to the progestin receptor in uterine cytosol from adult mice of different endocrine statuses. 10 nM [^{3}H]R5020 was incubated for 1 hr at 25°C with cytosol prior to layering on the gradient.

This observation was confirmed by sucrose density gradient analysis (Fig. 12). A marked peak was detected at proestrus. It was somewhat closer to the 4S serum peak than was the 7S peak observed in the cytosol of estradiol-primed castrated mice. Similar but less-marked peaks were found at estrus and diestrus 1 and 2.

PREGNANT MICE

The variations in progestin binding site concentration and in serum progesterone levels throughout pregnancy are illustrated in Fig. 13. At the beginning of pregnancy, the progestin receptor level was about the same as at estrus. It decreased to a minimum at midpregnancy, and, although it

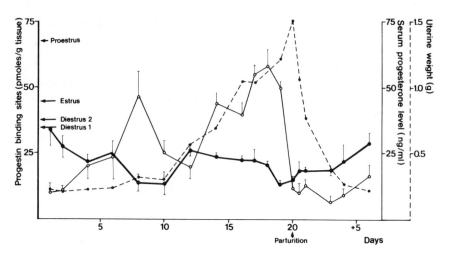

FIG. 13. Variations in tissue progestin receptor (●——●) and serum progesterone levels (○——○) during pregnancy.

increased afterward, it never reached values as high as those recorded at proestrus. In general, receptor levels were three times lower than at proestrus. At delivery a simultaneous decrease in serum progesterone and increase in receptor level were found. No direct relationship between plasma progesterone and progestin receptor levels was evident.

DISCUSSION

The detection and assay of the progesterone receptor have met with several difficulties in many species, in particular in the rat (8) and human myometrium (9), on account of the specific binding of progesterone to plasma corticosteroid binding globulin (CBG) and the rapid dissociation of progesterone from the tissue receptor. The synthetic radioligand R5020 has two advantages over progesterone: first, it is not specifically bound by CBG; second, it dissociates about six to eight times slower than progesterone from the uterine cytosol receptor in the mouse (*this chapter*), rat, rabbit (*unpublished observations*), and human myometrium (10). In the present studies, these properties of R5020 have enabled the visualization of uterine cytosol progestin binding in a sucrose gradient 30 min after *in vivo* injection of steroid. This R5020-receptor complex sediments in exactly the same region (~7S) as the complex obtained after *in vitro* incubation of [^{3}H]R5020 or [^{3}H]progesterone with cytosol. The complex formed between the receptor and progesterone, although relatively easily detectable *in vitro* in the presence of a stabilization agent such as glycerol in the gradient buffer, was just latent when cytosol obtained immediately (5 min) after injection of [^{3}H]-progesterone was layered on the gradient.

Compared to that in many species [hamster (11), man (10,12), and rabbit (*unpublished observation*)], the dissociation rate of the progesterone-receptor complex is very fast ($K_d = 70 \times 10^{-3}$ min^{-1}) in mouse uterine cytosol (1). The intrinsic association constant, as measured by a DCC adsorption technique and calculated from a Scatchard plot ($0.15 \pm 0.02 \ 10^9$ M^{-1}), is also lower in the mouse than in other species [man (12–14), sheep (15), guinea pig (16), rabbit (17), and hamster (11)]. Since, however, unfilled progestin binding sites are unstable even at 4°C, the validity of the measurement of a parameter such as the intrinsic association constant of binding (K_a) is doubtful. For most methods used to measure this parameter, there is no guarantee that the receptor level remains constant during measurement. It is highly probable that at low bound ligand concentrations, a fair proportion of sites are degraded, this proportion varying according to method. Physicochemical binding parameters such as K_a are of interest only insofar as they enable the selection of suitable ligands for further study on the basis of affinity and specificity; but in an investigation of the dynamics of different physiological states, a parameter such as number of binding sites is far more meaningful.

Since R5020 dissociates slowly from the receptor, it was considered a particularly suitable compound for the measurement of total binding sites by an exchange assay. Appropriate conditions for a valid assay were chosen by checking the stability of the steroid-receptor complex during incubation. The fact that empty sites were labile at 0°C whereas filled sites were stable over at least 24 hr stressed the importance of using a saturating radioligand concentration for the assay. By artificially saturating all available binding sites with exogenous progesterone in cytosol from castrated mice, it was shown that it is possible to exchange all bound hormone at 0°C within 4 hr or more by addition of excess [^{3}H]R5020. During exchange, the effect of the isotopic dilution resulting from the dissociation of endogenous progesterone from the receptor was minimal.

The exchange assay was applied to the measurement of progestin binding sites in uterine cytosol from mice of different endocrine statuses to confirm the validity of several well-known observations. Indeed, castration decreased binding site concentration, whereas priming by estradiol or estrogen analogues induced binding sites as already observed in the hamster (11,18) and guinea pig (19,20). This priming effect was specific to estrogens and could be counteracted by progestins. In adult animals, maximum binding site concentrations were recorded at proestrus with no significant variations at other times during the cycle. The value at proestrus, which was also the maximum value recorded after induction with a single dose of estrogen, was not exceeded even during pregnancy. In pregnant mice, tissue progestin receptor levels were low, in fact, much lower than in the pregnant rat (21,22) and somewhat lower than in the pregnant rabbit (23), whereas serum progesterone levels were high except at midpregnancy. No direct correlation was, however, apparent between tissue and serum levels. The low concentrations in the cytosol could be explained by the translocation of most of the tissue receptor into the nucleus by the high circulating progesterone levels.

ACKNOWLEDGMENT

The competent technical assistance of S. Viet is gratefully acknowledged.

REFERENCES

1. Feil, P. D., Glasser, S. R., Toft, D. O., and O'Malley, B. W. (1972): Progesterone binding in the mouse and rat uterus. *Endocrinology,* 91:738–746.
2. Philibert, D., and Raynaud, J. P. (1973): Progesterone binding in the immature mouse and rat uterus. *Steroids,* 22:89–98.
3. Philibert, D., and Raynaud, J. P. (1974): Progesterone binding in the immature rabbit and guinea pig uterus. *Endocrinology,* 94:627–632.
4. Philibert, D., and Raynaud, J. P. (1974): Binding of progesterone and R5020, a highly potent progestin, to human endometrium and myometrium. *Contraception,* 10:457–466.
5. Raynaud, J. P., Philibert, D., and Azadian-Boulanger, G. (1975): Progesterone-progestin receptors. In: *The Physiology and Genetics of Reproduction, Part A,* edited by E. M. Coutinho and F. Fuchs, pp. 143–160. Plenum Press, New York.
6. Katzenellenbogen, J. A., Johnson, H. J., Jr., and Carlson, K. E. (1973): Studies on the uterine, cytoplasmic estrogen binding protein. Thermal stability and ligand dissociation rate. An assay of empty and filled sites by exchange. *Biochemistry,* 12:4092–4099.
7. Lowry, O. H., Rosebrough, N. J., Farr, A. L., and Randall, R. J. (1951): Protein measurement with the Folin phenol reagent. *J. Biol. Chem.,* 193:265–275.
8. Milgrom, E., and Baulieu, E. E. (1970): Progesterone in uterus and plasma. I. Binding in rat uterus 105,000 g supernatant. *Endocrinology,* 87:276–287.
9. Batra, S. (1973): Binding of progesterone in vitro by the cytoplasmic components of the human myometrium. *J. Endocrinol.,* 57:561–562.
10. Jänne, O., Kontula, K., and Vihko, R. (1976): Kinetic aspects in the binding of various progestins to the human uterine progesterone receptor. *Abstracts,* Vth International Congress of Endocrinology, Hamburg, July 1976, p. 151, No. 367.
11. Leavitt, W. W., Toft, D. O., Strott, C. A., and O'Malley, B. W. (1974): A specific progesterone receptor in the hamster uterus: Physiologic properties and regulation during the estrous cycle. *Endocrinology,* 94:1041–1053.
12. Young, P. C. M., and Cleary, R. E. (1974): Characterization and properties of progesterone-binding components in human endometrium. *J. Clin. Endocrinol. Metab.,* 39:425–439.
13. Kontula, K., Jänne, O., Luukkainen, T., and Vihko, R. (1973): Progesterone-binding protein in human myometrium. Ligand specificity and some physicochemical characteristics. *Biochim. Biophys. Acta,* 328:145–153.
14. Rao, B. R., Wiest, W. G., and Allen, W. M. (1974): Progesterone "receptor" in human endometrium. *Endocrinology,* 95:1275–1281.
15. Kontula, K. (1975): Progesterone-binding proteins from endometrium and myometrium of sheep uterus: A comparative study. *Acta Endocrinol. (Kbh.),* 78:593–603.
16. Atger, M., Baulieu, E. E., and Milgrom, E. (1974): An investigation of progesterone receptors in guinea pig vagina, uterine cervix, mammary glands, pituitary and hypothalamus. *Endocrinology,* 94:161–167.
17. Rao, B. R., Wiest, W. G., and Allen, W. M. (1973): Progesterone "receptor" in rabbit uterus. I. Characterization and estradiol-17β augmentation. *Endocrinology,* 92:1229–1240.
18. Reel, J. R., and Shih, Y. (1975): Oestrogen-inducible uterine progesterone receptors. Characteristics in the ovariectomized immature and adult hamster. *Acta Endocrinol. (Kbh.),* 80:344–354.

19. Milgrom, E., Atger, M., Perrot, M., and Baulieu, E. E. (1972): Progesterone in uterus and plasma: VI. Uterine progesterone *receptors* during the estrus cycle and implantation in the guinea pig. *Endocrinology,* 90:1071-1078.
20. Freifeld, M. L., Feil, P. D., and Bardin, C. W. (1974): The in vivo regulation of the progesterone "receptor" in guinea pig uterus: dependence on estrogen and progesterone. *Steroids,* 23:93-103.
21. Davies, I. J., and Ryan, K. J. (1973): The modulation of progesterone concentration in the myometrium of the pregnant rat by changes in cytoplasmic "receptor" protein activity. *Endocrinology,* 92:394-401.
22. Haukkamaa, M. (1974): Binding of progesterone by rat myometrium during pregnancy and by human myometrium in late pregnancy. *J. Steroid Biochem.,* 5:73-79.
23. Davies, I. J., Challis, J. R. G., and Ryan, K. J. (1974): Progesterone receptors in the myometrium of pregnant rabbit. *Endocrinology,* 85:165-173.

*Progesterone Receptors in Normal and
Neoplastic Tissues,* edited by W. L. McGuire
et al. Raven Press, New York © 1977.

Changes in Progesterone and Estrogen Receptors in the Rat Uterus During the Estrous Cycle and the Puerperium

*Fulgencio Gómez, **Heinz G. Bohnet, and Henry G. Friesen

Department of Physiology, University of Manitoba, Winnipeg, Manitoba R3E OW3, Canada

Hormonal actions are believed to be mediated by an interaction between hormone and specific tissue receptors; in the case of steroid hormones, these receptors are located in the cytosol, and upon binding to the hormone, the complex is translocated to the nucleus, where an action on the genome is triggered (4,8). The uterus is a target organ for estrogen and progesterone, and the marked changes it undergoes throughout life are modulated to a great extent by these as well as other hormones. In order to better correlate hormonal influences on uterine changes, we have studied peripheral blood hormone levels and changes in uterine receptors for estradiol (E_2) and progesterone.

The potent progestational action of the synthetic steroid R5020(17,21-dimethyl-19-nor-4,9-pregnandiene-3,20-dione) is likely to be exerted through binding to the cytosol progesterone receptors, to which it binds with a higher affinity than progesterone itself (18). In addition, R5020 does not bind to glucocorticoid binding proteins as progesterone does. Therefore, R5020 is particularly useful in studying progesterone receptors without interference by other binding proteins of unknown physiological significance, which are present in crude subcellular preparations. Taking advantage of these characteristics, investigators have studied progesterone receptors in a number of tissues using R5020 (1,10,17–19).

In the present study we have examined changes in uterine weight, cytosol protein concentrations, and estrogen and progesterone receptor content during the estrous cycle and the puerperium of the rat. Correlations with serum E_2 and progesterone levels during the puerperium also have been made.

---

* Fellow of the Medical Research Council of Canada; and
** Fellow of the German Research Foundation (DFG).

MATERIALS AND METHODS

Animals

Young adult female to mid-to-late pregnant Sprague-Dawley rats were obtained from Canadian Breeding Farms and Laboratory, Montreal. Throughout the experiments, all rats were kept in individual plastic cages under standard conditions of temperature and light-dark cycles. The phase of the estrous cycle was assessed by daily vaginal smears (23). Only animals having two consecutive 4-day cycles were entered in this study and were killed in the morning of different days of the third cycle. Pregnant rats were examined every 8 hr around the time of parturition, and at birth litters were adjusted to eight pups per mother or removed completely. Thereafter, rats were killed at specified intervals in the puerperium, day 0 denoting the time when litters were first observed. Some rats were killed on day 16 of pregnancy and referred to as day −5. All the animals were killed by cervical dislocation under light ether anesthesia; blood was collected from the trunk for hormone assays; uteri were dissected from surrounding fat and connective tissue (when applicable, fetuses and placentas were removed), washed with ice cold saline, blotted, and snap frozen on dry ice, the whole operation lasting 15 to 20 min/rat. The frozen tissues were stored at −70°C until assayed for steroid binding 2 to 7 days later.

Hormone Measurements

Serum prolactin (PRL) concentrations were determined by radioimmunoassay (RIA) with double antibody using the kit supplied by the NIAMDD (NIH), and expressed as nanograms per milliliter in terms of the rat PRL reference preparation RP-1. Progesterone and E_2 (estradiol-17β) were measured by RIA in extracted serum as previously reported (20).[1]

Tissue Preparation

All the operations were done at 0 to 4°C unless otherwise specified. Uteri were homogenized on ice in buffer [10:1 (vol/wt) 10 mM Tris-HCl, 1.5 mM EDTA, 0.5 mM dithiothreitol, pH 7.6, for E_2 binding studies and same buffer containing 10% glycerol for R5020 binding studies] with a Polytron homogenizer (three bursts of 5 sec each at medium speed set, with 20-sec cooling intervals). The homogenate was centrifuged 90 min at 100,000 × g and the protein concentration of the supernatant (cytosol) was measured by the method of Lowry et al. (14). Prior to binding assays, cytosol preparations

---

[1] Progesterone and E_2 RIA were performed by the Endocrine-Metabolic Section of the Health Sciences Centre, Winnipeg.

were adjusted to 0.25 mg protein/0.1 ml with the corresponding homogenization buffer.

E_2 Binding Assay

Aliquots of cytosol (0.1 ml) were added to glass tubes containing different amounts of ^{3}H-E_2 (6,7-^{3}H-estradiol-17β, 48 Ci/mmole, purchased from New England Nuclear). For the simple binding assay, 5 pg of the tracer was used, and in a parallel set of tubes a large excess (1 μg) of unlabeled diethylstilbestrol (DES) was used to obtain maximum displacement; for saturation analysis (Scatchard's plots), 5 to 60 pg of ^{3}H-E_2 was used, and in parallel tubes the corresponding tracer plus 1 μg unlabeled DES was added. Tracer and cold hormone dissolved in 10 μl ethanol were added to 0.05 ml of "assay buffer" (identical to homogenization buffer but containing five times more concentrated dithiothreitol) prior to addition of the cytosol fraction. After 16 hr incubation at 4°C with constant shaking, 1.0 ml of a dextran-coated charcoal (DCC) suspension (0.25% Norit A charcoal, 0.0025% dextran 80, in 10 mM Tris-HCl, pH 8.0, buffer) was added, and shaking was continued for an additional 20 min at 4°C. Centrifugation at 1,000 × g for 10 min was carried out and 0.9 ml of the supernatant was transferred into scintillation vials containing 10 ml Bray's solution. DCC is thought to remove unbound and loosely bound steroid from the supernatant.

R5020 Binding Assay

In the simple binding assay, 10 pg of ^{3}H-R5020[2] was used, with excess (1 μg) unlabeled R5020 in parallel sets of tubes to obtain maximum displacement. For saturation analysis, increasing amounts of ^{3}H-R5020 (10 to 1,000 pg) with 1 μg unlabeled R5020 in parallel sets at each tracer concentration were used. The assays were performed as described for E_2 binding, except that "assay buffer" contained 10% glycerol and the incubation time was shortened to 4 hr at 4°C. As can be seen in Fig. 1, specific (displaceable) binding (SB) was found to be maximal between 2 and 6 hr at 4°C. In contrast to the relatively slow decay at 4°C, at 22° SB decreased rapidly, and at 37°C no significant SB could be observed at any time. Similar time-temperature courses were observed in different tissues such as human breast cancer (Fig. 2) and are also similar to the observation reported by Hsueh et al. (11) on ^{3}H-progesterone binding to rat uterine nuclear preparations. Our method could not distinguish between an accelerated dissociation rate and alteration of the cytosol binders at higher temperature. Under the assay conditions used, R5020 appeared on a molar basis to be approximately five times more

---

[3] H-R5020, 51.4 Ci/mmole, and R5020 were a gift from Dr. J. P. Raynaud, Roussel-Uclaf, France.

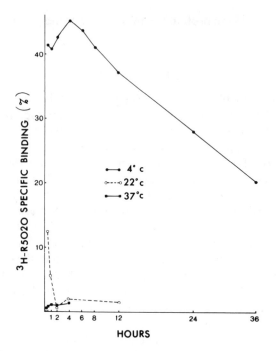

FIG. 1. ³H-R5020 specific binding to cytosol from the rat uterus. Influence of incubation time and temperature.

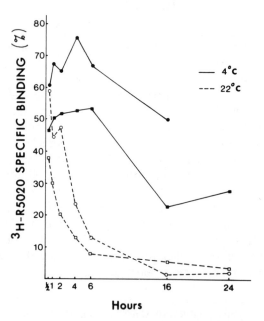

FIG. 2. ³H-R5020 specific binding to human breast cancer cytosol. Influence of incubation time and temperature. ●———● and ○---○, patient 1; ■———■ and □---□, patient 2.

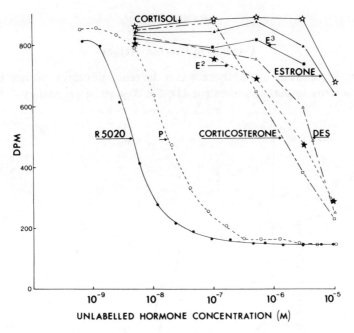

FIG. 3. Displacement of ³H-R5020 from its binding to rat uterine cytosol by increasing concentrations of steroid hormones and synthetic estrogen. Incubation 4 hr at 4°C; bound and free separated with DCC as described in the text.

potent than progesterone (Fig. 3), whereas nonprogestational steroids were several orders of magnitude less potent; similar relative potencies have been described in cytosol from DMBA-induced rat mammary tumors (1).

Sucrose Density Gradients

Cytosol aliquots (0.2 ml) were incubated with 10^{-8} M ³H-R5020 during 1 hr at 4°C; parallel samples were incubated in the presence of 1 µg unlabeled R5020, progesterone, or cortisol. The incubates were applied on 5 to 20% sucrose density gradient (SDG) freshly prepared with homogenization buffer containing 10% glycerol and centrifuged at 192,000 × g (Av.) in cellulose nitrate tubes in a Spinco SW-40 rotor (39,000 rpm) for 31 hr at 4°C. ¹⁴C-γ-globulin and ¹⁴C-bovine serum albumin (BSA) were used as reference standards for the 8S and 4S positions, respectively. Five-drop fractions were collected from the bottom and transferred into scintillation vials containing 10 ml Bray's solution.

B-Counting

Vials were counted 10 min in a B-counter (Nuclear Chicago) and counts were corrected for efficiency (31 to 33%) in each sample using the external standard ratio with the help of a desk computer.

RESULTS

Changes in Uterine Weight

During the estrous cycle, there was a decrease in uterine weight (Fig. 4) from proestrus to diestrus (diestrus II). On day 16 of pregnancy (−5 in the

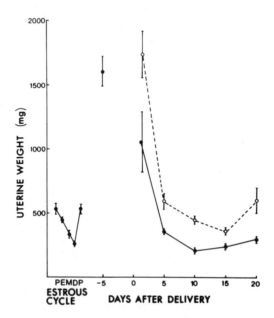

FIG. 4. Weight of rat uterus during the estrous cycle, pregnancy (day 16), and puerperium. P, proestrus; E, estrus; M, metestrus (diestrus I); D, diestrus (diestrus II). The difference in weight between D and P was significant ($p < 0.01$). Puerperium: ●———●, lactating animals (8 pups); ○---○, nonlactating animals (litter removed at birth). Each point represents mean ± SEM of 4 to 7 animals. Differences in weight between lactating and nonlactating animals were statistically significant at all times ($p < 0.05$).

figure), uterine weight was approximately three times as high as the maximum weight during the cycle (proestrus). After parturition it decreased rapidly and reached values similar to or lower than the minimal values seen in the cycle. However, when the litter was removed at birth, uterine involution took place at a slower pace and was less complete. A marked difference in uterine weight between lactating and nonlactating rats already was apparent as early as 36 hr post-partum.

Changes in Cytosol Protein Concentration

The lowest level of cytosol protein concentration (Fig. 5), was observed at diestrus, which coincides with the lowest uterine weight. Values 36 hr

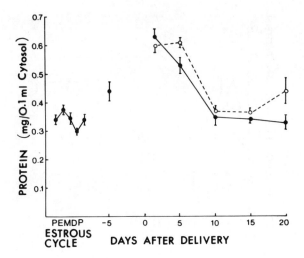

FIG. 5. Protein concentration in 0.1 ml cytosol from the rat uterus. Tissue homogenized with 10:1, vol/wt, buffer. Symbols as in Fig. 4. Statistical significance (lactating vs nonlactating), $p < 0.05$ days 5 and 20.

post-partum were twice as high as during the estrous cycle. By day 10 of the puerperium, cytosol protein concentration decreased to reach the range seen in the estrous cycle. There were no consistent differences between lactating and nonlactating animals.

E_2 Binding

The lowest specific binding of E_2 (Fig. 6) during the cycle was in proestrus. On day 16 of pregnancy, SB was lower than at proestrus. After parturition, at 36 hr SB had already started to increase toward the range seen in estrous in lactating animals, but in nonlactating animals it was extremely low. Thereafter, SB rose similarly in both groups, to reach a plateau by day 5 that was in the range seen in cycling rats. Binding capacity and affinity constant (K_a), as measured by saturation analysis, are shown in Fig. 7.

R5020 Binding

From proestrus to metestrus (diestrus I) there was a striking decrease in SB of R5020 (Fig. 8), followed by a subsequent rise at diestrus (diestrus II). On day 16 of pregnancy and 36 hr after parturition, SB was in the same low range as in metestrus. During normal lactation, SB rose progressively reaching the highest levels observed in this study on day 20. The same high level was attained more rapidly (by day 5) when the litter was removed at birth. These high levels were above the SB observed during the cycle. Binding capacity calculated from Scatchard analysis is shown in Fig. 9.

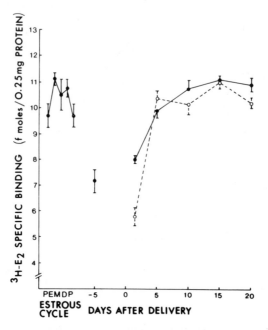

FIG. 6. E_2 specific binding to rat uterine cytosol. Total activity, 18.4 fmoles. Symbols as in Fig. 4. Statistical significance (lactating vs nonlactating), $p < 0.05$ at 36 hr.

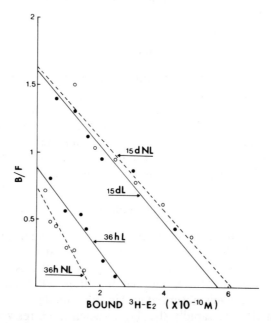

FIG. 7. Scatchard's analysis of ^3H-E_2 binding to rat uterine cytosol at different times of puerperium, in lactating (L) and nonlactating (NL) animals. K_a ($\times 10^9$ M^{-1}): 36 hr (NL) = 4.6; 36 hr (L) = 3.2; 15 days (L) = 2.8; 15 days (NL) = 2.7. Capacity (fmoles/0.25 mg protein): 36 hr (NL) = 27; 36 hr (L) = 46; 15 days (L) = 94; 15 days (NL) = 101.

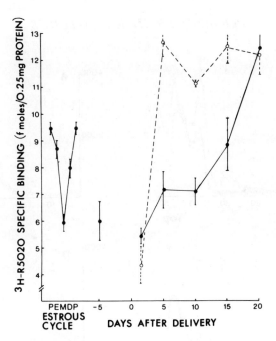

FIG. 8. ³H-R5020 specific binding to cytosol from the rat uterus. Total activity, 28 fmoles. Symbols as in Fig. 4. Statistical significance (lactating vs nonlactating): $p < 0.05$ at 5, 10, and 15 days; estrous cycle: $p < 0.05$ at M vs D (etc.).

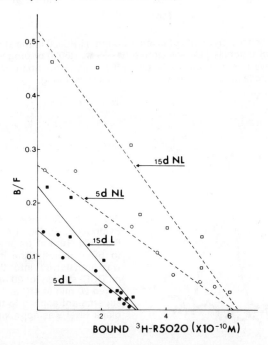

FIG. 9. Scatchard's analysis of ³H-R5020 binding to cytosol from the rat uterus at different times of puerperium in lactating (L) and nonlactating (NL) animals. K_a ($\times 10^9$ M^{-1}); 5 days (L) = 0.5; 15 days (L) = 0.75; 5 days (NL) = 0.45; 15 days (NL) = 0.85. Capacity (fmoles/ 0.25 mg protein): 5 days (L) = 50; 15 days (L) = 52; 5 days (NL) = 100; 15 days (NL) = 103.

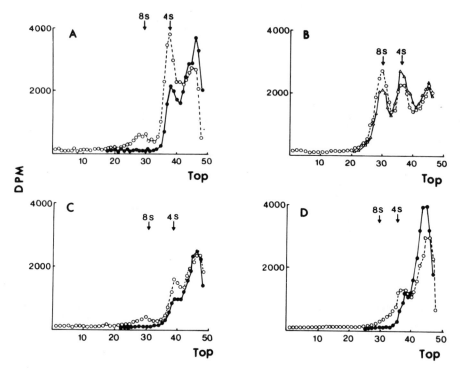

FIG. 10. Sucrose density gradients of cytosol-bound ³H-R5020 in the absence (O---O) or in the presence of unlabeled excess steroid (●——●, R5020 or progesterone; △——△, cortisol). **A:** 3 days post-partum nonlactating. **B:** 13 days post-partum nonlactating. **C:** 5 days post-partum lactating. **D:** 10 days post-partum lactating.

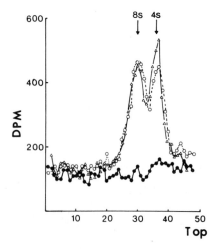

FIG. 11. Sucrose density gradient. Symbols as in Fig. 10. Same cytosol as **B** in Fig. 10, but treated with DCC prior to SDG analysis; the incubate (0.2 ml) was transferred into the pellet of 1.0 ml DCC, shaken for 20 min at 4°C, centrifuged 20 min at 4°C at $1,000 \times g$, and the supernatant applied to the gradient. The blank level was approximately 100 dpm.

TABLE 1. Serum levels of prolactin, estradiol, and progesterone in rats

				Time after parturition (days)				
	−5	12 (hr)	36 (hr)	2	5	10	15	20
Prolactin (ng/ml)								
Lactating	10 ± 2(4)	32 ± 12(4)	84 ± 10(4)	114 ± 9(4)	99 ± 26(4)		85 ± 6(5)	34 ± 13(3)
Nonlactating			28 ± 5(5) (*)	32 ± 8(4) (*)	21 ± 2(4) (*)		21 ± 4(4) (*)	54 ± 12(3)
Estradiol (ng/100 ml)								
Lactating	13 ± 1(4)	7 ± 1(4)	6 ± 2(4)			5 ± 1(4)	7 ± 0.5(4)	6 ± 0.5(3)
Nonlactating			8 ± 1(5)			8 ± 0.5(4) (*)	8 ± 0.5(4)	10 ± 2(3) (*)
Progesterone (ng/ml)								
Lactating	75 ± 23(4)	6.6 ± 1(4)	17 ± 3(4)		55 ± 9(4)	105 ± 21(4)	57 ± 21(4)	28 ± 21(3)
Nonlactating			8 ± 1(5) (*)		17 ± 7(4) (*)	20 ± 3(4) (*)	13 ± 3(4) (*)	7 ± 4(3)

Mean ± SEM; in parentheses, number of animals. Asterisk (*) indicates statistical significance ($p < 0.05$) in lactating vs nonlactating groups.

Sucrose density gradient centrifugation revealed that bound ^{3}H-R5020 migrated in the 8S and the 4S regions. The 8S binding proved to be specific since it was completely displaced by unlabeled R5020 or progesterone, whereas the 4S binding peak was only partly displaced by excess unlabeled progesterone or R5020. Cortisol did not compete for binding with ^{3}H-R5020 (Fig. 10). When the incubated cytosol was treated with DCC prior to SDG analysis, the binding in both peaks was substantially reduced (Fig. 11), but the residual binding was completely displaceable.

Serum Hormone Levels

From the low levels at day 16 of pregnancy, serum PRL concentrations increased rapidly during the first days of lactation and remained elevated at least until day 15 post-partum (Table 1). However, when lactation was prevented by removal of the litter, PRL levels were low throughout the observation period.

Mean serum E_2 levels tended to be lower in lactating than in nonlactating animals, but the differences were not consistently significant. The overall levels were below those observed on day 16 of pregnancy.

Serum progesterone levels showed a marked decrease between day 16 of pregnancy and 12 hr post-partum. When lactation was allowed to proceed normally, serum progesterone reached maximal concentrations by day 10 and thereafter declined progressively. Upon removal of the litter, the puerperal rise of progesterone was abolished.

DISCUSSION

The saturable, high-affinity binding of ^{3}H-E_2 and ^{3}H-R5020 in cytosol obtained from the rat uterus most likely represents specific steroid receptors. The specificity of R5020 binding is apparent from the relative potencies of the different steroids in the displacement curves, and the sedimentation pattern indicates binding to a specific 8S protein as well as to a 4S protein. The latter findings are similar to those of studies using human breast cancer (10) and rat DMBA-induced mammary tumor tissue (1). Excess progesterone displaced ^{3}H-R5020 in an identical manner to excess R5020, whereas cortisol failed to compete for binding sites.

The changes in uterine weight that we observed during the estrous cycle confirm previous observations by other authors (5,20) and are probably related to the sequential secretion of E_2 and progesterone by the rat ovary. The sequence of these changes is as follows: on the evening of diestrus (diestrus II) and during most of proestrus, E_2 secretion reaches a maximum (3,23); the rapid increase of uterine weight that ensues results from the uterotrophic action of E_2; this phase has been shown to be associated with an increased synthesis of proteins in the uterus (2,5,21,22) and an increase in the cytosol protein content (see also Fig. 5). At the same time, progesterone receptors in the cytosol increase, possibly under the regulatory (stimulatory) effect of estrogens (15). The concomitant decrease in cytosol estrogen receptors may merely be due to occupancy, since estrogen receptors seem also to increase under a stimulatory estrogenic action (12,16). Progesterone increases on the evening of proestrus (7,13), whereas E_2 decreases to low levels. The combined factors of decreased uterotrophic action of less circulating E_2 and increased antiuterotrophic action of elevated progesterone result in a fall of uterine weight (nadir on the morning of diestrus). Simultaneously there is a rapid decrease of progesterone receptors, which probably is in part due to occupancy by high circulating levels of progesterone but could also be due to the inhibitory effect on progesterone receptors [increased inactivation rate, according to Milgrom et al. (15)] of progesterone itself, and to lack of stimulation by E_2.

The changes in E_2 receptors, however, cannot be interpreted on the basis of changes in regulatory mechanisms but seem to reflect circulating E_2 levels.

After parturition ovulation occurs, and if lactation follows normally the "corpus luteum of lactation" continues to secrete progesterone (6) under the luteotrophic stimulus of PRL, the levels of which are notably elevated. However, if the litter is removed at birth, the increases in PRL and eventually in progesterone are aborted (9) (Table 1); under these circumstances, uterine regression is relatively incomplete and retarded. Since estrogen levels are practically identical, the likely explanation would be the lack of appropriate antiuterotrophic action of progesterone; similarly, the high levels of progesterone receptors in nonlactating rats, which are above values seen in cycling animals, could be due to lack of negative control by progesterone. A similar rapid rise of progesterone receptors in normally lactating rats was prevented during the first 10 days post-partum, a period during which serum progesterone rose to a peak. As soon as progesterone levels declined, there was an especially steep rise in SB. It is therefore tempting to correlate progesterone levels and SB and to explain a substantial part of the results of binding in terms of occupancy. However, some of the changes observed cannot easily be interpreted in that way. Soon after parturition serum progesterone levels are low and yet SB is at its lowest point; and between 36 hr and 10 days post-partum both serum levels and specific binding are increasing. Therefore, there must be additional factors that participate in this regulation during the puerperium. One of the possible factors is oxytocin, which is released in response to suckling and also has an effect on the contractility of the uterus.

In conclusion, correlation of circulating hormonal changes and tissue hormone receptors may serve to elucidate the mechanisms responsible for the structural and functional changes observed in the uterus in the puerperium.

ACKNOWLEDGMENTS

This research was supported by the Medical Research Council of Canada and the USPHS Child Health and Human Development Institute, HD 07843-03. The authors wish to thank Mrs. Jocelyne Gomez for skillful technical assistance and Mrs. Linda Holeman for typing the manuscript.

REFERENCES

1. Asselin, J., Labrie, F., Kelly, P. A., Philibert, D., and Raynaud, J. P. (1976): Specific progesterone receptors in dimethylbenzanthracene (DMBA)-induced mammary tumors. *Steroids*, 27:395–404.
2. Blahna, D. G., and Yochim, J. M. (1975): Protein and RNA in endometrium of the rat uterus during early progestation. *Biol. Reprod.*, 13:527–534.
3. Brown-Grant, K., Exley, D., and Naftolin, F. (1970): Peripheral plasma estradiol and luteinizing hormone concentrations during the estrous cycle of the rat. *J. Endocrinol.*, 48:295–296.
4. Buller, R. E., and O'Malley, B. W. (1976): The biology and mechanism of steroid hormone receptor interaction with the eukaryotic nucleus. *Biochem. Pharmacol.*, 25:1–12.
5. Clark, J. H., Anderson, J., and Peck, E. J. (1972): Receptor-estrogen complex in the nuclear fraction of rat uterine cells during the estrous cycle. *Science*, 176:528–530.
6. Ford, J. J., Takahashi, M., Yoshinaga, K., and Greep, R. O. (1975): Progestin levels after inhibition of post-partum ovulation in rats. *Biol. Reprod.*, 12:584–589.
7. Freeman, M. C., Dupe, K. C., and Croteau, C. M. (1976): Extinction of the estrogen-induced daily signal for LH release in the rat: a role for the proestrous surge of progesterone. *Endocrinology*, 99:223–229.
8. Gorski, J., and Gannon, F. (1976): Current models of steroid hormone action; a critique. *Annu. Rev. Physiol.*, 38:425–450.
9. Grota, L. J., and Eik-Nes, K. B. (1967): Plasma progesterone concentrations during pregnancy and lactation in the rat. *J. Reprod. Fertil.*, 13:83–91.
10. Horwitz, K. B., and McGuire, W. L. (1975): Specific progesterone receptors in human breast cancer. *Steroids*, 25:497–505.
11. Hsueh, A. J. W., Peck, E. J., and Clark, J. H. (1974): Receptor progesterone complex in the nuclear fraction of the rat uterus: demonstrations by 3H-progesterone exchange. *Steroids*, 24:599–611.
12. Hsueh, S. J. W., Peck, E. J., and Clark, J. H. (1976): Control of uterine estrogen receptor levels by progesterone. *Endocrinology*, 98:438–444.
13. Kennedy, T. G., and Armstrong, D. T. (1975): Loss of uterine luminal fluid in the rat: Relative importance of changing peripheral levels of estrogen and progesterone. *Endocrinology*, 97:1379–1385.
14. Lowry, O. H., Rosebrough, N. D., Farr, A. L., and Randall, R. J. (1951): Protein measurement with the Folin phenol reagent. *J. Biol. Chem.*, 193:265.
15. Milgrom, E., Thi, L., Atger, M., and Baulieu, E. E. (1973): Mechanisms regulating the concentration and the conformation of progesterone receptor(s) in the uterus. *J. Biol. Chem.*, 218:6366–6374.
16. Pavlik, E. J., and Coulson, P. B. (1976): Modulation of estrogen receptors in four different target tissues: Differential effects of estrogen vs. progesterone. *J. Steroid Biochem.*, 7:369–376.
17. Philibert, D., and Raynaud, J. P. (1973): Progesterone binding in the immature mouse and rat uterus. *Steroids*, 22:89–98.
18. Philibert, D., and Raynaud, J. P. (1974a): Progesterone binding in immature rabbit and guinea-pig uterus. *Endocrinology*, 94:627–632.
19. Philibert, D., and Raynaud, J. P. (1974b): Binding of progesterone and R5020, a highly potent progestin, to human endometrium and myometrium. *Contraception*, 10:457–466.
20. Reyes, F. J., Winter, J. S. D., Faiman, C., and Hobson, W. C. (1975): Serial serum levels of gonadotropins, prolactin and sex steroids in the nonpregnant and pregnant chimpanzee. *Endocrinology*, 96:1447–1455.

21. Schwartz, N. B. (1964): Acute effects of ovariectomy on pituitary LH, uterine weight and vaginal cornification. *Am. J. Physiol.,* 207:1251–1257.
22. Shain, S. A., and Barnea, A. (1971): Some characteristics of the estradiol binding proteins of the mature, intact rat uterus. *Endocrinology,* 89:1270–1279.
23. Yoshinaga, K., Hawkins, R. A., and Stocker T. F. (1969): Estrogen secretion by the rat ovary in vivo during the estrous cycle and pregnancy. *Endocrinology,* 85:103–112.

Use of [³H]R5020 for the Assay of Cytosol and Nuclear Progesterone Receptor in the Rat Uterus

E. Milgrom, M. T. Vu Hai, and F. Logeat

Groupe de Recherches sur la Biochimie Endocrinienne et la Reproduction (U 135 INSERM), Faculté de Médecine Paris-Sud, 94 Bicêtre, France

It is now well established that the concentration of hormonal receptors is modulated in various physiological and pathological states. However, in most such situations endogenous hormone is secreted by the animal and this renders considerably more difficult the assay of the receptor. On the one hand, receptor is distributed between two subcellular compartments: cytosol and nuclei. On the other hand, part of the cytosol and all of the nuclear receptor sites are occupied by endogenous hormone. Since binding of *in vitro* added radioactive steroid is prerequisite for receptor assay, it is necessary to establish experimental conditions in which this radioactive hormone is exchanged with endogenous steroid without modification of receptor concentration. Such exchange assays must be devised for both cytosol and nuclei.

In the case of estrogen receptor, well-documented assays have been described (1,2). For progesterone receptor only partial or preliminary work has been published (3,4). Such an assay would be especially useful in the rat, a species whose reproductive physiology has been most thoroughly studied. Studies on progesterone receptor in the rat have been hampered by the fact that receptor-progesterone complexes are very unstable (5) and also by the presence in the uterus of a corticosteroid binding globulin (CBG)-like protein which binds progesterone with high affinity (6). The use of R5020 has been shown to overcome both of these difficulties since this steroid binds to the progesterone receptor with high affinity and does not interact with CBG (7).

In this chapter are described experimental conditions for assaying cytosol and nuclear progesterone receptors in the rat uterus. Variations of the receptors during the estrous cycle and pregnancy are also reported.

ASSAY OF CYTOSOL RECEPTOR

Before R5020 could be used for the assay of the cytosol progesterone receptor, various conditions had to be fulfilled:

1. R5020 Must bind only to the progesterone receptor. Various experiments confirmed the previously reported fact (7) that R5020 was bound to a macromolecule having the characteristics of a receptor (sedimentation coefficients at low and high ionic strength, inhibition by SH-blocking agents, precipitation by ammonium and protamine sulfate, etc.). This macromolecule had the hormonal specificity and tissular distribution expected from the progesterone receptor. It was also established that R5020 did not bind to the two other saturable steroid-binding proteins present at appreciable concentrations in the uterine cytosol: the estrogen receptor and the CBG-like protein. Up to 10^{-6} M of unlabeled R5020 did not compete with [^{3}H]estradiol for the estrogen receptor, but concentrations over 10^{-7} M of unlabeled R5020 competed with [^{3}H]cortisol for CBG. However, progesterone receptor is saturated at concentrations one order of magnitude lower, thus the use of such high concentrations of R5020 is not necessary.

2. The receptor should be saturated by the hormone. It was established that binding equilibrium was attained at 1 or 20 nM steroid concentrations in less than 90 min. The affinity of the receptor for R5020 was, at 0°C, $K_a = 8.10^8$ M^{-1}. A concentration of 10 nM [^{3}H]R5020 was sufficient to saturate all specific sites.

3. Endogenous hormone should not interfere in the assay. Usually this problem is solved by exchanging endogenous hormone with added radioactive steroid. However, in the present case the assay had to be carried out

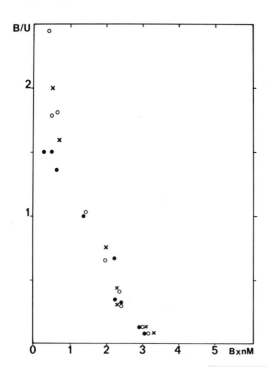

FIG. 1. Assay of cytosol receptor in the presence or absence of unlabeled progesterone. ○, absence of unlabeled progesterone; △, 1 nM unlabeled progesterone; X, 10 nM unlabeled progesterone. Cytosol was incubated 90 min at 0°C with unlabeled progesterone (1 or 10 nM). One volume of dextran (0.05%)-coated charcoal (0.5%) was then added and the suspension agitated for 60 min at 0°C. The supernatant was obtained and incubated with various concentrations of [^{3}H]-R5020 for 2 hr at 0°C. Bound radioactivity was then measured.

during pregnancy when very high concentrations of progesterone are present in the uterine cytosol and can dilute to an unknown extent the added [^{3}H]R5020. For this reason it was decided to try to remove most of the endogenous progesterone before incubation with the synthetic progestin. It was shown that over 90% of progesterone present in the uterine cytosol could be removed by a 60-min contact at 0°C with a dextran-coated charcoal suspension. There was no change in receptor concentration during this time.

TABLE 1. *Assay of cytosol receptor*

1. Each uterus is homogenized in 4 volumes of buffer (Tris 0.01 M, EDTA 1.5 mM, β-mercaptoethanol 1 mM, glycerol 10%, pH 7.4.).
2. The homogenate is centrifuged at 800 × g for 10 min.
3. The supernatant is centrifuged at 105,000 × g for 90 min.
4. The cytosol is diluted to obtain an optical density at 280 nM equal to 4.
5. One volume of dextran (0.05%)-coated charcoal (0.5%) is added. Agitation 60 min. Centrifugation 3,000 × g for 10 min.
6. The supernatant is incubated for 2 hr at 0°C either with a saturating concentration of [^{3}H]R5020 (10 nM) or with various concentrations of [^{3}H]R5020.
   Parallel incubations in the presence of 2 μM unlabeled progesterone are also performed.
7. One volume of dextran-coated charcoal is added. The suspension is agitated for 10 min. The supernatant of a 3,000 × g for 3 min centrifugation is counted for radioactivity.
8. Receptor concentration is obtained either by Scatchard plot analysis or as the difference between total and nonspecific binding at hormone saturation. The temperature is kept at 0°C throughout the procedure.

To further assess this technique, we divided cytosol in three aliquot portions: one was incubated with 1 nM unlabeled progesterone, a second with 10 nM of the hormone, and the third kept without added steroid. All incubations were of 90 min at 0°C. After the two incubates containing progesterone were treated with dextran-coated charcoal (60 min at 0°C), all three solutions were further incubated with various concentrations of [^{3}H]R5020. As shown in Fig. 1, no difference in binding of radioactivity could be observed between cytosol previously incubated with and without unlabeled progesterone.

4. The assay of the cytosol receptor was established according to these findings and is summarized in Table 1.

ASSAY OF THE NUCLEAR RECEPTOR

Nuclear progesterone-receptor complexes were measured by an exchange technique.

In most of the experiments nuclei were prepared from castrated, estradiol-primed (5μg × 2 days) animals to which progesterone (1 mg) was injected subcutaneously 1 hr prior to sacrifice.

Time Course of Exchange

Nuclei containing progesterone-receptor complexes were incubated for various times and at 0°C with 10 nM [^{3}H]R5020 or 10 nM [^{3}H]R5020 plus 2 μM unlabeled progesterone. The latter incubation measures nonspecific, nonsaturable binding and is subtracted from the first incubation (total binding) to yield specific binding.

As shown in Fig. 2, the specific binding increased up to the fifth hour, then remained on a plateau and decreased only after very long incubations. However, the nonspecific binding increased progressively even after the fifth hour. For this reason a 6-hr incubation at 0°C was chosen in further experiments to minimize nonspecific binding.

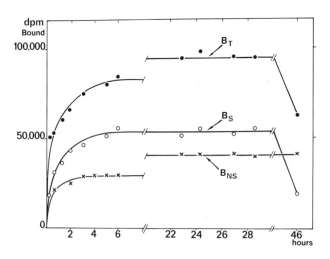

FIG. 2. Time course of exchange of [^{3}H]R5020 with unlabeled progesterone in nuclear progesterone-receptor complexes. Nuclei were incubated for various periods of time at 0°C with either 10 nM [^{3}H]R5020 (B_T) or 10 nM [^{3}H]R5020 and 2 μM unlabeled progesterone (B_{NS}). Specific binding $B_S = B_T - B_{NS}$.

Concentration of [^{3}H]R5020

Exchange was performed at various concentrations of [^{3}H]R5020. Saturation of specific binding was obtained at 10 nM. Scatchard plot analysis gave a single line ($K_a = 4.10^8$ M^{-1}).

Specificity of the Exchange Reaction

Specificity of the exchange reaction was studied by competition with unlabeled steroids. R5020, progesterone, and norethisterone competed effectively, whereas estradiol, corticosterone, and testosterone did not.

Washing Conditions, Decrease of Nonspecific Binding

It was established that a single washing of nuclei before incubation with [^{3}H]R5020 and four washings after incubation realized optimal conditions. Nonspecific binding was also decreased by the use of albumin (10%) in the washing buffer, the use of polystyrene tubes for incubations and washings, and other factors.

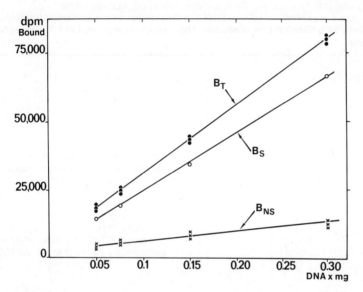

FIG. 3. Assay of nuclear progesterone-receptor complexes at different concentrations of nuclei. Nuclear progesterone receptor was assayed as described in Table 2 but at different concentrations of nuclei.

All these precautions led to a markedly decreased nonspecific binding ($\simeq$ 15% of total binding as shown in Fig. 3).

Assay of Nuclear Receptor

Assay of nuclear receptor is summarized in Table 2. The linearity of this assay was verified at various concentrations of nuclei (Fig. 3). All further measurements were performed at a concentration of 0.15 to 0.2 mg DNA per milliliter.

Dependence of Nuclear Receptor Concentration on Endogenous Progesterone

In castrated, estradiol-primed animals the concentration of nuclear receptors was very small (Fig. 4). After adrenalectomy a slight decrease was observed. The concentration of receptor increased markedly after proges-

TABLE 2. *Assay of nuclear receptor*

1. The precipitate of the 800 g centrifugation (see Table 1) is washed with 6 ml of buffer. It is then resuspended in 6 ml of buffer.
2. Aliquots (1 ml) of the suspension are then added to 1 ml of buffer containing the steroids. Final steroid concentration is 10 nM [^{3}H]R5020 or 10 nM [^{3}H]R5020 plus 2 μM unlabeled progesterone. Incubation is 6 hr at 0°C.
3. Centrifugation at 3,000 g for 10 min. The precipitate is washed three times with 5 ml of buffer containing bovine serum albumin (10 mg/ml) and once with 5 ml of buffer.
4. The precipitate is suspended in 1.5 ml of buffer and transferred into a counting vial.
5. After radioactivity counting, DNA is measured in the nuclear pellet.
6. Receptor concentration equals total binding minus nonspecific binding (incubation in presence of unlabeled progesterone). The temperature is kept at 0°C throughout the procedure.

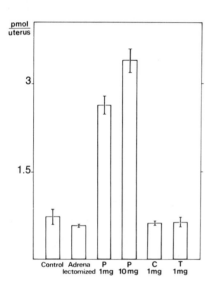

FIG. 4. Effect of steroid administration on nuclear progesterone receptor concentration. Ovariectomized estrogen-primed rats were used. Adrenalectomy was performed 3 days before the experiment. Progesterone (P) (1 or 10 mg), corticosterone (C) (1 mg), and testosterone (T) (1 mg) were injected subcutaneously 60 min before sacrifice.

terone injection, whereas the injections of corticosterone and testosterone were without effect.

RECEPTOR VARIATIONS DURING THE ESTROUS CYCLE AND PREGNANCY

The methods described above were used for the assay of both cytosol and nuclear receptors during the estrous cycle and pregnancy.

Receptor Variations During the Estrous Cycle

Cytosol receptor (Fig. 5) peaked at proestrus ($\simeq$ 30,000 binding sites per cell), decreased at estrus ($\simeq$ 11,000 sites per cell) and metestrus ($\simeq$ 6,000 sites per cell), and increased again at diestrus ($\simeq$ 17,000 sites per cell). This pattern was thus similar to that described in the guinea pig (3).

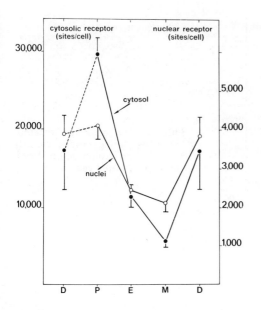

FIG. 5. Progesterone receptor variations during the estrous cycle.

In the nuclei the highest concentration was obtained at proestrus and diestrus ($\simeq$ 4,000 sites per cell), whereas the lowest was observed at estrus and metestrus ($\simeq$ 2,000 sites per cell). However, the highest ratio of nuclear receptor/cytosol receptor was seen at metestrus. The concentration of nuclear receptor is related, as expected, to both the concentration of cytosol receptor and plasma progesterone concentration. A correlation coefficient of $r = 0.78$ was obtained when comparing nuclear receptor concentration with the product of cytosol receptor concentration times the plasma progesterone concentration.

Receptor Variations During Pregnancy

Cytosol receptor was low on day 3 of pregnancy (Fig. 6). There was a further decrease at day 5 (corresponding probably to a transfer of progesterone-receptor complexes to the nuclei due to increased plasma progesterone). Then similar low concentrations were observed between days 6 and 12 of pregnancy. Thereafter, receptor concentration started to increase until the end of pregnancy. At day 22 cytosol receptor concentration was 26,000 binding sites per cell, similar to that observed at proestrus. This increase of cytosol receptor was probably due on one hand to the decreasing progesterone concentration and on the other hand to a peak of estrogen secretion at the end of pregnancy.

The nuclear receptor pattern was completely different: concentration was low on day 3, a sharp increase was observed at day 5, then the concentration dropped again on day 6. A plateau of highest concentration of nuclear

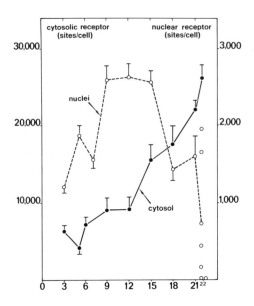

FIG. 6. Progesterone receptor variations during pregnancy. Day 1 = day on which spermatozoa were found in the vagina.

receptor ($\simeq$ 2,600 sites per cell) was observed between days 9 and 15. Then nuclear receptor concentration sharply decreased. At day 22 the mean concentration was very low ($\simeq$ 700 sites per cell); some animals, probably on the verge of parturition, had no detectable nuclear receptor.

DISCUSSION

The relatively low affinity of rat uterine receptor for progesterone and the fast dissociation rate of the progesterone-receptor complexes allow either removal of endogenous progesterone or its exchange with [^{3}H]R5020, even at 0°C. This low temperature prevents alteration of receptor during the assay period.

During both the estrous cycle and pregnancy only a small fraction of total receptor was found in the nuclei. The explanation of this phenomenon is unknown. It may be because progesterone is bound by other proteins in the rat uterus (albumin and CBG-like protein) and the unbound fraction is insufficient to saturate the receptor (the relatively high K_d of the interaction further preventing this saturation). Alternatively, it is also possible that only a relatively small fraction of receptor-steroid complexes are bound to the nuclei.

Receptor variations during the estrous cycle essentially confirmed the results obtained previously in guinea pigs. During pregnancy in the rat, peculiar receptor modifications were observed at two privileged periods. At implantation the overall receptor concentration in the uterus was low, but a peak of plasma progesterone provoked a transfer of receptor from cytosol to nuclei. At parturition the concentration of receptor-progesterone complexes in the nuclei was quite low. Since these nuclear complexes are supposed to mediate hormone action, this probably means that progesterone's biological action is markedly diminished shortly before parturition. The reason for this decrease of nuclear receptor appears to lie in the falling plasma progesterone concentration and not in a modification of receptor concentration since the latter is high in the cytosol at the end of pregnancy. It is possible that the increase of cytosol receptor at the end of pregnancy is necessary to compensate for a falling plasma progesterone concentration and to prevent a premature onset of parturition. In other species (women, for instance) there is no marked decrease in hormonal level prior to parturition. It will be of interest to learn how receptor concentrations vary in these cases.

The methods of assay described in this chapter could also be used in various experimental situations to correlate receptor concentrations with biological responses to progesterone.

SUMMARY

[^{3}H]R5020 was used to measure the concentration of progesterone receptor in rat uterine cytosol and nuclei even in the presence of endogenous progesterone. In the cytosol, a 60-min treatment of dextran-coated charcoal was sufficient to remove over 90% of endogenous progesterone without harming the receptor. Receptor concentration was then measured with [^{3}H]R5020. A 6-hr incubation of nuclei at 0°C in a saturating concentration (10 nM) of [^{3}H]R5020 was sufficient to completely exchange the progesterone bound to receptor with [^{3}H]R5020.

Using these methods, we studied variations in receptor concentration in both cytosol and nuclei during the estrous cycle and pregnancy.

ACKNOWLEDGMENTS

We thank J. P. Raynaud and D. Philibert (Roussel-Uclaf) for the gift of R5020. This work was supported by the INSERM and the CNRS. M. L. Brenot typed the manuscript.

REFERENCES

1. Anderson, J., Clark, J. H., and Peck, E. J. (1972): Oestrogen and nuclear binding sites. Determination of specific sites by [^{3}H]oestradiol exchange. *Biochem. J.*, 126:561–567.
2. Katzenellenbogen, J. A., Johnson, H. J., and Carlson, K. E. (1973): Studies on the uterine, cytoplasmic estrogen binding protein. Thermal stability and ligand dissociation rate. An assay of empty and filled sites by exchange. *Biochemistry*, 12:4092–4099.
3. Milgrom, E., Perrot, M., Atger, M., and Baulieu, E. E. (1972): Progesterone in uterus and plasma. V. An assay of the progesterone cytosol receptor of the guinea pig uterus. *Endocrinology*, 90:1064–1070.

4. Hsueh, A. J. W., Peck, E. J., and Clark, J. A. (1974): Receptor progesterone complex in the nuclear fraction of the rat uterus: demonstration by [^{3}H]progesterone exchange. *Steroids,* 24:599–611.
5. Feil, P. D., Glaser, S. R., Toft, D. O., and O'Malley, B. W. (1972): Progesterone binding in the mouse and rat uterus. *Endocrinology,* 91:738–746.
6. Milgrom, E., and Baulieu, E. E. (1970): Progesterone in uterus and plasma. I. Binding in rat uterus 105,000 g supernatant. *Endocrinology,* 87:276–287.
7. Philibert, D., and Raynaud, J. P. (1973): Progesterone binding in the immature mouse and rat uterus. *Steroids,* 22:89–98.

> Progesterone Receptors in Normal and
> Neoplastic Tissues, edited by W. L. McGuire
> et al. Raven Press, New York © 1977.

Nuclear and Cytoplasmic Progesterone Receptors in the Rat Uterus: Effects of R5020 and Progesterone on Receptor Binding and Subcellular Compartmentalization

Marian R. Walters and James H. Clark

Department of Cell Biology, Baylor College of Medicine, Houston, Texas 77030

The presence of a specific progesterone receptor has been clearly demonstrated in uterine cytosols of several mammals, i.e., hamster (16,17,32), rabbit (9,10,19,20,31,33,36), guinea pig (7,8,11,13,18,25–27), and mouse (12). In contrast to these studies, no reliable technique has been described for the quantitation of progesterone receptors in rat uterine cytosol, although a few reports have described a progesterone receptor in qualitative terms (9,12,35). The principal problems encountered by these investigators have been the instability of the progesterone receptor and interference in the assays by the binding of ^{3}H-progesterone to lower affinity proteins, especially a corticosteroid binding globulin (CBG)-like protein described by Milgrom, Baulieu, and co-workers (2,24,25).

Previous measurements of nuclear progesterone receptors in mammalian models other than the rat were based on the extraction of receptors by KCl (11,18,27,31), which may exclude significant portions of the nuclear populations from the assays (5). One report of a nuclear exchange assay (NEA) for the rat uterine progesterone receptor has appeared (14). However, the method is tedious (incubation at 15°C for 5 hr was necessary to avoid degradation) and the nonspecific binding was high.

The fact that specific progesterone receptors have been detected in the rat uterus suggested the possibility of optimizing the procedures to afford sensitive and reliable cytosol and nuclear exchange assays. The availability of R5020 (17,21-dimethyl-19-nor-4,9-pregnadiene-3,20-dione), a compound which is relatively specific for progesterone receptors (28–30), presented a unique opportunity to develop these assays and to validate their specificity for the rat uterine progesterone receptor. In this chapter we report the characteristics of the progesterone receptors of rat uterine cytosol and nuclei, their binding to R5020, and preliminary data on their hormonal control.

METHODS

Female rats of the Sprague-Dawley strain (60 days of age) were obtained from Texas-Inbred Mice Co., Houston, Texas. They were ovariectomized under ether anesthesia on day 0 and received 1 μg E_2 per rat s.c. on days 3 and 4. The rats were sacrificed by decapitation on day 5. In experiments in which nuclear receptors were measured, rats received 500 μg progesterone s.c. in 0.5 ml 30% ethanol in saline 1 hr prior to sacrifice.

Uteri were homogenized (50 mg/ml) on ice in Tris buffer (10 mM), pH 7.4, containing 30% vol/vol glycerol (TG). Nuclei were prepared by filtering the homogenate through a single layer of organza, pelleting at 860 g for 20 min, and washing the pellet in TG buffer three times. Cytosol was prepared by centrifuging the homogenate or the 860 g supernatant at 25,000 g for 30 min. Overnight incubation (18 to 24 hr) at 4°C of 250λ cytosol with 20 nM ^{3}H-progesterone (specific activity 50 Ci/mmole = 29,668 cpm/pmole) ± 2 μM progesterone, cortisol, or R5020 was followed by exposure to 250λ dextran-coated charcoal for about 30 sec, centrifugation at 2,400 g for 6 min, and counting of a 250λ aliquot of the supernatant. The 30-sec exposure to charcoal represents the shortest time possible between the addition of the charcoal and the beginning of centrifugation. Next 500λ of the resuspended nuclear pellet was incubated under similar conditions, washed three times with TG, and extracted with ethanol for counting. DNA determinations (4) of the initial homogenate and the washed nuclear pellet prior to incubation allowed calculation of the percentage yield of nuclei.

The amount of ^{3}H-progesterone that was specifically bound to the receptor was determined by subtracting the amount nonspecifically bound (^{3}H-progesterone bound in the presence of excess unlabeled progesterone) from the total ^{3}H-progesterone bound (in the absence of excess unlabeled progesterone). Occasionally in this study we observed a sporadic, minor cortisol-compatible component of ^{3}H-progesterone binding. Correction for this component (noted in the appropriate experiments) was accomplished by subtracting the cortisol-compatible ^{3}H-progesterone bound from the specifically bound ^{3}H-progesterone (progesterone competible). In the future, this correction will be simplified by calculating the specifically bound ^{3}H-progesterone from the nonspecifically bound ^{3}H-progesterone in the presence of excess unlabeled R5020 to determine the R5020-compatible component of bound ^{3}H-progesterone.

A comprehensive validation of the above assays will appear elsewhere (Walters and Clark, *in preparation*).

RESULTS

Progesterone Binding and Exchange in Cytosol

The following experiments were done to establish the time of saturation and exchange for the ^{3}H-progesterone exchange assay in rat uterine cytosol.

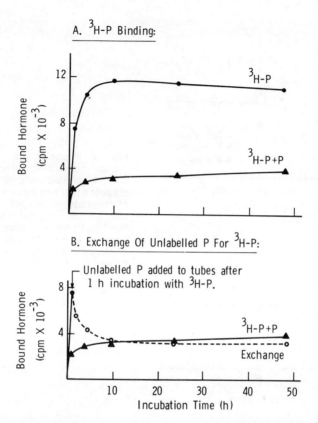

FIG. 1. Progesterone binding and exchange in cytosol. **A:** Cytosol was incubated at 4°C for varying times in the presence of 20 nM ³H-progesterone ± 2 µM progesterone. **B:** After 1 hr incubation in the presence of 20 nM ³H-progesterone, tubes received 2 µM unlabeled progesterone and the exchange phenomenon was measured. Symbols: ●—— total ³H-progesterone bound; △—— nonspecifically bound; ○---- exchange of unlabeled progesterone for ³H-progesterone.

Cytosol was incubated for varying lengths of time with 20 nM ³H-progesterone. The system reached saturation by 4 to 10 hr and then remained stable for at least 38 hr thereafter (Fig. 1A). Nonspecific (lower curve, Fig. 1A) and specific binding (difference between curves in Fig. 1A) followed a similar pattern. Exchange of unlabeled progesterone for ³H-progesterone was demonstrated by adding unlabeled progesterone to tubes previously incubated for 1 hr with ³H-progesterone alone (Fig. 1B). Exchange was rapid and reached levels which were equivalent to nonspecific binding by 9 hr.

Saturation Analysis of ³H-R5020 and ³H-Progesterone Binding to Cytosol

In order to compare the binding parameters of ³H-progesterone and ³H-R5020, we ran saturation analysis by incubating cytosol overnight with

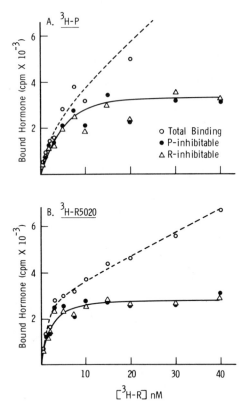

FIG. 2. Progesterone receptor of cytosol: saturation analysis. Cytosol was incubated overnight in the presence of varying concentrations of ^{3}H-progesterone **(A)** or ^{3}H-R5020 **(B)** ± 2 μM progesterone or R5020. Symbols: ○---- total ^{3}H-steroid bound, ●—— specifically bound (P-inhibitable), △—— specifically bound (R-inhibitable).

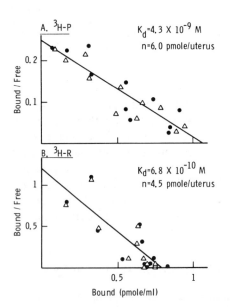

FIG. 3. Progesterone binding to cytosol: Scatchard analysis. Data for the amount of specifically bound ^{3}H-progesterone **(A)** and ^{3}H-R5020 **(B)** from Fig. 2 were plotted by the method of Scatchard (34). Lines were fitted and K_d and N calculated by linear regression analysis. Symbols: ●—— specifically bound (P-inhibitable); △—— specifically bound (R-inhibitable).

varying concentrations of ^{3}H-progesterone $\pm$ 2 μM progesterone or R5020 (Fig. 2A). The specific binding component saturated at 15 nM, and nonspecific binding did not equal specific until 20 nM. The curves of progesterone-inhibitable and R5020-inhibitable ^{3}H-progesterone binding were identical in this experiment. ^{3}H-R5020 binding to the progesterone receptor was similar to that of ^{3}H-progesterone, although the amount bound was slightly less. Additionally, the total binding component of ^{3}H-R5020 was lower and less variable than that of ^{3}H-progesterone.

Scatchard analysis of these specific binding curves yielded $K_d = 4.3$ nM and $N = 6.0$ pmoles/uterus for ^{3}H-progesterone (Fig. 3A), and $K_d = 0.68$ nM and $N = 4.5$ pmoles/uterus for ^{3}H-R5020 (Fig. 3B).

Relative Binding Activities of Steroids for the Cytosol Progesterone Receptor

The hormone binding specificity of the cytosol assay was established by competition analysis. A typical curve obtained when 20 nM ^{3}H-R5020 or ^{3}H-progesterone was incubated with cytosol in the presence of several steroids over a wide concentration range (2 nM to 2×10^4 nM) is shown in Fig. 4. The variation in the zero point is typical of these curves and is not well understood. In the curves for ^{3}H-progesterone (*not shown*) the corticoids routinely increased the binding when present at low levels. One interpretation of this phenomenon is that the corticoids displace ^{3}H-progesterone from nonspecific components, effectively increasing the free ^{3}H-progesterone concentration exposed to the receptor.

Table 1 summarizes the relative binding affinities (RBAs) for the steroids tested (determined from the concentration of competitor at the 50% level

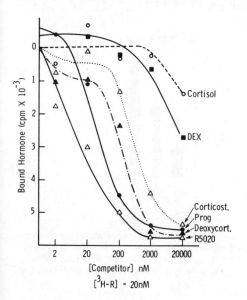

FIG. 4. Competition of ^{3}H-R5020 binding in the cytosol. Cytosol was incubated overnight in the presence of 20 nM ^{3}H-R5020 and 2 nM to 2×10^4 nM unlabeled competitors. The amount of competible binding of ^{3}H-progesterone (specifically bound) is plotted along the ordinate as a function of competitor concentration for each test steroid.

TABLE 1. *Relative binding activities of steroids for rat uterine cytosol*

Steroid	RBA vs ^{3}H-progesterone[a]	RBA vs ^{3}H-R5020[b]
Progesterone	100%	30%
R5020	1,600	100
Deoxycorticosterone	55	8
Corticosterone	8	3
E_2	5	—
Dexamethasone	5	<0.1
Cortisol	<1	<0.1
Nafoxidine	0	—

Cytosol from E_2-primed ovariectomized rats was incubated with 20 nM ^{3}H-progesterone in the presence of competing steroids (2 nM–2 × 10^4 nM). The relative binding activities were calculated from the concentrations of the competing steroids at the 50% level of competition.

[a] Two to three determinations.
[b] One determination.

of competition). The specificity of the progesterone receptor for these steroids is typical of data obtained for progesterone receptors in other animal models (9,11,15,17,31). The specificities agree very well between ^{3}H-progesterone and ^{3}H-R5020 when one takes into account the change in reference steroid.

Effect of Estradiol on the Quantity of Progesterone Receptors in the Rat Uterus

Previous reports have indicated that the quantity of cytosol progesterone receptors is increased by E_2 pretreatment (9,13,17,21,27,32). As a part of

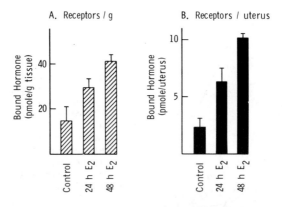

FIG. 5. Effect of estradiol-17β on cytosol progesterone receptors. Rats were ovariectomized for 3 days and then received 0, 1, or 2 injections (1 μg/rat s.c.) spaced 24 hr apart. Uterine cytosol was prepared for 24 hr after the last E_2 injection and was incubated at 4°C overnight with 20 nM ^{3}H-progesterone ± 2 μM progesterone. Data are expressed as the mean ± SEM of 4 observations.

the validation of the assay, we have confirmed this finding in rat uterine cytosol. Cytosol progesterone receptors were measured in 3-day castrates 24 hr after zero, one, or two E_2 injections (1 μg/rat s.c. spaced 24 hr apart) by incubating overnight in the presence of 20 nM ^{3}H-progesterone. As shown in Fig. 5, there was a significant ($p < 0.05$) increase in the receptor numbers whether expressed as per gram uterus or per uterus.

Exchange of ^{3}H-Progesterone for Nuclear Bound Progesterone

The characteristics of ^{3}H-progesterone exchange for nuclear receptor sites were assessed in uteri from rats which were injected with 500 μg

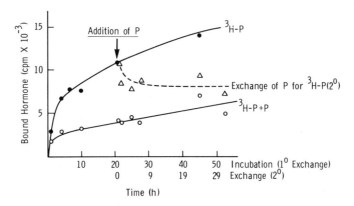

FIG. 6. Progesterone exchange in nuclei. Nuclei from rats treated 1 hr *in vivo* with progesterone were incubated for varying times with 20 nM ^{3}H-progesterone ± 2 μM unlabeled progesterone. Data are plotted for total ^{3}H-progesterone bound (●———) and nonspecifically bound ^{3}H-progesterone (○———). Additionally, a secondary exchange of unlabeled progesterone for ^{3}H-progesterone was examined. After 24 hr of incubation, 2 μM unlabeled progesterone was added to tubes previously incubated only with ^{3}H-progesterone and the incubation was continued for 24 hr (△---).

progesterone 1 hr before sacrifice. Nuclei were incubated at 4°C for varying times in the presence of 20 nM ^{3}H-progesterone ± 2 μM progesterone. Exchange of ^{3}H-progesterone was complete by about 10 hr and remained stable for at least 38 hr thereafter (Fig. 6). In order to confirm that this binding of ^{3}H-progesterone was due to an exchange reaction for unlabeled progesterone bound to nuclear receptors, we performed an additional experimental manipulation. After specific ^{3}H-progesterone binding/exchange was complete (24 hr), unlabeled progesterone (2 μM) was added to tubes previously incubated only with ^{3}H-progesterone. The exchange reaction was then measured for varying times (Fig. 6). Exchange of unlabeled progesterone for ^{3}H-progesterone was complete at 8 hr and remained relatively stable for at least 20 hr thereafter.

Saturation Analysis of ^{3}H-R5020 and ^{3}H-Progesterone Binding to Nuclei

Nuclear pellets from progesterone-treated rats were incubated with varying concentrations of ^{3}H-progesterone $\pm$ 2 μM progesterone (or ^{3}H-R5020 $\pm$ R5020). ^{3}H-progesterone and ^{3}H-R5020 displayed similar satura-

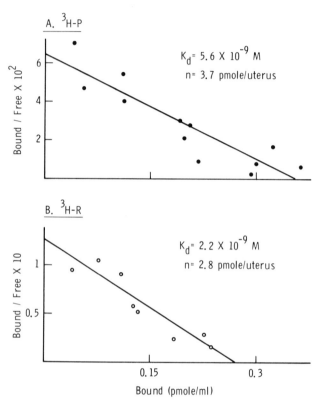

FIG. 7. Progesterone binding to nuclei: Scatchard analysis. Data for the amount of specifically bound ^{3}H-progesterone **(A)** or ^{3}H-R5020 **(B)** were plotted according to Scatchard (34). Lines were fitted and K_d and N calculated by linear regression analysis. Symbols: ●——^{3}H-progesterone; ○——^{3}H-R5020.

tion characteristics (*not shown*). Scatchard analysis of the specific components yielded $K_d = 5.6$ nM and $N = 3.7$ pmoles/uterus for ^{3}H-progesterone (Fig. 7A) and $K_d = 2.2$ nM and $N = 2.8$ pmoles/uterus for ^{3}H-R5020 (Fig. 7B).

Competition Analysis of the Nuclear Progesterone Receptor

Steroid specificity was measured in the nuclear preparation (Fig. 8) as described previously for the cytosol. The ability of R5020, progesterone, and deoxycorticosterone to compete for ^{3}H-progesterone binding sites is similar to that observed for the cytosol receptor. Cortisol and E_2 show little ability to compete over most of the concentration range; however, cortisol does act as a competitor for a small number of high-affinity sites. This component of the curve is probably due to the presence of CBG-like material described earlier. This assumption is further substantiated by the observation that R5020 does not compete for a number of ^{3}H-progesterone binding sites that are quantitatively similar to the high-affinity cortisol-compatible sites. The relative binding affinities were determined in the area of the graph between the CBG-like component and the tailing-off of the R5020-inhibitable component, where the competition curves were parallel.

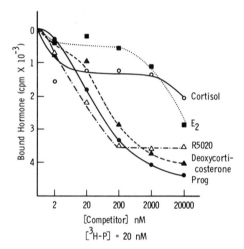

FIG. 8. Competition of ^{3}H-progesterone binding in nuclei. Nuclei were incubated overnight in the presence of 20 nM ^{3}H-progesterone and 2 nM–2×10^4 nM unlabeled competitors. The amount of compatible binding (i.e., specifically bound) is plotted along the ordinate as a function of competitor concentration for each test steroid.

The specificity of the nuclear receptor for these steroids was similar to that of the cytosol: R5020 > progesterone > deoxycorticosterone > corticosterone >> dexamethasone or cortisol. Although not shown here, the competition curves for ^{3}H-R5020 did not include a high-affinity cortisol-inhibitable component, supporting the conclusion that this component is the CBG-like protein.

Cytosol and Nuclear Receptors: Translocation and Hormonal Dependence

Cytosol and nuclear progesterone receptor levels were measured at various times after injection of 500 μg progesterone or the vehicle. Data are expressed as percentage of control binding per uterus at 24 hr post-E_2 (second 24-hr injection).

One hour after injection of the vehicle, there was no change in cytosol receptors; but by 9 to 24 hr the cytosol receptor had decreased to 30 to 50% of the control levels (Figs. 9A and 10A). No changes in the level of nuclear receptor were observed in this control group. These results suggest that

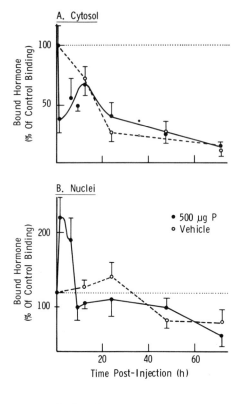

FIG. 9. Progesterone receptors of rat uterine cytosol and nuclei after progesterone injection *in vivo*: ^{3}H-progesterone binding. Receptors were quantitated at varying times after the injection of 500 μg progesterone or the vehicle *in vivo*. Thus, cytosol **(A)** and nuclei **(B)** were separated and incubated overnight in the presence of 20 nM ^{3}H-progesterone alone or with 2 μM progesterone or cortisol. Receptor binding was measured as the magnitude of specifically bound ^{3}H-progesterone and was corrected for the cortisol-inhibitable component. Data were expressed as pmoles receptor per uterus and normalized by dividing individual values by the average zero-time control value (cytosol: 12.9 ± 1.5 pmoles/uterus; nuclei: 3.1 ± 0.6 pmoles/uterus). Points and bars, respectively, represent the mean ± SEM.

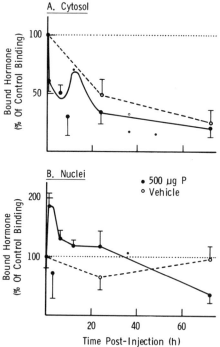

FIG. 10. Progesterone receptors of rat uterine cytosol and nuclei after progesterone injection *in vivo*: ^{3}H-R5020 binding. Receptors were quantitated at varying times after the injection of 500 μg progesterone or the vehicle *in vivo*. Thus, cytosol **(A)** and nuclei **(B)** were separated and incubated overnight in the presence of 20 nM ^{3}H-R5020 ± 2 μM R5020. Receptor binding was measured as the magnitude of specifically bound ^{3}H-R5020. Data were expressed as pmoles receptor per uterus and normalized by dividing individual values by the average control value (cytosol: 11.5 ± 1.8 pmoles/uterus; nuclei: 3.2 ± 0.6 pmoles/uterus). Points and bars, respectively, represent the mean ± SEM.

rat uterine progesterone receptor levels are maintained by E_2, and in its absence the receptors are rapidly degraded and/or not resynthesized.

Progesterone injection resulted in depletion of cytosol receptor and concomitant accumulation of nuclear receptor (presumably by translocation) by 1 hr (Figs. 9 and 10). The level of nuclear receptor returned to the control value by 3 to 6 hr after the injection. During this time the quantity of cytosol receptor increased to a level comparable to that of the vehicle-injected control and then declined at a rate which did not differ from that of the control.

DISCUSSION

The data presented in this study indicate that reliable measurements of the specific progesterone receptor can be obtained in cytosol and nuclei of rat uteri. Association (binding) and dissociation-reassociation (exchange) curves indicate that an exchange reaction is feasible between occupied receptor sites and free steroid in the incubation medium. Thus, it is possible to assess occupied and unoccupied receptors in the cytosol and nuclei. Proper measurement of these occupied and unoccupied receptors is an absolute necessity in experiments concerning nuclear translocation and cytosol replenishment. In our experiments, the concomitant decrease in cytosol receptors and increase in nuclear receptors indicates that translocation follows progesterone injection. Additionally, there is a subsequent cytosol receptor replenishment of a transient nature.

The characteristics of the progesterone receptor of rat uterine cytosol are similar to those of other species (3,12,17,19,21). Thus deoxycorticosterone has a relatively high affinity for the receptor, whereas E_2 and cortisol do not (Table 1). Any further question as to the identity of this progesterone receptor should be resolved by its high affinity for R5020 and negligible affinity for dexamethasone. The similarity between the nuclear and cytosol receptors (Figs. 3 through 5, 9 through 11, and Table 1) agrees with data for the E_2 receptors (1) but has not been previously demonstrated for mammalian progesterone receptors. Additionally, the data here confirm the specificity of R5020 for the progesterone receptor. The higher affinity of R5020 for both the cytosol and nuclear receptors has been observed both in Scatchard analysis (Figs. 3 and 7) and in competition assays (Figs. 4 and 8). These data also verify that R5020 does not bind significantly to cortisol-inhibitable proteins in this low concentration range. We have also shown that R5020 induces translocation of the progesterone receptor to the nucleus after R5020 injection *in vivo* (to appear elsewhere).

Charcoal was chosen as an absorbent for free hormone because it appears to diminish the interference due to the CBG-like component in a manner similar to that observed for low-affinity proteins in the estrogen receptor assay (6). Although we have used other methods (to be published elsewhere), the charcoal method has proven highly reproducible, with little intraassay variability (Fig. 1). However, in this chapter we have observed interassay variations in receptor numbers. This variation may be due to the animal model chosen. Ovariectomy was performed randomly during the estrous cycle and thus a variable receptor population was present 3 days later when E_2 was injected. Regardless of the source of variation, a cortisol-independent binding component (i.e., progesterone receptor) has been observed in rats in every physiological state tested (immature with or without E_2, adults during the estrous cycle, and ovariectomized adults with or without E_2) in these experiments (to appear elsewhere).

The rapid kinetics of the exchange phenomena at 4°C in both cytosol (Fig. 1) and nuclei (Fig. 6) corresponds to that observed in guinea pig cytosol (22). This characteristic may help to account for previous problems with detection of the rat uterine progesterone receptor and emphasizes the necessity of maintaining ice cold conditions and avoiding experimental delays.

The E_2-induced increase in the rat uterine progesterone receptor has been demonstrated previously (9) and is well described in other species (9,13,17, 21,23,27,32). However, in general the loss of receptors after E_2 withdrawal (13,17,27) is much slower than that observed herein (Figs. 9 and 10). The decrease in rat progesterone receptors at 24 hr post-progesterone reported here appears to be more correctly associated with E_2 withdrawal than with progesterone injection as observed in other species. For example, progesterone injection in the guinea pig resulted in a rapid loss of the receptor population by 24 hr, which was clearly different from the slow degradation observed after E_2 withdrawal (13,27). Cytosol receptors were examined at short intervals post-progesterone in the guinea pig by only one laboratory (13), and there was no apparent replenishment phenomenon as observed in the rat. An apparent progesterone-dependent receptor loss has also been indirectly inferred in the hamster from observations during the estrous cycle and after ovariectomy (17). Unfortunately, nuclear receptors were not evaluated in that study, making direct comparisons difficult. Thus, further studies on hormonal control of the rat progesterone receptor are needed to resolve the differences.

In conclusion, the progesterone receptor of the rat uterus shares similar characteristics with those of other species. Differences in hormonal control of these receptors may be due to species differences in reproductive physiology, but further studies are clearly necessary to distinguish between the mechanisms and their implications.

SUMMARY

Existing exchange assay procedures for the measurement of occupied and unoccupied progesterone receptors have been modified and extended for the evaluation of progesterone receptors in the rat uterus. Estrogen-primed uteri were homogenized in 10 mM Tris buffer containing 30% vol/vol glycerol, and the nuclear and cytosol fractions were prepared by centrifugation. Total cytosol binding of ^{3}H-progesterone or ^{3}H-R5020 was determined by incubation at 4°C overnight. Nonspecifically bound ^{3}H-steroid was estimated by competition analysis as described previously (*Steroids,* 24:599, 1974). Separation of free and bound steroid was accomplished by brief (30 sec) exposure to dextran-coated charcoal. Nuclear receptor was assayed under similar conditions and the quantity of bound steroid was measured by ethanol extraction from thoroughly washed nuclear pellets. Exchange of ^{3}H-progesterone with cytosol and nuclear receptors exhibited half-lives of 2.2 and 2.4 hr, respectively, and remained stable for 24 to 48 hr. The relative binding activities of steroids to ^{3}H-progesterone or ^{3}H-R5020 were R5020 > progesterone > deoxycorticosterone > corticosterone >> dexamethasone or cortisol for the cytosol receptor. Similar hormone specificity was observed for ^{3}H-progesterone exchange in the nuclear fractions. Thus, the steroid specificity of the cytosol receptor is similar to that of other previously described progesterone receptors, and ^{3}H-R5020 appears to bind in a manner consistent with ^{3}H-progesterone binding. Scatchard analysis confirms the presence of high-affinity ($K_d = 4.3 \times 10^{-9}$), low-capacity ($N = 6.0$ pmoles/uterus) ^{3}H-progesterone binding to the cytosol receptor. ^{3}H-R5020 binds with a higher affinity ($K_d = 6.8 \times 10^{-10}$) to the same number of sites ($N = 4.5$ pmoles/uterus). Injections of either progesterone or R5020 result in the depletion of cytosol receptor and concomitant accumulation of nuclear receptor. Progesterone injection and the resultant nuclear translocation is followed by partial cytosol replenishment by 3 to 9 hr. Estradiol injection increases the number of cytosol progesterone receptors per gram uterus or per uterus 24 hr after the injection. This increase is followed by a relatively rapid decrease in the number of progesterone binding sites after withdrawal of estrogen treatment. Thus, the methods described in this chapter should permit evaluation of the relationship between progesterone receptor binding and progesterone-induced responses in the rat uterus.

ACKNOWLEDGMENTS

The authors wish to thank Dr. J. P. Raynaud of the Centre de Recherches Roussel-Uclaf for the gifts of ^{3}H-R5020 and R5020. This work was supported by NIH grant HD-8436. The competent technical assistance of James Kovar is gratefully appreciated.

REFERENCES

1. Anderson, J., Clark, J. H., and Peck, E. J. (1972): Oestrogen and nuclear binding sites: determination of specific sites by [^{3}H]oestradiol exchange. *Biochem. J.,* 126:561–567.
2. Baulieu, E.-E., Alberga, A., Jung, I., Lebeau, M.-C., Mercier-Bodard, C., Milgrom, E., Raynaud, J. P., Raynaud-Jammet, C., Rochefort, H., Truong, H., and Robel, P. (1971):

Metabolism and protein binding of sex steroids in target organs: an approach to the mechanism of hormone action. *Recent Prog. Horm. Res.,* 27:351–412.
3. Baulieu, E.-E., Raynaud, J. P., and Milgrom, E. (1970): Measurement of steroid binding proteins. *Acta Endocrinol. (Kbh.)* [*Suppl.*], 147:104–118.
4. Burton, K. (1956): A study of the conditions and mechanism of the diphenylamine reaction for the colorimetric estimation of deoxyribonucleic acid. *Biochem. J.,* 62:315–323.
5. Clark, J. H., and Peck, E. J. (1976a): Nuclear retention of receptor-oestrogen complex and nuclear acceptor sites. *Nature,* 260:635–637.
6. Clark, J. H., and Peck, E. J. (1976b): Steroid receptors: biology, function and measurement. In: *Hormone Action and Molecular Biology Workshop,* edited by B. W. O'Malley and W. T. Schrader. Houston Biological Association, Houston.
7. Corval, P., Falk, R., Freifeld, M., and Bardin, C. W. (1972): In vitro studies of progesterone binding proteins in guinea pig uterus. *Endocrinology,* 90:1464–1469.
8. Faber, L. E., Sandmann, M. L., and Stavely, H. E. (1972a): Progesterone binding in uterine cytosols of the guinea pig. *J. Biol. Chem.,* 247:8000–8004.
9. Faber, L. E., Sandmann, M. L., and Stavely, H. E. (1972b): Progesterone-binding proteins of the rat and rabbit uterus. *J. Biol. Chem.,* 247:5648–5649.
10. Faber, L. E., Sandmann, M. L., and Stavely, H. E. (1973): Progesterone and corticosterone binding in rabbit uterine cytosols. *Endocrinology,* 93:74–80.
11. Feil, P. D., and Bardin, C. W. (1975): Cytoplasmic and nuclear progesterone receptors in the guinea pig uterus. *Endocrinology,* 97:1398–1407.
12. Feil, P. D., Glasser, S. R., Toft, D. O., and O'Malley, B. W. (1972): Progesterone binding in the mouse and rat uterus. *Endocrinology,* 91:738–746.
13. Freifeld, M., Feil, P. D., and Bardin, C. W. (1974): The in vivo regulation of the progesterone "receptor" in guinea pig uterus: dependence on estrogen and progesterone. *Steroids,* 23:93–103.
14. Hseuh, A. J. W., Peck, E. J., and Clark, J. H. (1974): Receptor progesterone complex in the nuclear fraction of the rat uterus: demonstration by ^{3}H-progesterone exchange. *Steroids* 24:599–611.
15. Kontula, K., Jänne, O., Vihko, R., de Jager, E., de Visser, J., and Zeelen, F. (1975): Progesterone-binding proteins: in vitro binding and biological activity of different steroidal ligands. *Acta Endocrinol. (Kbh.),* 78:574–592.
16. Leavitt, W. W., and Blaha, G. C. (1972): An estrogen-stimulated, progesterone-binding system in the hamster uterus and vagina. *Steroids,* 19:263–274.
17. Leavitt, W. W., Toft, D. O., Strott, C. A., and O'Malley, B. W. (1974): A specific progesterone receptor in the hamster uterus: physiologic properties and regulation during the estrous cycle. *Endocrinology,* 94:1041–1053.
18. Luu Thi, M. T., Baulieu, E. E. and Milgrom, E. (1975): Comparison of the characteristics and of the hormonal control of endometrial and myometrial progesterone receptors. *J. Endocrinol.,* 66:349–356.
19. McGuire, J. L., and Bariso, C. D. (1972): Isolation and preliminary characterization of a progestogen specific binding macromolecule from the 273,000 g supernatant of rat and rabbit uteri. *Endocrinology,* 90:496–506.
20. McGuire, J. L., Bariso, C. D., and Shroff, A. P. (1974): Interaction between steroids and a uterine progestogen specific binding macromolecule. *Biochemistry,* 13:319–322.
21. Milgrom, E., Atger, M., and Baulieu, E.-E. (1970): Progesterone in uterus and plasma:IV. Progesterone receptor(s) in guinea pig uterus cytosol. *Steroids,* 16:741–764.
22. Milgrom, E., Atger, M., and Baulieu, E.-E. (1972): Progesterone in uterus and plasma: V. An assay of the progesterone cytosol receptor of the guinea pig uterus. *Endocrinology,* 90:1064–1070.
23. Milgrom, E., Atger, M., Perrot, M., and Baulieu, E.-E. (1972): Progesterone in uterus and plasma: VI. Uterine progesterone receptors during the estrus cycle and implantation in the guinea pig. *Endocrinology,* 90:1071–1078.
24. Milgrom, E., and Baulieu, E.-E. (1970a): Progesterone in uterus and plasma: I. Binding in rat uterus 105,000 g supernatant. *Endocrinology,* 87:276–287.
25. Milgrom, E., and Baulieu, E.-E. (1970b): Progesterone in the uterus and plasma: II. The role of hormone availability and metabolism on selective binding to uterus protein. *Biochem. Biophys. Res. Commun.,* 40:723–730.
26. Milgrom, E., Luu Thi, M., Atger, M., and Baulieu, E.-E. (1973): Mechanisms regulating

the concentration and the conformation of progesterone receptor(s) in the uterus. *J. Biol. Chem.*, 248:6366–6374.
27. Milgrom, E., Luu Thi, M., and Baulieu, E.-E. (1973): Control mechanisms of steroid hormone receptors in the reproductive tract. *Acta Endocrinol. (Kbh.) [Suppl.]*, 180:380–397.
28. Philibert, D., and Raynaud, J. P. (1973): Progesterone binding in the immature mouse and rat uterus. *Steroids*, 22:89–98.
29. Philibert, D., and Raynaud, J. P. (1974a): Binding of progesterone and R5020, a highly potent progestin, to human endometrium and myometrium. *Contraception*, 10:457–466.
30. Philibert, D., and Raynaud, J. P. (1974b): Progesterone binding in the immature rabbit and guinea pig uterus. *Endocrinology*, 94:627–632.
31. Rao, B. R., Wiest, W. G., and Allen, W. M. (1973): Progesterone "receptor" in rabbit uterus. I. Characterization and estradiol-17β augmentation. *Endocrinology*, 92:1229–1240.
32. Reel, J. R., and Shih, Y. (1975): Oestrogen-inducible uterine progesterone receptors: Characteristics in the ovariectomized immature and adult hamster. *Acta Endocrinol. (Kbh.)*, 80:344–354.
33. Reel, J. R., Van Dewark, S. D., Shih, Y., and Callantine, M. R. (1971): Macromolecular binding and metabolism of progesterone in the decidual and pseudo-pregnant rat and rabbit uterus. *Steroids*, 18:441–461.
34. Scatchard, G. (1949): The attractions of proteins for small molecules and ions. *Ann. N.Y. Acad. Sci.*, 51:660–672.
35. Terenius, L. (1972): Specific progestogen binder in uterus of normal rats. *Steroids*, 19:787–794.
36. Terenius, L. (1974): Affinities of progesterone and estrogen receptors in rabbit uterus for synthetic progestogens. *Steroids*, 23:909–919.

Progesterone Receptors in Normal and Neoplastic Tissues, edited by W. L. McGuire et al. Raven Press, New York © 1977.

Measurement of the Progesterone Receptor in Human Endometrium Using Progesterone and R5020

F. Bayard, B. Kreitmann, and B. Derache

Laboratoire d'Endocrinologie Expérimentale, C. H. U. de Rangueil, Toulouse, 31052 Cedex, France

The presence of a specific progesterone receptor in the human endometrial as well as myometrial tissue is now well established (1–6). The detection of this receptor and estimation of its physicochemical binding parameters and of the concentration of binding sites have been carried out using either progesterone (2,5) or synthetic progestins and particularly R5020 (3,4). The use of progesterone is complicated by the constant contamination of the tissular cytosol by the cortisol binding globulin (CBG), which binds progesterone with high affinity (7), and by a high dissociation rate of the progesterone receptor complexes (8). The interest in R5020 appeared to reside mainly in its lack of binding to CBG and in its pronounced affinity for the progesterone receptor (4,9). However, no systematic study has ever been carried out on the comparative analysis of the progesterone receptor in human endometrium comparing the two methods. It is the purpose of the present chapter to report such a study.

MATERIALS AND METHODS

Steroids

Radioinert and tritiated R5020 (specific activity 51.4 Ci/mmole, ^{3}H-R5020) were a generous gift from Roussel-Uclaf Laboratories.

Tritiated progesterone (specific activity 96 Ci/mmole ^{3}H-progesterone) was obtained from the Radiochemical Center, Amersham, England. All these steroids were kept in benzene-ethanol (85/15, vol/vol) at 4°C and checked regularly for purity by thin-layer chromatography. Nonradioactive cortisol and progesterone were obtained from Sigma Chemical Company (St. Louis, Mo.) and not further purified. All solvents used were obtained from Merck and were Uvasol grade.

Preparation of Tissue Samples

Endometrial curettages or biopsies were washed in ice cold saline to remove mucus and blood and processed immediately as described elsewere (6,8). Briefly, homogenization was carried out at 0°C using a Teflon-glass homogenizer in 2 volumes (ml/g of tissue) of cold 0.02 M Tris-HCl buffer, pH 7.8, containing 1 mM $MgCl_2$ (Merck), 1 mM dithiothreitol (Sigma), and 0.25 M sucrose (Merck) (buffer A). The volume of the homogenate was accurately measured and an aliquot was taken for DNA determination. The homogenate was centrifuged at 600 g for 10 min at 0 to 4°C. The 600 g supernatant, again accurately measured, was diluted one to two with ice cold 0.02 M Tris-HCl buffer, pH 7.8, containing 3 mM EDTA and 1 mM dithiothreitol (buffer B), centrifuged at 105,000 g for 1 hr at 4°C, and the supernatant (cytosol) collected.

Incubations

Fractions of 0.1 ml cytosol were incubated at 0°C for variable periods of time (precised below) in presence of various concentrations of radioactive (0.25 to 20 nM) and radioinert (1 to 3 μM) steroids and varying concentrations of glycerol. An excess of 100-fold cortisol was added in the experiments studying binding of the labeled progesterone.

Binding Measurements

The free or loosely bound hormone was removed using an equal volume of an ice cold dextran-coated charcoal solution (0.05/0.5 g%, wt/vol, in buffer B containing 30 to 60% glycerol) for 30 min at 0°C. After centrifugation at 3,500 g for 5 min, the radioactivity was measured in the supernatant.

The nonspecific low-affinity binding, measured in presence of a 100-fold excess of nonradioactive hormone, was subtracted from the total binding. The number of cytoplasmic receptor sites was expressed in femto-moles per milligram of deoxyribonucleic acid in the homogenate.

RESULTS

Association Rate Constant

The rate of association (K_a) was determined using the method described in detail by Best-Belpomme et al. (10). In absence of glycerol in the incubation medium (but in presence of 30% glycerol during the dextran-coated charcoal treatment), the second-order rate constant for the association of progesterone, measured between 0 and 2 min, when the dissociation rate can still be neglected, was slightly lower (2.5 × 10^5 M^{-1} s^{-1}) than the association rate constant (2.8 × 10^5 M^{-1} s^{-1}) for R5020 (Fig. 1). In

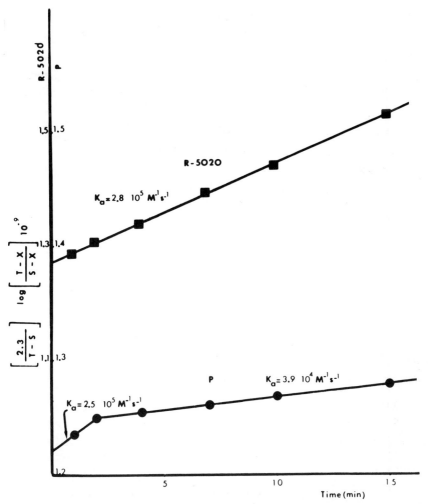

FIG. 1. Association rate constant determination between the progesterone receptor and progesterone or R5020 in absence of glycerol (protein concentration = 1.5 mg/ml).

another series of experiments, the incubations of ^{3}H-R5020 were conducted in absence or in presence of 30% glycerol. The association rate constant was 4.4×10^4 M^{-1} s^{-1} in absence and 1.11×10^5 M^{-1} s^{-1} in presence of 30% glycerol in the incubation medium (Fig. 2). It can be noted that the association rate constant decreased when the protein concentration of the cytosol increased (1.5 mg/ml in the first experiment and 3.2 mg/ml in the second experiment).

Dissociation Rate Constant

The rate of dissociation (K_d) was estimated by adding a 1,000-fold molar excess of unlabeled steroid to tubes containing aliquots of the cytosol which

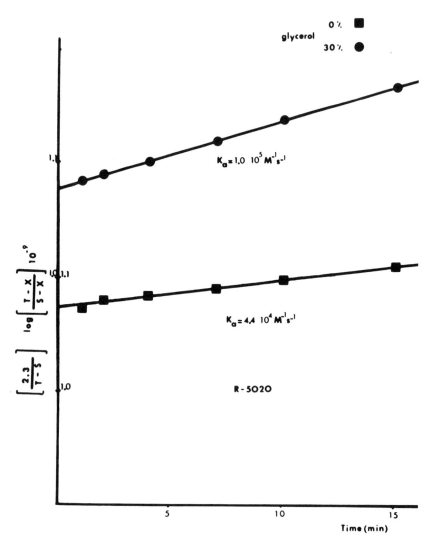

FIG. 2. Effect of glycerol on the association rate constant between the progesterone receptor and R5020 (protein concentration = 3.2 mg/ml).

had been labeled previously with 1×10^{-9} M ^{3}H-steroid. At various times after the addition of the unlabeled steroid, the receptor-bound ^{3}H-steroid was determined by the dextran charcoal assay. Semilogarithmic plots of bound labeled hormone versus time present two parts in the time interval considered (Fig. 3). Between 0 and 60 min the dissociation rate constant of the pseudo-first-order dissociation reaction was determined from the slope of the curve. For progesterone, this dissociation rate constant was 1.3×10^{-3} s^{-1} in absence of glycerol in the incubation medium (but in presence of 30% glycerol during the dextran charcoal treatment), 7.4×10^{-4} s^{-1} in presence of 15% glycerol, and 3.1×10^{-4} s^{-1} in presence of 30% glycerol in the incubation medium (*data not shown*). These data will be published in more detail elsewhere (8). For R5020, the dissociation rate constant was 5.9×10^{-4} s^{-1} in absence of glycerol, and 2.3×10^{-4} s^{-1} and 1.4×10^{-4} s^{-1} in presence of 15 and 30% glycerol in the incubation

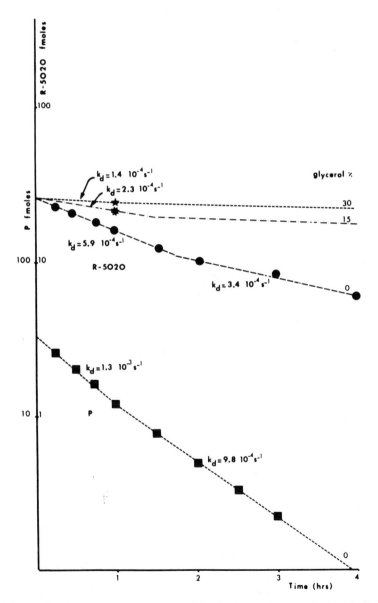

FIG. 3. Dissociation rate constant determination of the progesterone-receptor complexes and the R5020-receptor complexes in presence or absence of glycerol.

medium. These decay curves represent dissociation rather than binding site denaturation since preparations containing no added unlabeled hormone showed no loss of bound hormone for at least 4 hr in presence or absence of glycerol.

Scatchard Plot Estimation (11) of Equilibrium Constant of Dissociation and Binding Site Concentration

In a first experiment, both labeled progesterone and R5020 were incubated in absence of glycerol for 4 hr at 0 to 4°C and bound and free fractions were

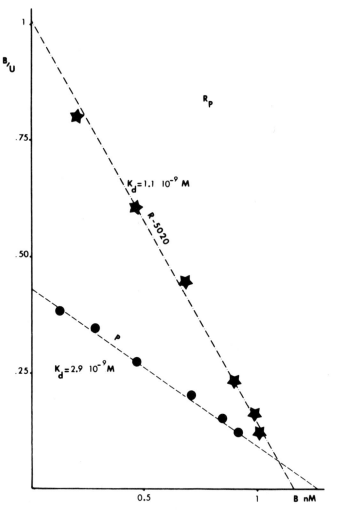

FIG. 4. Estimation of the binding equilibrium constants of the progesterone receptor for progesterone and R5020.

separated using a dextran charcoal solution in presence of 30% glycerol. As shown in Fig. 4, R5020 binds to endometrial cytosol with a higher affinity than progesterone, but the number of binding sites is the same. It has been repeatedly observed that a concentration of 20×10^{-9} M of labeled steroid

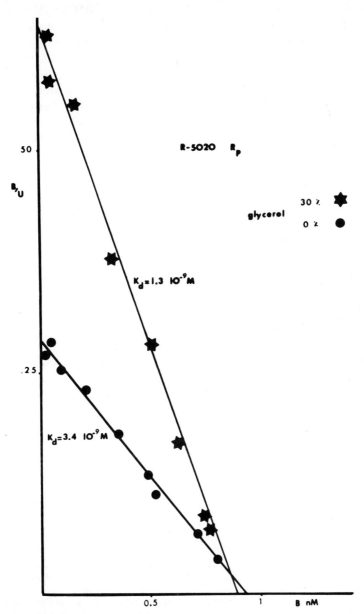

FIG. 5. Effect of glycerol on the binding equilibrium constant of the progesterone receptor using R5020.

in the incubation medium was a saturating concentration for the measurement of the receptor sites in the conditions of our study.

In a second experiment, the effect of the presence or absence of glycerol in the incubation medium on the equilibrium constant of dissociation of the hormone-receptor complexes was studied using R5020. A cytosol containing no endogenous progesterone measurable by radioimmunoassay, to prevent variation of the equilibrium constant of dissociation due to the process of exchange (12), was used. Incubations were carried out for 4 hr in a medium containing 0 or 30% glycerol. Bound and free fractions were separated using protamine sulfate precipitation of the progesterone receptor complexes. Both types of pellets were then washed with buffer B containing 10% glycerol. The results are presented in Fig. 5. The equilibrium dissociation constant was 3.4×10^{-9} M in absence and 1.3×10^{-9} M in presence of glycerol.

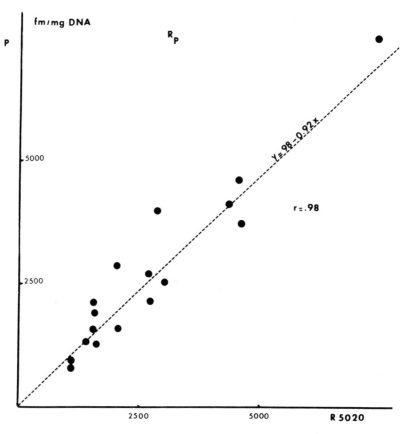

FIG. 6. Comparison of the concentration of progesterone receptors in endometrium measured using ^{3}H-progesterone or ^{3}H-R5020. (Measurements using ^{3}H-progesterone are corrected for the isotopic dilution by endogenous progesterone. Measurements using ^{3}H-R5020 are not corrected.)

Comparative Measurements of the Progesterone Receptor Content of Endometrial Cytosol

Using saturating concentrations of labeled hormones and conditions of complete exchange between the labeled and the endogenous hormone (8), we measured the cytosolic content of progesterone receptor in 17 endometrial samples taken by biopsies from normal women at different times during the menstrual cycle. When labeled progesterone was used for this measurement, an isotopic dilution by the endogenous hormone had to be considered (8). When the data obtained in these conditions were compared with those obtained on the same samples using labeled R5020 without consideration of any isotopic dilution, a good agreement was found between the two series of measurements ($r = 0.98$) as shown in Fig. 6.

DISCUSSION

From the rate constant data, it appears that the equilibrium constants for dissociation reactions calculated as the ratio K_d/K_a are in the range of the equilibrium constant estimated by analysis of saturation binding data by the method of Scatchard: 5.2×10^{-9} M for progesterone, 2.1×10^{-9} M for R5020. The higher affinity of R5020 for the progesterone receptor appears to result mainly from the slower dissociation rate of the R5020-receptor complexes. Finally, glycerol increases the association rate constant and decreases the dissociation rate constant, so that the calculated equilibrium constant for dissociation reactions decreases and is again in good agreement with the equilibrium constant estimated by saturation analysis.

When R5020 is used for the measurement of the progesterone receptor concentration in human endometrial tissues, it offers three advantages. It does not bind to CBG, and the slow dissociation rate of the R5020-receptor complexes allows the practical use of dextran-coated charcoal to separate bound and free fractions provided there is a minimum of 15% glycerol in the incubation buffer at the time of separation. This slow rate of dissociation will have to be considered when the progesterone receptor concentration is measured in tissue samples of patients treated with this drug. The usual conditions of exchange defined when progesterone is the only circulating progestin (8) will have to be corrected accordingly. Finally, when working with tissues of human origin, investigators should consider the endogenous progesterone since it inhibits to a variable extent the binding of the labeled hormone used for the measurement of the progesterone receptor concentrations. The degree of inhibition depends on the concentration of the labeled hormone (S) used for the measurement, the concentration of the endogenous progesterone (I), and the respective equilibrium constant of dissociation of the labeled hormone-receptor complexes (K_d) and of the endogenous progesterone-receptor complexes (K_I). The percentage of inhibition (13) is described by the equation:

$$\% \text{ inhibition} = 1 - \frac{K_d + (S)}{K_d \text{ app.} + (S)},$$

with $K_d \text{ app.} = K_d \left(1 + \frac{(I)}{K_I}\right).$

Using the respective values for K_d and K_I determined in Fig. 4, and for a concentration of the labeled hormone of 20×10^{-9} M, one can evaluate the degree of inhibition when progesterone or R5020 is used for the measurement of the progesterone receptor concentration:

$$\% \text{ inhibition} = 1 - \frac{22.9 \times 10^{-9}}{22.9 \times 10^{-9} + (I)}$$

when using labeled progesterone;

$$\% \text{ inhibition} = 1 - \frac{21.1 \times 10^{-9}}{21.1 \times 10^{-9} + 0.38 \, (I)}$$

when using labeled R5020.

Using a radioimmunoassay, we have measured the concentration of endogenous progesterone in 0.1 ml (volume of the aliquot used for the progesterone receptor measurement) of endometrial cytosol of 42 samples taken from normal women at different stages of their menstrual cycle or at 6 to 8 weeks gestation (voluntary abortion). These concentrations were compared with the corresponding plasma concentrations. It has been observed that there is a linear relationship between the plasma and the cytosolic concentrations, an observation already reported (14). Moreover, for most of the endometrial samples taken during the menstrual cycle, the amount of progesterone measured was less than 1 pmole in the aliquot of cytosol, which means that the concentration of the endogenous hormone was usually in the range of 0 to 10×10^{-9} M. In these conditions, it results from the above equations that the underestimation of the total number of progesterone receptor binding sites would vary from 0 to 30% using labeled progesterone but only from 0 to 15% using labeled R5020. This is probably the explanation for the good correlation coefficient observed between the measurements obtained using labeled progesterone and a correction for the isotopic dilution and those obtained using labeled R5020 without any correction. It appears, then, that R5020 has a net advantage over progesterone for the measurement of the progesterone receptor concentration in tissues taken from postmenopausal patients and most premenopausal women at different stages of their menstrual cycle. During pregnancy, however, the interference by the endogenous hormone cannot be ignored even when using R5020.

ACKNOWLEDGMENT

This work was supported by Grant No. 741–406 AU from the Institut National de la Santé et de la Recherche Médicale.

REFERENCES

1. Wiest, W. G., and Rao, B. R. (1971): Progesterone binding proteins in rabbit uterus and human endometrium. In: *Advances in the Biosciences, Vol. 7*, edited by G. Raspe. Pergamon Press, Oxford.
2. Kontula, K., Jänne, O., Luukkainen, T., and Vihko, R. (1973): Progesterone-binding protein in human myometrium. *Biochem. Biophys. Acta*, 328:145–153.
3. Murugesan, K., and Laumas, K. R. (1973): Binding of 6,7 ³H-norethynodrel to the rat uterine cytosol and to human endometrium and myometrium. *Contraception*, 8:451–470.
4. Philibert, D., and Raynaud, J. P. (1974): Binding of progesterone and R5020, a highly potent progestin, to human endometrium and myometrium. *Contraception*, 10:457–466.
5. Haukkamaa, M., and Luukkainen, T. (1974): The cytoplasmic progesterone receptor of human endometrium during the menstrual cycle. *J. Steroid Biochem.*, 5:447–452.
6. Bayard, F., Damilano, S., Robel, P., Magnier, J. C., and Baulieu, E. E. (1976): The concentration of estradiol and progesterone "receptors" in normal endometrium during the menstrual cycle in human. In: *Research on Steroids, Vol. 7*, edited by A. Vermeulen. Elsevier, Amsterdam (*in press*).
7. Westphal, U. (1971): Steroid protein interactions. In: *Monographs on Endocrinology, Vol. 4*. Springer-Verlag, Berlin.
8. Bayard, F., Damilano, S., Robel, P., and Baulieu, E. E. (1976): *In preparation*.
9. Philibert, D., and Raynaud, J. P. (1973): Progesterone binding in the immature mouse and rat uterus. *Steroids*, 22:89–98.
10. Best-Belpomme, M., Fries, J., and Erdos, T. (1970): Interactions entre l'oestradiol et des sites récepteurs utérins. Données cinétiques et d'équilibre. *Eur. J. Biochem.*, 17:425–432.
11. Scatchard, G. (1949): The attractions of proteins for small molecules and ions. *Ann. N.Y. Acad. Sci.*, 51:660–672.
12. Milgrom, E., Perrot, M., Atger, M., and Baulieu, E. E. (1972): Progesterone in uterus and plasma: V. An assay of the progesterone cytosol "receptor" of the guinea pig uterus. *Endocrinology*, 90:1064–1070.
13. Segel, I. H. (1968): *Biochemical Calculations*. Wiley, New York.
14. Bayard, F., Louvet, J. P., Monroziès, M., Boulard, A., and Pontonnier, G. (1975): Endometrial progesterone concentrations during the menstrual cycle. *J. Clin. Endocrinol. Metab.*, 41:412–414.

Progestin Receptor in Endometrial Carcinoma

Jan-Åke Gustafsson, * Nina Einhorn, Gunilla Elfström,
* Bo Nordenskjöld, and Örjan Wrange

*Department of Chemistry, Karolinska Institutet, and
* Radiumhemmet, Karolinska Sjukhuset, S-104 01 Stockholm 60, Sweden*

Administration of progestins constitutes the primary choice of therapy in cases of advanced endometrial adenocarcinoma no longer within the reach of radiotherapy or surgery (2). It is now generally agreed that a temporary or sometimes complete remission from tumor disease will occur in about one-third of patients treated with progestins (13). Furthermore, histologically well-differentiated tumors are more responsive to progestin therapy than poorly differentiated ones (7,19). However, objective remissions have been reported in patients with anaplastic tumors (1), and at the present time it is not possible to predict with certainty from clinical or histological data whether a patient will respond to progestin therapy.

Nordqvist (13) has developed a method to study effects of steroids *in vitro* on rate of DNA synthesis in specimens of endometrial carcinoma, and he has suggested that this method might be of value in predicting a given patient's response to progestin therapy. Another way to create a predictive test for evaluation of responsiveness to progestin therapy could be to analyze progestin receptor contents in tumor specimens. This is an obvious possibility since several groups of investigators have shown that the presence of an appreciable concentration of cytoplasmic estrogen receptor in breast cancer is predictive of a response to estrogen therapy (for references, see 11). Several laboratories have reported on the occurrence of specific cytoplasmic progesterone receptors in normal human endometrium (12,15,21). Although initial studies indicated the absence of progesterone receptors in endometrial adenocarcinoma (18,21), subsequent investigations have unequivocally demonstrated that some endometrial carcinomas retain their ability to bind progesterone (9,17,22).

The possibility to measure progestin receptors conveniently without interference from corticosteroid binding globulin that also binds progesterone with high affinity and low capacity has increased considerably since Philibert and Raynaud (14) synthesized the highly active progestin $17\alpha,21$-dimethyl-19-nor-4,9-pregnadiene-3,20-dione (R5020) in tritium-labeled form. This steroid binds to the progestin receptor in immature mouse, rat, rabbit, and guinea pig uterus and to human endometrium and myometrium but not to corticosteroid binding globulin (14–16). In the present investigation we have used tritium-labeled R5020 to measure the progestin receptor concentration in cytosol from tumor specimens from patients with endometrial adenocarcinoma. We have also investigated the characteristics of the R5020-receptor complex in cytosol from normal human endometrial tissue in order to define the right conditions for measuring the complex in tumor tissue. The method used for receptor quantitation has been a micromodification of isoelectric focusing that we have previously found to be a versatile, accurate, and sensitive method for quantitation of estradiol receptor in breast cancer (20).

MATERIALS AND METHODS

Steroids

$17\alpha,21$-Dimethyl-19-nor-[6,7-^{3}H]pregna-4,9-diene-3,20-dione (R5020) (specific radioactivity, 51 Ci/mmole) as well as unlabeled R5020 were generously supplied by Dr. J. P. Raynaud, Roussel-Uclaf, Romainville, France. Unlabeled cortisol, progesterone, testosterone, and estradiol were kind gifts from Dr. J. Babcock, Upjohn Co., Kalamazoo, Michigan. Unlabeled dexamethasone was purchased from Sigma Chemical Co., St. Louis, Missouri.

Labeling of Progestin Receptor in Normal Endometrial Tissue

Normal endometrial tissue was obtained by hysterectomy. Four grams of tissue were minced and homogenized in 20 ml of 10 mM Tris-HCl, pH 7.4, 10 mM KCl, 1.5 mM EDTA, 12 mM thioglycerol, 10% (vol/vol) glycerol. The tissue was first homogenized using a few strokes with an Ultra-Turrax homogenizer (Janke und Kunkel, Staufen i. Br., W. Germany) and then using a Potter-Elvehjem homogenizer. All manipulations were carried out at 4°C, and during homogenization the sample was cooled with an ice jacket. The cytosol was prepared by centrifuging the homogenate at 43,000 rpm for 45 min in an SW 50.1 rotor in an L3–50 Beckman ultracentrifuge. Labeling of the progestin receptor was routinely performed by incubating 1-ml aliquots of the cytosol with 10 nM [^{3}H]R5020 in the presence or absence of 1,000 nM unlabeled R5020 for 2 hr at 0°C. Free and protein-bound radioactivity were separated by treatment with dextran-coated charcoal. The cytosol sample was thoroughly mixed with 0.3 ml of a suspension of dextran-coated charcoal [3.8% (wt/vol) Norit A and 0.38% (wt/vol) dextran T 500 in homogenization buffer (see above)], incubated for 10 min at 4°C, and centrifuged at 3,200 × g for 10 min. The supernatant was removed and treated with dextran-coated charcoal a second time according to the same procedure. Aliquots of the cytosol before and after treatment with dextran-coated charcoal were taken for measurement of radioactivity (see below). Aliquots of the labeled cytosol after treatment with dextran-coated charcoal were also taken for sucrose gradient analysis and isoelectric focusing (see below). Protein determinations were carried out according to Lowry et al. (8).

Labeling of Progestin Receptor in Endometrial Cancer Specimens

Carcinomatous endometrial tissue was obtained from 11 patients by diagnostic curettage and simultaneously submitted for analysis of progestin receptor contents and for histological examination. Endometrial carcinomas were graded I through III (5). The grade I endometrial adenocarcinomas were well organized and were either papillary or adenomatous. The papillary cords might be covered with a single layer or a few layers of uniform cells with little if any polymorphism. Adenomatous tumors were distinguished by slightly irregular structures, often tightly packed, with a minimum of stroma. The grade II endometrial adenocarcinomas were adenomatous and less well organized than grade I tumors, and solid areas were common. The epithelial proliferations were irregular and there was obvious polymorphism. The grade III endometrial adenocarcinomas were predominantly solid with more or less marked polymorphism and numerous mitotic figures.

Endometrial cancer tissue (0.04 to 0.50 g) was homogenized in 2 ml of homogenization buffer as described above. Cytosol was prepared by centrifuging the homogenate in an SW 40 rotor at 40,000 rpm for 70 min. Labeling of cytosol receptor and separation of free and protein-bound radioactivity were performed as described above. After aliquots were removed for scintillation counting (see below), the two radioactive samples (labeled with [^{3}H]R5020 in the presence or absence of 100-fold excess of unlabeled ligand) were analyzed by isoelectric focusing.

Isoelectric Focusing

Isoelectric focusing was performed essentially as described by Katsumata and Goldman (6). Electrofocusing was performed in 6-ml columns stabilized by a 15 to 60% (wt/vol) sucrose gradient using 1% (vol/vol) Ampholine® of pH 3.5 to 10 (LKB-Produkter, Stockholm, Sweden). A maximum of 0.3 W was applied for 16 hr at 4°C. The gradient was fractionated into plastic counting vials and pH was measured at ambient temperature using a surface electrode (type 403–30, Ingold, Zurich, Switzerland). Contaminating hemoglobin was used as internal pH standard and had a reproducible pI of 7.6.

Sucrose Density Gradient Centrifugation

Next, 0.2-ml samples were layered on 5-ml linear gradients of 5 to 20% (wt/vol) sucrose in homogenization buffer (see above). Centrifugation was performed at 45,000 rpm for 16 hr at 4°C in an SW 50.1 rotor. After centrifugation the gradients were fractionated using a Metrohm pisan burette E 274 (Metrohm AG, CH-9100 Herisan, Switzerland). Bovine serum albumin (4.6S) was used as standard.

Radioactivity Measurement

Radioactivity was measured in an Intertechnique SL 30 liquid scintillation spectrometer. Then 5 ml of Instagel® (Packard Instrument Co., Inc., Warrenville, Downess Grove, Ill.) was added to each sample for counting. Quenching measurements were carried out using the external standard technique.

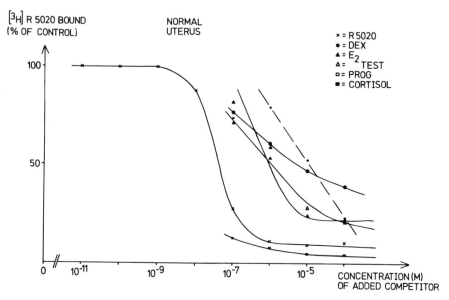

FIG. 1. Competition between different ligands for binding to R5020 binding sites in normal human endometrial cytosol. [^{3}H]R5020, 1 nM, was incubated at 0°C for 2 hr with cytosol (2 mg/ml protein) from normal human endometrial tissue in the presence of increasing concentrations of unlabeled R5020, progesterone, testosterone, estradiol, cortisol, or dexamethasone. Protein-bound [^{3}H]R5020 was measured as the radioactivity resistant toward treatment with dextran-coated charcoal (see Materials and Methods).

RESULTS

Characterization of the R5020 Binding Component in Human Endometrial Cytosol

Human endometrial cytosol contained a limited number of binding sites for R5020 (Fig. 1). Competition experiments with increasing concentrations of unlabeled ligands other than R5020 showed that progesterone

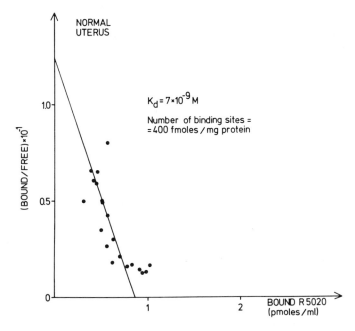

FIG. 2. Scatchard plot of [^{3}H]R5020 binding in cytosol from normal human endometrial tissue. Protein-bound radioactivity was determined after treatment of incubated cytosol with dextran-coated charcoal. The Scatchard plot was corrected for nonspecific binding as indicated by Chamness and McGuire (3). A K_d of 7×10^{-9} M and a number of binding sites of 400 fmoles/mg of cytosol protein were calculated.

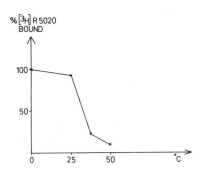

FIG. 3. Heat inactivation of the R5020-receptor complex in normal human endometrial cytosol. Cytosol from normal human endometrium (2 mg of protein per ml) was labeled with [^{3}H]R5020 by incubation for 2 hr at 0°C, and the formed [^{3}H]R5020-receptor complex was incubated for 10 min at 0°, 25°, 37°, and 50°C. Protein binding was assayed by treatment with dextran-coated charcoal and the results are expressed as a percentage of the control incubation.

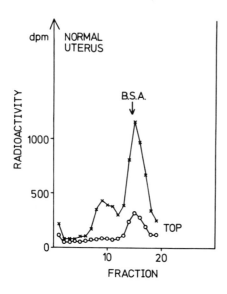

FIG. 4. Sucrose density gradient centrifugation analysis of the [^{3}H]R5020-receptor complex in normal human endometrial cytosol in the absence or presence of a 100-fold excess of unlabeled R5020. The cytosol (2 mg of protein per ml) was incubated at 0°C for 2 hr with 1 nM [^{3}H]R5020 in the absence or presence of 100 nM unlabeled R5020. The labeled cytosol samples were treated with dextran-coated charcoal prior to sucrose density gradient centrifugation. For further experimental details, see Materials and Methods.

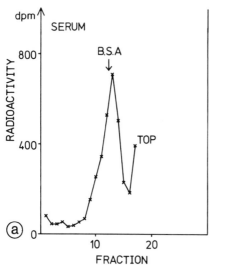
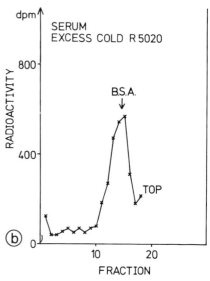

FIG. 5. Sucrose density gradient centrifugation analysis of [^{3}H]R5020 binding components in serum pooled from three patients with endometrial adenocarcinoma in the absence **(a)** or presence **(b)** of a 100-fold excess of unlabeled R5020. Serum samples from patients W. O., I. P., and B. N. (see Table 1) were pooled. Then 1-ml aliquots of the pool (7 mg of protein per ml) were incubated at 0°C for 2 hr with 1 nM [^{3}H]R5020 in the absence or presence of 100 nM unlabeled R5020. The labeled serum samples were treated with dextran-coated charcoal prior to sucrose density gradient centrifugation. For further experimental details, see Materials and Methods.

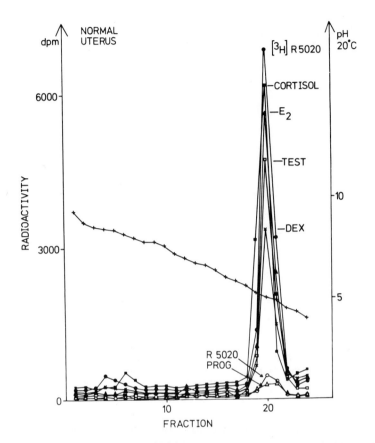

FIG. 6. Isoelectric focusing of the [^{3}H]R5020-receptor complex in normal human endometrial cytosol in the absence or presence of a 100-fold excess of R5020, progesterone, dexamethasone, testosterone, estradiol (E$_2$), or cortisol. Some 1-ml aliquots of cytosol (2 mg of protein per ml) were incubated at 0°C for 2 hr with 1 nM [^{3}H]R5020 in the absence or presence of 100 nM unlabeled R5020, progesterone, dexamethasone, testosterone, estradiol, or cortisol. The labeled cytosol samples were treated with dextran-coated charcoal prior to isoelectric focusing. Electrofocusing was carried out using a sucrose density gradient and Ampholine® with a pH range of 3.5 to 10. +−+−+, pH. For further experimental details, see Materials and Methods.

displaced [^{3}H]R5020 efficiently from its binding sites in the cytosol, whereas testosterone and estradiol were less-efficient competitors and dexamethasone and cortisol the least-efficient competitors of the steroids tested (Fig. 1). Scatchard analysis (Fig. 2) indicated a dissociation constant of the R5020-receptor complex of about 7×10^{-9} M and a number of binding sites of about 400 fmoles/mg of cytosol protein.

The R5020-receptor complex was stable during incubation for 10 min at 25°C but not at 37°C (Fig. 3). Sucrose density gradient analysis showed that the R5020-receptor complex sedimented as a major 4S and a smaller 8S peak (Fig. 4). Only the latter peak was completely eliminated following incubation of [^{3}H]R5020 in the presence of a 100-fold excess of un-

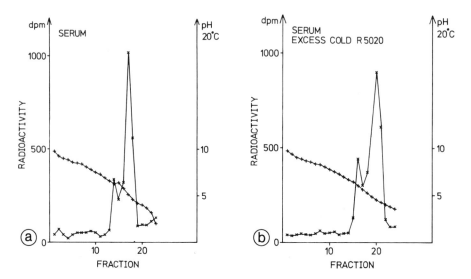

FIG. 7. Isoelectric focusing of [³H]R5020 binding components in serum pooled from three patients with endometrial adenocarcinoma in the absence (a) or presence (b) of a 100-fold excess of unlabeled R5020. Serum samples from patients W. O., I. P., and B. N. (see Table 1) were pooled. Then 1-ml aliquots of the pool (7 mg of protein per ml) were incubated at 0°C for 2 hr with 1 nM [³H]R5020 in the absence or presence of 100 nM unlabeled R5020. The labeled serum samples were treated with dextran-coated charcoal prior to electrofocusing. Isoelectric focusing was carried out using a sucrose density gradient and Ampholine ® with a pH range of 3.5 to 10. +−+−+, pH. For further experimental details, see Materials and Methods.

labeled R5020 (Fig. 4). In order to investigate whether contaminating serum proteins could contribute to the binding of R5020 in endometrial cytosol, we labeled with [³H]R5020 serum samples pooled from three patients according to the same procedure as described above for endometrial cytosol. Sucrose gradient analysis showed that the R5020-serum protein complex had a sedimentation coefficient of about 4S (Fig. 5a). However, it was not possible to displace [³H]R5020 from binding sites on serum proteins by addition of a 100-fold excess of unlabeled R5020 during incubation (Fig. 5b). These results indicate that the binding sites for R5020 in serum have a high capacity and are different from the high-affinity, low-capacity binding sites in endometrial cytosol.

Isoelectric focusing of the R5020-receptor complex showed that the complex had an isoelectric point of about 5.0 (Fig. 6). After unlabeled R5020 or progesterone was added during the incubation of cytosol with [³H]R5020, the radioactive peak of pI 5.0 was almost totally eliminated (Fig. 6). Dexamethasone, testosterone, estradiol, and cortisol were much less efficient competitors. The [³H]R5020-serum protein complex was also focused at pH 5.0 (Fig. 7a) but was not displaceable by an excess of unlabeled R5020 (Fig. 7b).

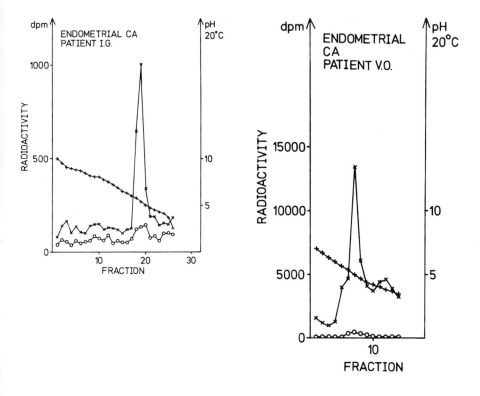
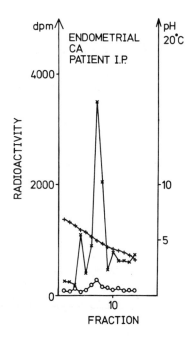

FIG. 8. Isoelectric focusing of [^{3}H]R5020-receptor complex in specimens of endometrial adenocarcinoma. Cytosol from carcinomatous tissue was labeled with 1 nM [^{3}H]R5020 in the absence (x - x - x) or presence (o - o - o) of 100 nM unlabeled R5020 by incubation at 0°C for 2 hr. The samples were treated with dextran-coated charcoal and analyzed by isoelectric focusing. Electrofocusing was carried out using a sucrose density gradient and Ampholine® with a pH range of 3.5 to 10. +−+−+, pH. For further experimental details, see text.

Binding of R5020 in Human Endometrial Carcinomatous Tissue

Since the method of isoelectric focusing allowed application of larger sample volumes than the method of sucrose density gradient analysis, and since, accordingly, isoelectric focusing under the conditions used was a more sensitive technique for detecting the [^{3}H]R5020-receptor complex, this method was chosen as the routine method for demonstration and quantitation of the progestin receptor in endometrial cancer specimens.

The receptor-positive specimens contained [^{3}H]R5020-receptor complexes focusing at about pH 5.0, although in some cases slightly higher pI values were observed (Fig. 8). The specific R5020 binding sites were calculated as the areas under the pI 5 peak that were displaceable with a 100-fold excess of unlabeled R5020. The results are summarized in Table 1. The range of R5020 binding sites in the specimens of endometrial adenocarcinoma ranged from an undetectable level up to 62 fmoles/mg of protein. The number of binding sites in the two specimens of grade III tumors (patients G. S. and E. C.) did not exceed 10 fmoles/mg protein, whereas all three specimens of grade I tumors (patients W. O., G. A., and B. F.) had more than 10 fmoles of binding sites per milligram of protein. On the other hand, the number of binding sites in grade II tumors showed great variation and therefore there was no absolute correlation between histological type of tumor and number of R5020 binding sites.

TABLE 1. *Isoelectric focusing analysis of progestin receptor in 11 specimens of endometrial adenocarcinoma, classified with respect to degree of differentiation and histological type*

Patient	Age (yr)	Weight (g) of specimen	No. of specific R5020 binding sites (fmoles/mg protein)	Pathology Degree of differentiation	Grade
W. O.	86	0.37	62	Relatively high	I,II
G. A.	51	0.04	23	High	I
A. H.	70	0.50	19	Suspected carcinosarcoma	II
I. P.	75	0.50	15	Medium	II
B. F.	58	0.18	14	High-medium	I
B. N.	56	0.20	9.8	Medium	II
G. S.	77	0.04	8.7	Relatively low	III
E. C.	69	0.50	1.8	Low, anaplastic	III
K. M.	53	0.10	1.4	High-medium	II
I. K.	58	0.09	<1	Medium	II
R. L.	69	0.01	<1	Medium	II

In all cases the tumor tissue was stored at −70°C prior to analysis. Control experiments showed that the progestin receptor contents were the same in frozen as in fresh specimens.

DISCUSSION

The results presented in this investigation show the occurrence in normal human endometrial cytosol of a high-affinity, low-capacity progestin-specific binding component. Its temperature instability indicates that it is a progestin receptor. The present study therefore confirms previous reports (9,12,15,21) on the existence of a progestin receptor in human endometrium. The characteristics of the R5020-receptor complex found in the present investigation also resemble those described for the same complex in human myometrium and endometrium by Philibert and Raynaud (15).

The nondisplaceable (nonspecific) binding of R5020 was found to be less than 10% of total binding in normal human endometrial cytosol. Cortisol was inefficient in displacing R5020 from its binding sites in the cytosol, indicating that R5020 did not bind to corticosteroid binding globulin, in agreement with previous findings of Philibert and Raynaud (15), later confirmed by Horwitz and McGuire (4). Serum also bound R5020, but this binding was of nonspecific character and, consequently, did not disturb receptor measurements in endometrial cytosol.

Quantitation of specific progestin binding sites by sucrose density gradient analysis is complicated by the fact that the R5020-receptor complex seems to sediment as two distinct peaks, an 8S peak and a 4S peak, the latter of which is often contaminated by nonspecific R5020 binding proteins (4,15). In view of this fact, it was of interest to examine the possible advantages of isoelectric focusing as an alternative method for quantitation of the progestin receptor.

In 1973 Mainwaring and Irving (10) introduced isoelectric focusing in steroid receptor analysis and found that the partially purified 5α-dihydrotestosterone-receptor complex in rat prostate focused at pH 5.8. Using a micromethod for electrofocusing, Katsumata and Goldman (6) described the occurrence of multiple receptors for 5α-dihydrotestosterone in unfractionated rat prostate cytosol. These authors pointed out the great advantages of the micromethod as a sensitive method for detection of steroid receptor proteins allowing characterization of receptor components in tissue specimens containing less than 1 mg of protein. We have been interested in using this method as a potentially more specific and sensitive method than sucrose density gradient centrifugation in characterization and measurement of estradiol receptor in human breast cancer specimens (20). In these studies the estradiol receptor was found to focus at pH 6.6, whereas this region of the pH gradient was free from nonspecific and specific nonreceptor binding of estradiol. The present investigation has shown that isoelectric focusing is applicable also to the analysis of the R5020-receptor complex. Generally little nonspecific binding occurred at pH 5, where the R5020-receptor complex focused. However, nonspecific R5020 binding components of serum also focused in the pH 5 region, and a few of the endometrial cancer specimens were also characterized by some nonspecific binding in this region. It is therefore evident that electrofocusing of the R5020-receptor complex does not offer the same great advantages as electrofocusing of the estradiol-receptor complex. On the other hand, isoelectric focusing seems to be superior to sucrose density gradient centrifugation due to its greater capacity and, consequently, higher sensitivity. This is of special significance in the case of endometrial carcinoma, where routinely only small specimens of tumor tissue are available after diagnostic curettage.

The finding of specific binding sites for R5020 in cytosol from several specimens of endometrial carcinoma focusing at pH 5 indicates that this tissue may contain a similar or identical progestin receptor as normal endometrial cytosol. At the present time it is not possible to say whether the presence of high-affinity R5020 binding sites in endometrial adenocarcinoma indicates that the tumor in question is sensitive to treatment with progestin. It is evident that there exists no absolute correlation between histological grading of endometrial adenocarcinoma and number of specific R5020 binding sites. Consequently, measurement of progestin receptor contents in tumor specimens may give additional information about the individual biology of the tumors. A study is in progress to correlate quantity of R5020 binding sites in tumor specimens to responsiveness of tumors to progestin treatment.

SUMMARY

High-affinity, low-capacity binding of $[^3H]R5020$ has been studied in normal human endometrial tissue and in 11 specimens of human endometrial adenocarcinoma. In normal tissue R5020 bound specifically to protein that sedimented at 8S and 4S and that had an isoelectric point of approximately 5. Progesterone competed efficiently with $[^3H]R5020$ for binding sites, whereas dexamethasone, cortisol, estradiol, and testosterone were weak competitors. Binding disappeared after incubation of the R5020-protein complex at 37°C for 10 min. Scatchard analysis indicated that the R5020 binding protein had a K_d of 7×10^{-9} M and that normal endometrial tissue contained about 400 fmoles of binding sites per milligram of protein. Serum also bound R5020 and the R5020-serum protein complex sedimented at 4S and had an isoelectric point of about 5. However, it was not possible to displace $[^3H]R5020$ from its binding sites on serum protein with an excess of unlabeled R5020, and it is therefore concluded that serum binding of R5020 represented nonspecific, high-capacity binding.

Isoelectric focusing showed that 9 of the 11 specimens of endometrial adenocarcinoma bound $[^3H]R5020$ specifically. There was no absolute correlation between histological grading of tumor and contents of specific

R5020 binding sites, although grade I tumors tended to have a larger number of binding sites than grade III tumors. It is suggested that the amount of specific progestin binding sites in endometrial adenocarcinoma may be of predictive value when contemplating progestin treatment as an alternative to other forms of therapy.

ACKNOWLEDGMENTS

Dr. J. P. Raynaud is gratefully acknowledged for providing R5020. This work was supported by grants from Cancerföreningen in Stockholm and from Lotten Bohmans foundation.

REFERENCES

1. Andersson, D. G. (1965): Management of advanced endometrial adenocarcinoma with medroxyprogesterone acetate. *Am. J. Obstet. Gynecol.*, 92:87–98.
2. Bloomfield, R. D. (1971): Current cancer chemotherapy in obstetrics and gynecology, *Am. J. Obstet. Gynecol.*, 109:487–528.
3. Chamness, G. C., and McGuire, W. L. (1975): Scatchard plots: common errors in correction and interpretation. *Steroids*, 26:538–542.
4. Horwitz, K. B., and McGuire, W. L. (1975): Specific progesterone receptors in human breast cancer. *Steroids*, 25:497–505.
5. Karlstedt, K. (1968): Carcinoma of the uterine corpus. *Acta Radiol. [Suppl.] (Stockh.)*, 282:1–98.
6. Katsumata, M., and Goldman, A. S. (1974): Separation of multiple dihydrotestosterone receptors in rat ventral prostate by a novel micromethod of electrofocusing. Blocking action of cyproterone acetate and uptake by nuclear chromatin. *Biochim. Biophys. Acta*, 359:112–129.
7. Kelley, R. M., and Baker, W. H. (1965): The role of progesterone in endometrial cancer. *Cancer Res.*, 25:1190–1196.
8. Lowry, O. H., Rosebrough, N. J., Farr, A. L., and Randall, R. J. (1951): Protein measurement with the Folin phenol reagent. *J. Biol. Chem.*, 193:265–275.
9. MacLaughlin, D. T., and Richardson, G. S. (1976): Progesterone binding by normal and abnormal human endometrium. *J. Clin. Endocrinol. Metab.*, 42:667–678.
10. Mainwaring, W. I. P., and Irving, R. (1973): The use of deoxyribonucleic acid-cellulose chromatography and isoelectric focusing for the characterization and partial purification of steroid-receptor complexes. *Biochem. J.*, 134:113–127.
11. McGuire, W. L., Carbone, P. P., and Vollmer, E. P. (Eds.) (1975): *Estrogen Receptors in Human Breast Cancer*. Raven Press, New York.
12. Murugesan, K., and Laumas, K. R. (1973): Binding of [6,7-^{3}H]norethynodrel to the rat uterine cytosol and to human endometrium and myometrium. *Contraception*, 8:451–470.
13. Nordqvist, S. R. B. (1974): In vitro effects of progestins on DNA synthesis in metastatic endometrial carcinoma. *Gynecol. Oncol.*, 2:415–428.
14. Philibert, D., and Raynaud, J. P. (1973): Progesterone binding in the immature mouse and rat uterus. *Steroids*, 22:89–98.
15. Philibert, D., and Raynaud, J. P. (1974a): Binding of progesterone and R5020, a highly potent progestin, to human endometrium and myometrium. *Contraception*, 10:457–466.
16. Philibert, D., and Raynaud, J. P. (1974b): Progesterone binding in the immature rabbit and guinea pig uterus. *Endocrinology*, 94:627–632.
17. Pollow, K., Lubbert, H., Boquoi, E., Kreuzer, G., and Pollow, B. (1975): Characterization and comparison of receptors for 17β-estradiol and progesterone in human proliferative endometrium and endometrial carcinoma. *Endocrinology*, 96:319–328.
18. Rao, B. R., Wiest, W. G., and Allen, W. M. (1974): Progesterone "receptor" in human endometrium. *Endocrinology*, 95:1275–1281.

19. Varga, A., and Henriksen, E. (1965): Histological observations on the effect of 17-alpha-hydroxyprogesterone-17-n-caproate on endometrial carcinoma. *Obstet. Gynecol.*, 26:656–664.
20. Wrange, Ö., Nordenskjöld, B., Silferswärd, D., Granberg, P. O., and Gustafsson, J.-Å. (1976): Isoelectric focusing of estradiol receptor protein from human mammary carcinoma—a comparison to sucrose gradient analysis. *Eur. J. Cancer*, 12:695–700.
21. Young, P. C. M., and Cleary, R. E. (1974): Characterization and properties of progesterone-binding components in human endometrium. *J. Clin. Endocrinol. Metab.*, 39:425–439.
22. Young, P. C. M., Ehrlich, C. E., and Cleary, R. E. (1976): Progesterone binding in human endometrial carcinomas. *Am. J. Obstet. Gynecol.*, 125:353–358.

Progesterone Receptors in Normal Human Endometrium and Endometrial Carcinoma

K. Pollow, M. Schmidt-Gollwitzer, and J. Nevinny-Stickel

Institut f. Molekularbiologie u. Biochemie der Freien Universität Berlin, 1000 Berlin 33, Germany; and Universitäts-Frauenklinik und -Poliklinik Charlottenburg der Freien Universität Berlin, 1000 Berlin 19, Germany

Studies on the mechanism of action of steroid hormones in various cell systems have led to the discovery of the steroid hormone receptors. These receptors bind the appropriate hormone with high affinity and low capacity and are saturable at or below physiological concentrations of the circulating steroid. The sensitivity of any tissue for steroid hormones seems to be directly related to the content of receptors in the target cells.

The most significant progress in this field of research has been made by Jensen and his co-workers (1,2). Their pioneering work eventually led to new knowledge about the molecular events induced by steroid hormones in normal and neoplastic tissue. Endocrine aspects of carcinogenesis and tumor growth were studied in many laboratories, especially in human breast cancer (1–15) and to a lesser degree in normal human endometrium and endometrial carcinoma (16–28).

Clinical experience has shown that endometrial carcinoma is significantly more frequent in women with an "abnormal" endocrine constellation, for example, as associated with polycystic ovarian syndrome, severe obesity, liver disease, and a variety of ovarian tumors (29–31). It also has been shown that neoplastic proliferation of the tumor can often be reduced by high doses of progestins (32–34). The above observations have raised the question whether some alteration of the human uterine estrogen or progesterone receptor system could be involved in the genesis of endometrial carcinoma. From therapeutic effects it may be expected that tumor cells containing progesterone receptors will regress after appropriate progestin therapy, whereas cells which have lost their complex endocrine regulatory unit will not regress.

In view of the above hypothesis, the following investigation was undertaken, firstly in an attempt to further characterize the progesterone receptor in normal endometrial cytosol during the menstrual cycle, and secondly to present a comparison of biochemical and physical properties of this receptor in normal and neoplastic endometrium. Such a comparison may reveal differences between normal and neoplastic cells which can be used for the discrimination between two populations of tumors, with or without progesterone receptors.

METHODS

Uterine Tissue

Normal human endometrium (68 cases) from normal menstruating patients or those with endometrial carcinoma (30 cases) was obtained after hysterectomy or diagnostic curettage on the following indications: myomata uteri, prolapsus uteri, and endometrial carcinoma. Immediately after hysterectomy the endometrium was scraped from the uterine cavity. Adequate samples were sent for histological examination and the rest divided for extraction of steroids and for determination of receptors. For the receptor assay the specimens were chilled in ice cold buffer composed of 50 mM Tris-HCl, 1.25 mM EDTA, 12 mM mercaptoethanol, 20% glycerol, pH 7.0 (TEM-glycerol buffer). Further processing of the samples was initiated within the next 30 min. Blood samples were withdrawn simultaneously with the biopsies or the hysterectomy.

The menstrual age of the mucosa was based on a 28-day cycle according to the method of Noyes et al. (35). As there is little morphological change during the ovulation period, dating can be made with confidence except 2 to 3 days before or after this period. The carcinomas were graded from "well differentiated" to "undifferentiated" by conventional methods.

Preparation of Subcellular Fractions

Tissue processing and experiments were done at 4°C. The tissue was washed several times in ice cold TEM-glycerol buffer, weighed, and homogenized in 4 volumes (wt/vol) of the same buffer by five strokes with a Potter-Elvehjem-type homogenizer (B. Braun, Melsungen, Germany). The homogenate was centrifuged for 15 min at 850 × g to remove the nucleus fraction. The nuclear pellet was purified as described by Pollow et al. (36–40). The supernatant was centrifuged at 5,500 × g for 15 min. After the floating fat was removed, the supernatant was carefully decanted. The pellet was suspended in TEM-glycerol buffer containing 0.25 M sucrose and recentrifuged at 5,500 × g for 10 min. This procedure was repeated one to three times. The sediment obtained after the final centrifugation was designated as "washed mitochondria." The 5,500 × g supernatant was centrifuged at 12,000 × g; the pellet obtained was not used in this study. The supernatant was centrifuged in a Spinco L2-65B ultracentrifuge (Beckman Instruments) using a Ti-50 rotor at 105,000 × g for 90 min. The final supernatant was designated "cytosol." Microsomes were obtained from the resulting pellet after washing twice in TEM-glycerol buffer.

The cytosol fraction was treated with a dextran-coated charcoal pellet (2.5 mg charcoal, dextran 0.025 mg/ml cytosol) for 20 min to remove the endogenous, unbound steroids and centrifuged again at 20,000 × g for 20 min to remove the charcoal.

Density Gradient Centrifugation

The cytosol (prepared in TEM buffer without glycerol) was incubated with 2 nM ^{3}H-progesterone or ^{3}H-R5020 with or without excess unlabeled steroids (50-fold excess relative to ^{3}H-labeled steroids) for 4 hr at 4°C. Next, 0.2 ml was layered on a linear sucrose gradient (5 to 20% sucrose in 50 mM Tris-HCl, pH 7.0, 1.25 mM EDTA, 12 mM mercaptoethanol, with or without 0.3 M KCl and 10% glycerol, vol/vol) and centrifuged for 16 hr at 234,000 × g (rotor SW 50.1, Beckman Instruments).

Bovine serum albumin was run simultaneously as the reference protein in the estimation of sedimentation coefficients (41). Fractions were collected from the top of the gradient by means of the Isco piercing unit and drop counter. Radioactivity was determined in 0.1-ml aliquots of undiluted fractions. Quenching of radioactivity by sucrose did not vary significantly.

Agarose Gel Filtration

Bio-Gel A-0.5 m was packed in columns (1.5 cm i.d.) to a height of 100 cm and washed with the eluting buffer solution (TEM-glycerol buffer). The gel exclusion characteristics of the column were measured with standard proteins. Samples were applied in 2-ml volumes and elution was carried out at a flow rate of 8.5 ml/hr. Then 1.5-ml fractions were collected and aliquots were counted for radioactivity.

DEAE-Sephadex Chromatography

Columns (2 × 5 cm) of DEAE-Sephadex were equilibrated with TEM-glycerol buffer until the conductivity of the effluent had stabilized. Samples were washed into the columns with 30 ml of equilibrating buffer and eluted with linear gradient of KCl (200 ml, 0 to 0.3 M); 1.5-ml fractions were collected and aliquots were assayed for radioactivity.

Isoelectric Focusing

According to the method of Vesterberg and Svensson (42), 2 ml of ^{3}H-progesterone- or ^{3}H-R5020-labeled cytosol was subjected to isoelectric focusing in a 110-ml LKB 8101 column with double-cooling jackets containing a linear glycerol gradient (10 to 50%, vol/vol), which contained 3% ampholine (pH 4 to 6). The column was prefocused at 300 V for 24 hr, then the sample in 50 mM Tris-HCl, pH 7.0, 1.25 mM EDTA, 12 mM mercaptoethanol, and 20% glycerol was applied in the upper third of the column. Focusing lasted for approximately 48 hr until milliamperage fell to a constant value. Fractions of 2 ml were collected at a flow rate of 1 ml/min; pH was measured in each fraction; aliquots of each fraction were counted for radioactivity.

Polyacrylamide Gel Electrophoresis

Disc electrophoresis was performed with a Shandon disc electrophoresis apparatus; 7.5% acrylamide was used with a running pH of 9.5. Both gels and buffers contained the corresponding ^{3}H-labeled steroid according to Milgrom and Baulieu (43) (0.05 μCi/gel and 0.5 μCi/500 ml buffer).

A current of 1.25 mA/gel was applied for 1 hr to remove the ammonium persulfate ions in the gel. Then 200 μl cytosol was incubated with 2 nM ^{3}H-progesterone or ^{3}H-R5020 for 4 hr and 10 μl was applied on the polymerized gel. Electrophoresis was performed at 4°C, first for 30 min at 1.25 mA/gel, then 2.5 mA/gel was applied until the free bromophenol blue was 1 to 2 mm above the bottom of the gel. As a control, the ^{3}H-labeled steroids without cytosol were also electrophoresed simultaneously. After electrophoresis, the gel was removed, frozen at -35°C for 12 hr and then cut into 3-mm slices. The slices were dissolved in 30% H_2O_2 at 80°C for 18 hr and counted by adding scintillation fluid.

Equilibrium Dialysis

Cytosols were diluted with TEM-glycerol buffer to a final protein concentration of approximately 1 mg/ml. The equilibrium dialysis was performed according to the method described in detail by Davies (44). The diluted cytosols were incubated at 4°C for 16 hr with shaking, in the presence of increasing concentrations of ^{3}H-progesterone or ^{3}H-R5020 (0.2 to 2 nM) and a constant amount of unlabeled cortisol, 50-fold molar excess over the highest ^{3}H-steroid hormone level. Cortisol was added to eliminate binding by contaminating plasma corticosteroid binding globulin. The derivation of the number of binding sites as well as the association constant from the multiple dialysis were based on linearization of the data as described by Scatchard (45).

Determination of Estradiol Binding

The diluted cytosols (1 mg/ml) were incubated at 4°C for 16 hr with ^{3}H-estradiol over a 30-fold concentration range (0.5×10^{-9} M to 15×10^{-9} M).

Binding was then measured using the charcoal adsorption technique. Tubes containing 10^{-6} M unlabeled estradiol were used to correct for nonspecific binding. The equilibrium association constant (K_a) and the concentration of binding sites were calculated according to Scatchard (45).

Competitive Process Experiments

For 16 hr at 4°C, 0.2 ml of cytosol was incubated with 1 nM ^{3}H-progesterone or ^{3}H-R5020 and increasing amounts of unlabeled steroids. Separation of unbound and bound steroid was achieved by charcoal adsorption technique. Each incubation was carried out in duplicate. The radioactivity in the bound fraction was determined in dioxane-PPO-POPOP scintillation fluid using a Berthold liquid scintillation counter, BF 5000.

17β-Hydroxysteroid Dehydrogenase Activity

For the oxidation reaction the standard reaction mixtures (total volume 4.1 ml) contained 0.01 M Tris-HCl buffer, pH 7.4, 0.1 μCi [4-^{14}C]estradiol-17β (0.47 μg) plus 10 μM of unlabeled steroids (dissolved in 0.1 ml of propylene glycol, 2.4%, vol/vol), 500 μM NAD, and enzyme preparations.

The incubation temperature was 37°C. Reactions were started by addition of coenzymes and terminated (after 30 min of incubation) by addition of 5 ml ether/chloroform (3:1, vol/vol). The extracts of the reaction mixtures (3 × 5 ml ether/chloroform) were pooled and evaporated under nitrogen and dissolved in 0.5 ml of benzene. An aliquot (50 μl) was removed for liquid scintillation counting (Berthold BF 5000 liquid scintillation counter, Wildbad, Germany) in order to estimate the total amount of radioactive steroids present in the extract.

The benzene extracts were dried down under nitrogen and the dry residues were transferred with 0.2 ml chloroform/methanol (1:1, vol/vol) to thin-layer plates (Silica Gel, 0.25 mm, Woelm, Eschwege, Germany). Thin-layer chromatography and identification of reaction products were performed as described previously (36–40).

Protein Assay

Protein concentration was measured by the method of Lowry et al. (46) using bovine serum albumin (BSA) as standard.

Radioimmunoassays

EXTRACTION OF THE STEROIDS

The tissue samples (5 to 200 mg) were weighed, minced, and homogenized with 10 volumes of ice cold assay buffer (PBS buffer). Tritiated internal standards (approximately 1,000 cpm) were added to every specimen and kept for 30 min in an ice bath. After equilibration, 3 volumes of diethylether (Merck, Darmstadt, Germany) were added to the tissue samples. The contents were mixed for 15 sec. The diethylether was decanted and evaporated under nitrogen at 40°C. The residue was dissolved in 1 ml 0.1% gelantine PBS buffer, pH 7.0, and aliquoted for determination of recovery and radioimmunoassay of steroids.

The radioimmunoassays for estradiol-17β and progesterone were performed without chromatography by using specific antisera. Estradiol-17β was determined with an anti-estradiol-6-BSA serum donated generously by Dr. G. D. Niswender, Colorado State University (47). Progesterone was measured by specific radioimmunoassay using an 11-OH-progesterone-BSA antiserum (48). In brief, aliquots of the extracted serum or tissue specimen were incubated for 16 hr with 100 μl antiserum in appropriate dilution and 100 μl tritiated steroid (approximately 5,000 cpm of estradiol-1,4,6,7-^{3}H and progesterone-1,2,6,7-^{3}H purchased from Amersham Buchler, Braunschweig, Germany).

Bound and free fractions were separated using dextran-coated charcoal. The bound fraction was transferred to a liquid scintillation cocktail and counted in an automatic scintillation counter. Statistical analysis of the raw radioimmunoassay data was done by a computer program described by Schmidt-Gollwitzer and Pachaly (49). The recovery of the two steroids under investigation was assessed by adding known amounts of tritiated or untritiated steroids to the endometrial samples prior to homogenization. Mean percentage recoveries for both hormones were between 70 and 80 and over 95 for endometrial specimen and serum samples, respectively. Blank values in all assays were negligible if any. The detection levels at 95% probability were 5 and 15 pg for estradiol-17β and progesterone, respectively. The coefficient of within and between assay variation was for both hormones below 12%.

RESULTS

Density Gradient Centrifugation

Initial evidence for a progesterone binding in normal and neoplastic endometrial cytosol was provided by sucrose gradient centrifugation. As shown in Fig. 1 centrifugation of the ^{3}H-progesterone-labeled cytosol, prepared from either proliferative endometrium or undifferentiated endometrial carcinoma, through a 5 to 20% sucrose gradient (in the absence of KCl) revealed two binding components which sedimented in the 4S and 8S region.

Addition of unlabeled steroids displaced ^{3}H-progesterone bound to these two binding proteins to various extents: progesterone reduced both these peaks, R5020 shifted some of the radioactivity from the 8S to the 4S region. Cortisol competition was also observed but to a lesser extent.

When the endometrial cytosol was labeled *in vitro* with ^{3}H-R5020, a highly potent progestin, and centrifuged on a density gradient, a ^{3}H-steroid macromolecule complex with a sedimentation rate of about 8S in the

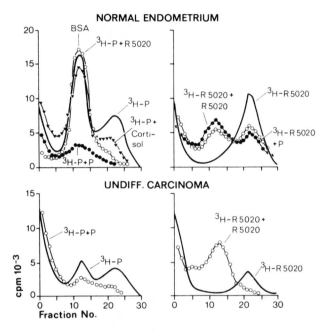

FIG. 1. Ultracentrifugation of normal and neoplastic endometrial cytosol preincubated with radioactive progesterone and R5020 on linear 5 to 20% sucrose density gradients in TEM buffer. Cytosols were labeled at 4°C for 4 hr, and 0.2-ml samples were centrifuged at 234,000 × g for 16 hr at 4°C. Thirty fractions were collected from the top of the gradients (abscissa) counted and the results expressed in cpm × 10^{-3}/fraction (ordinate). Normal endometrium, late proliferative endometrium; undiff. carcinoma, undifferentiated endometrial carcinoma. ^{3}H-P, 2 nM ^{3}H-progesterone; ^{3}H-P + P, 2 nM ^{3}H-progesterone plus 100 nM unlabeled progesterone; ^{3}H-P + cortisol, 2 nM ^{3}H-progesterone plus 100 nM cortisol; ^{3}H-R5020, 2 nM ^{3}H-R5020; ^{3}H-R5020 + R5020, 2 nM ^{3}H-R5020 plus 100 nM R5020; ^{3}H-R5020 + P, 2 nM ^{3}H-R5020 plus 100 nM progesterone.

hypoionic gradient was observed. Both nonlabeled R5020 and progesterone were capable of displacing ^{3}H-R5020 from the 8S binding protein, the radioactivity having been recovered in the 4S region. The sedimentation profiles revealed a striking similarity between receptors of normal endometrium and neoplastic tissue, but the amounts of ^{3}H-steroid in the 4S and 8S regions were reduced.

Sucrose gradient centrifugation analysis of the ^{3}H-progesterone-labeled cytosol of proliferative endometrium in 0.3 M KCl showed that the progesterone binding component migrated as a single 4S peak. In the absence

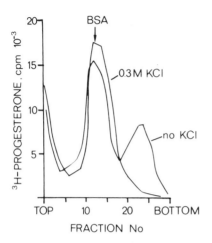

FIG. 2. Effect of KCl on the sedimentation rates of cytoplasmic ^{3}H-progesterone binding components of late proliferative endometrium. Portions of 0.2 ml of cytosol were centrifuged at 234,000 × g for 16 hr in a linear sucrose gradient (5 to 20%) with 0.3 M or without KCl. BSA was used as a reference for the calculation of sedimentation coefficients. Fractions were collected from the top of the gradients; radioactivity is expressed as cpm × 10^{-3}/fraction.

of KCl two peaks with sedimentation rates of 8S and 4S were noted on density gradient centrifugation (Fig. 2).

Agarose Gel Filtration

Fractionation of normal and neoplastic endometrial cytosol on Agarose gel in the absence of KCl resulted in three radioactive peaks (Fig. 3). The major ^{3}H-progesterone binding component of cytosol, chromatographed as peak II, has a mobility identical to that of human serum CBG (as shown by the superposition of chromatographs of ^{3}H-cortisol-labeled human serum and ^{3}H-progesterone-labeled human endometrial cytosol, *results not shown*). The first radioactive peak (peak I) was eluted close behind aldolase (M.W. 158,000). The third peak (peak III) eluted in front together with free ^{3}H-steroid. The ^{3}H-peaks were greatly reduced when unlabeled progesterone was present. It can also be seen in Fig. 3 that it is possible for cortisol to replace only to a certain extent the ^{3}H-progesterone where it is bound.

When ^{3}H-R5020 was used, the major part of steroid binding components of cytosol chromatographed in peak I. Addition of unlabeled R5020 reduced the binding activity of high molecular species as reflected in the small area of peak I followed by a displacement of the radioactivity toward the peak II area. The elution profile obtained with undifferentiated endometrial carcinoma was similar to that of normal endometrium except that binding of ^{3}H-progesterone or ^{3}H-R5020 was very low in peak I.

DEAE-Sephadex Chromatography

The elution profiles presented in Fig. 4 were obtained when cytoplasmic ^{3}H-progesterone or ^{3}H-R5020 binding components of normal and neo-

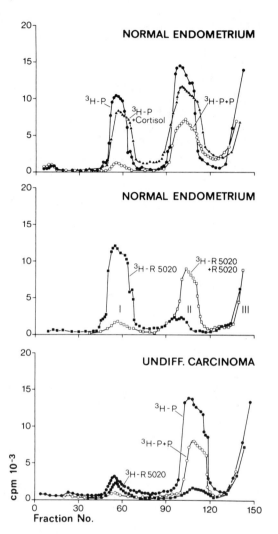

FIG. 3. Elution profiles obtained from Agarose gel filtration of cytoplasmic ³H-progesterone and ³H-R5020 binding components of late proliferative endometrium and undifferentiated endometrial carcinoma. Elution patterns were obtained when cytosol steroid incubates were eluted from the BioGel columns with TEM-glycerol buffer. Radioactivity is expressed as cpm × 10⁻³/fraction. Peaks of radioactivity are designated by roman numerals.

plastic endometrial tissue were chromatographed on DEAE-Sephadex. Free ³H-labeled steroid hormones were eluted with the washing buffer (peak I); when the KCl gradient elution was performed, two peaks of radioactivity were eluted. The same elution profiles were obtained from several different tissue preparations. The absolute amount of radioactivity

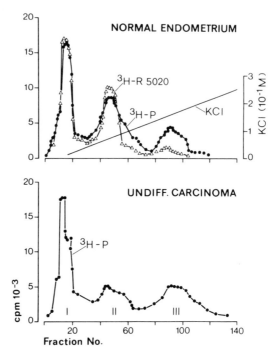

FIG. 4. Chromatographic patterns obtained from ion exchange chromatography of cytoplasmic ^{3}H-progesterone and ^{3}H-R5020 binding components of late proliferative endometrium and undifferentiated endometrial carcinoma. DEAE-Sephadex A-50 was equilibrated with TEM-glycerol buffer. After application of samples, columns were washed with TEM-glycerol buffer and eluted with a linear gradient of KCl (0 to 0.3 M KCl in TEM-glycerol buffer). Radioactivity is expressed as cpm $\times$ 10^{-3}/fraction. Free steroids were eluted with the washing buffer (fraction no. 0 to 18). Peaks of radioactivity are designated by roman numerals.

per fraction, however, varied significantly, suggesting different amounts of steroid binding activity per preparation. The bulk of the ^{3}H-steroid hormone binding proteins was eluted in the more acidic fractions (peak II).

Elution profiles of ^{3}H-progesterone and ^{3}H-R5020 were identical. No differences could be found in the elution profiles between ^{3}H-progesterone receptors of normal and neoplastic endometrium.

Isoelectric Focusing

When ^{3}H-progesterone- or ^{3}H-R5020-labeled human endometrial cytosol obtained from normal and neoplastic tissue was subjected to isoelectric focusing, only one major peak of radioactivity could be detected (Fig. 5), but some small peaks were also seen. The isoelectric point of the main peak was located around pH 4.8.

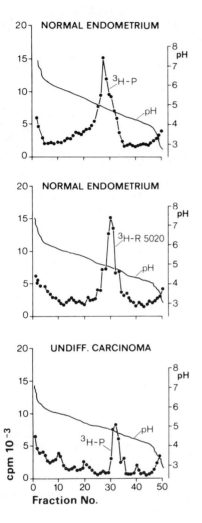

FIG. 5. Isoelectric focusing of cytoplasmic ^{3}H-progesterone and ^{3}H-R5020 binding components of late proliferative endometrium and undifferentiated endometrial carcinoma. Columns (110 ml, LKB 8101) contained a linear glycerol gradient (10 to 50%, vol/vol) with 3% (vol/vol) of carrier ampholytes (pH 4 to 6). Samples were applied in TEM-glycerol buffer in the upper third of the column. Focusing lasted for approximately 48 hr. Radioactivity is expressed as cpm × 10^{-3}/fraction.

Polyacrylamide Gel Electrophoresis

The electrophoretic patterns of ^{3}H-progesterone- or ^{3}H-R5020-labeled endometrial cytosol are shown in Fig. 6. The ^{3}H-labeled steroid hormones are associated with two proteins. One peak of radioactivity was detected near the top surface of the separation gel, the other peak had a relative mobility like that of human serum CBG binding only ^{3}H-progesterone. It was possible for nonlabeled progesterone to replace the bound ^{3}H-progesterone in the slower-moving region. When ^{3}H-progesterone binding components of cytosol from endometrial carcinoma were separated by polyacrylamide gel electrophoresis under the same conditions, the radioactivity

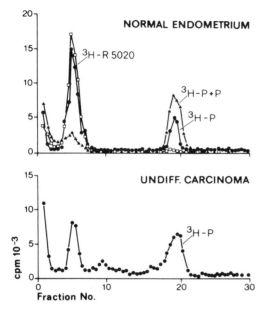

FIG. 6. Comparison of polyacrylamide gel electrophoresis pattern of the cytoplasmic ³H-progesterone and ³H-R5020 binding components of late proliferative endometrium and undifferentiated endometrial carcinoma. Total acrylamide concentration: 7.5%; pH 9.5; gels and buffers contained the corresponding ³H-labeled steroid according to the method described by Milgrom and Baulieu (43). Human serum CBG has the same mobility as the faster-moving peak; the slower-moving peaks of bound ³H-progesterone and ³H-R5020 are located near the top surface of the separation gel.

was located in the same distinct peaks. It should be pointed out that free steroid hormones do not enter the gel.

Steroid Specificity of Cytoplasmic Progesterone Binding Components

The competition between various progestational and nonprogestational steroids for the ³H-progesterone or ³H-R5020 binding sites at the cytoplasmic progesterone receptor of normal and neoplastic human endometrium was studied. The relative affinities of the steroids tested were calculated by the method described by Korenman (50) and are summarized in Table 1. It was observed that 5α-pregnane-3,20-dione and R5020 were bound to the progesterone receptor protein with an affinity comparable to that of ³H-progesterone. 5β-Pregnane-3,20-dione, 20α-hydroxy-4-pregnen-3-one, 3β-hydroxy-5-pregnen-20-one, and cortisol also showed some affinity to the progesterone receptor, whereas the binding of ³H-progesterone or ³H-R5020 was unchanged in the presence of estradiol-17β, estrone, testosterone, and 4-androstene-3,17-dione. There was no difference between the specificity of the progesterone binding components of normal and neoplastic endometrium. When the competitive effect of all these steroids on the binding of ³H-R5020 was measured, progesterone and (to a lesser extent) 5α-pregnane-3,20-dione competed effectively with ³H-R5020 for the binding sites on endometrial receptor protein.

TABLE 1. *Relative affinities of cytoplasmic progesterone (R5020) receptor of human late proliferative endometrium and undifferentiated endometrial carcinoma for different steroids*

	^{3}H-progesterone binding		^{3}H-R5020 binding	
	Normal endometrium	Endometrial carcinoma	Normal endometrium	Endometrial carcinoma
Steroid added	Relative (%)	affinity (%)	Relative (%)	affinity (%)
Progesterone	100	100	81.3	83.9
5α-Pregnane-3,20-dione	88.2	82.8	56.3	61.5
5β-Pregnane-3,20-dione	2.6	1.7	0.8	0.5
20α-Hydroxy-4-pregnen-3-one	3.1	1.3	0.1	0.1
3β-Hydroxy-5-pregnen-20-one	0.8	0.5	0.1	0.1
R5020	97.8	95.8	100	100
Cortisol	1.8	1.4	0.1	0.1
4-Androstene-3,17-dione	0.1	0.1	0.1	0.1
Testosterone	0.1	0.1	0.1	0.1
Estradiol-17β	0.1	0.1	0.1	0.1
Estrone	0.1	0.1	0.1	0.1

As measured by equilibrium dialysis according to the method of Korenman (50).

Level of the Cytoplasmic Progesterone Binding Proteins During the Menstrual Cycle

Figures 7 and 8 suggest that progesterone receptor levels in normal endometrial tissue are directly dependent on the stage of the menstrual cycle. The level of the ^{3}H-progesterone (^{3}H-R5020) binding sites was low in the early proliferative phase but increased toward the 14th day of the menstrual cycle having its highest value around the time of ovulation and declining sharply during the secretory phase.

The association constant (K_a) of the progesterone receptor in normal human endometrium was $1.1 \pm 0.6 \times 10^{-9}$ M (mean ± SD). A comparison of the K_a values of the progesterone receptor complex at each stage of the menstrual cycle indicated that there was no statistically significant difference.

Endocrine Parameters Which Regulate the Steroid Hormone Receptor Level in the Human Endometrial Cell

It has been shown in animal experiments that the level of the progesterone receptor is under steroidal control. Estrogen treatment increases the progesterone receptor concentration by inducing the *de novo* synthesis of

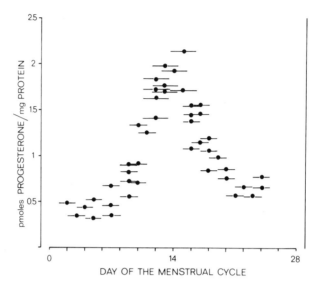

FIG. 7. Concentration of specific progesterone binding sites of normal human endometrial cytosol in relation to the day of the menstrual cycle. For the determination of the progesterone receptor concentration, endometrial cytosol was incubated with increasing amounts of ^{3}H-progesterone or ^{3}H-R5020 in the presence of an excess of unlabeled cortisol according to the method described by Davies (44). Binding site concentration was determined by the method of Scatchard (45).

receptor molecules, whereas progesterone seems to be a negative effector of its own receptor.

Although our small material does not allow any definitive conclusions, Fig. 9 shows that there seems to be correlation between the cytosol progesterone receptor level and the serum concentration of estradiol, pointing to

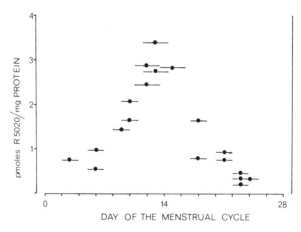

FIG. 8. Specific ^{3}H-R5020 binding concentration of endometrial cytosol in relation to the day of the menstrual cycle.

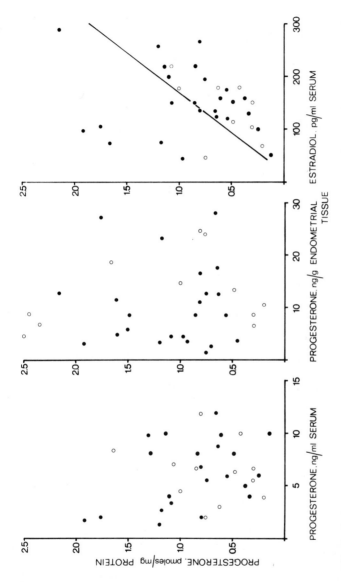

FIG. 9. Correlation of the specific progesterone or R5020 binding concentration of endometrial cytosol with serum and tissue levels of progesterone and estradiol. Binding site concentration was determined according to the method of Davies (44); for details see Methods. Estradiol and progesterone were analyzed by radioimmunoassay. ●, ^{3}H-progesterone; ○, ^{3}H-R5020.

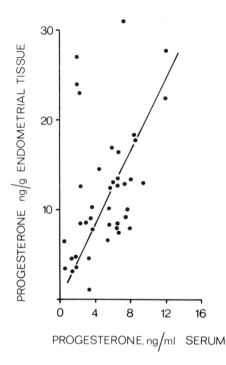

FIG. 10. Correlation of the serum progesterone levels with the endometrial tissue progesterone concentration. Progesterone was analyzed by radioimmunoassay as described in Methods.

a possible positive control by estrogen on the progesterone receptor level. In addition, no significant correlation was found between the progesterone receptor concentration and the serum or tissue levels of progesterone. The correlation between the endogenous progesterone levels in the endometrial tissue preparations and the serum progesterone levels is shown in Fig. 10. There is a strong correlation between these two parameters.

In Fig. 11 progesterone and estradiol cytoplasmic receptor levels as well as serum and endometrial tissue concentrations of progesterone and estradiol in the different phases of the menstrual cycle are depicted. The estradiol receptor level was highest during the early proliferative phase, whereas the highest progesterone receptor levels were observed at midcycle (as seen also in Figs. 7 and 8). There was an inverse correlation between the estradiol receptor level and serum concentration of estradiol. The midcycle peak of the progesterone receptor concentration correlated well with the first peak of the serum and tissue concentration of estradiol.

During the first half of the menstrual cycle the amount of progesterone bound by endometrial supernatant correlated well with the progesterone concentration in serum and endometrial tissue, whereas during the luteal phase the binding capacity seemed to be inversely correlated to both serum and tissue progesterone concentrations.

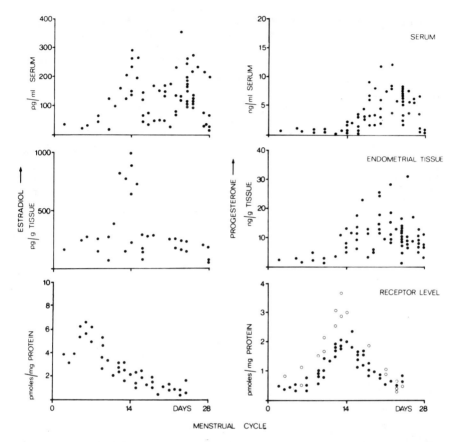

FIG. 11. Dependence of serum and endometrial tissue levels of progesterone and estradiol as well as progesterone and estradiol receptor levels on the phase of the menstrual cycle. ○, ³H-R5020

Level of the Cytoplasmic Progesterone and Estradiol Binding Proteins at Different Degrees of Tumor Differentiation

The cytosols of 30 endometrial carcinomas were examined for the presence of estradiol and progesterone binding components. The receptor levels correlated with the histological type of the tumor (Table 2). The level of estradiol binding sites was highest in undifferentiated carcinomas, whereas progesterone binding sites were lowest. In 6 of 11 undifferentiated tumor specimens, progesterone binding activity could not be detected. The affinity of estradiol and progesterone for their respective receptors in endometrial carcinoma as reflected by their apparent association constants seemed independent of the degree of tumor differentiation.

TABLE 2. *Progesterone and estradiol binding characteristics of late proliferative endometrial and undifferentiated endometrial carcinoma cytosol*

Patient	Age	Degree of differentiation	Progesterone binding pmoles/mg protein	K_a 10^{-9} M	Estradiol binding pmoles/ mg protein	K_a 10^{-9} M
K. A.	68	Well	0.63	0.8	1.28	3.4
A. P.	53	differentiated	0.38	1.1	0.98	2.8
B. A.	43		0.43	1.6	1.01	1.7
Ch. A.	50		0.31	0.7	1.08	3.1
B. S.	64		0.21	0.6	1.01	4.8
E. K.	68		0.78	1.3	0.98	6.1
A. K.	72		1.02	1.1	0.77	4.2
R. S.	51		0.41	1.3	1.01	3.8
R. Sch.	63		0.53	1.5	1.03	3.3
I. B.	61		0.98	0.9	1.12	3.5
S. B.	76	Moderately	0.48	0.8	1.58	6.8
L. P.	63	well	0.37	0.8	1.23	5.8
U. H.	63	differentiated	0.27	0.7	0.87	3.1
M. Sch.	66		0.29	1.1	1.54	2.8
K. B.	63		0.31	1.3	1.23	5.4
E. S.	58		0.18	1.1	0.98	3.1
R. L.	71		0.21	0.8	0.61	2.4
E. Ch.	68		0.13	0.6	1.12	3.1
B. Sch.	61		0.12	0.9	1.18	3.8
K. B.	61	Undifferentiated	0.09	0.8	1.38	2.4
B. Z.	72		nm	nm	1.45	5.1
I. A.	79		0.08	1.1	2.15	4.3
T. K.	73		0.13	1.3	1.69	4.4
L. H.	69		0.03	1.8	1.96	5.1
L. K.	68		nm	nm	1.81	1.8
A. L.	61		nm	nm	1.12	2.7
K. H.	71		0.07	1.1	1.78	2.8
L. C.	72		nm	nm	0.98	2.1
K. P.	68		nm	nm	1.18	2.4
W. R.	65		nm	nm	1.12	2.3

nm, not measurable.

17β-Hydroxysteroid Dehydrogenase

Figure 12 shows that the specific activity of 17β-hydroxysteroid dehydrogenase (17β-HSD) in normal human endometrium is dependent on the phase of the menstrual cycle. This is particularly evident in particulate fractions: the specific activity of the 17β-HSD in microsomes and mitochondria was approximately 10-fold higher during the early secretory than during the proliferative phase.

The specific 17β-HSD activities of microsomal fractions of normal human endometrium are plotted against the cytosol estradiol and progesterone binding capacity in Fig. 13. It was interesting to observe that the cytosol progesterone receptor level highly correlated to the specific 17β-HSD activity of microsomes and that there was a significant inverse corre-

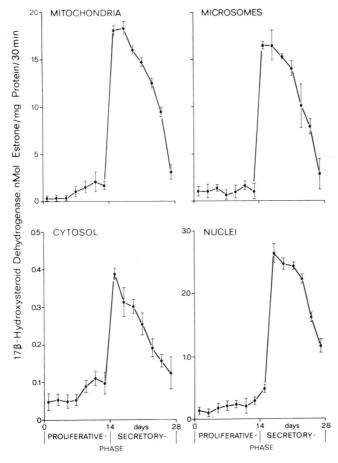

FIG. 12. Dependence of specific 17β-HSD activity in microsomes, mitochondria, cytosol, and nuclear fraction on the phase of the menstrual cycle. Activities of the 17β-HSD were determined as initial velocities by measuring the rate of oxidation of [4-^{14}C]estradiol to [4-^{14}C]estrone in 4.1 ml Tris-HCl buffer containing 0.1/μCi [4-^{14}C]estradiol (0.47 μg), 40 nmoles unlabeled steroid (added in 100 μl propylene glycol), 2 μmoles NAD, 2.4% propylene glycol, and various amounts of enzyme preparation. The temperature of incubation was 37°C. After 30 min of incubation the reaction was stopped by the addition of 2 volumes of ether/chloroform (3:1). The subsequent extraction procedure and separation of the radiometabolites by thin-layer chromatography were as described by Pollow et al. (36). Points are means of 5 to 7 different determinations.

lation between the cytosolic estradiol receptor level and 17β-HSD activity.

Figure 14 shows that in subcellular fractions of endometrial carcinomas the 17β-HSD activity decreased with decreasing differentiation of the tumor. After patients were treated with gestagens, only the well-differentiated carcinomas increased significantly in their 17β-HSD activity, which shows that the hormonal stimulus leads to a similar effect on the 17β-HSD activity as in normal endometrium.

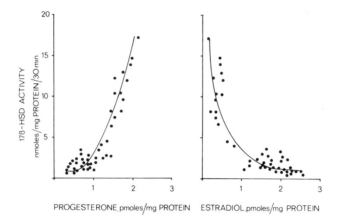

FIG. 13. Correlation of specific 17β-HSD activity in microsomes of normal endometrium with the specific progesterone and estradiol cytoplasmic receptor concentration. Progesterone and estradiol were analyzed by radioimmunoassay as described in Methods.

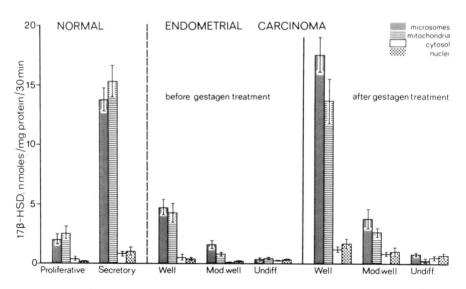

FIG. 14. 17β-HSD activity in various subcellular fractions of proliferative and secretory endometrium (different patients) as compared with carcinomas before and after gestagen treatment (same patients). Gestagens used for oral treatment were chlormadinone acetate or medroxyprogesterone acetate (doses 100 to 500 mg/day for 1 to 2 weeks). Normal proliferative and secretory endometrium: 18 cases each. Well, moderately well, and undifferentiated carcinomas: 4 cases each (before and after gestagen treatment). Vertical lines indicate ± SD.

DISCUSSION

The present studies show that normal and neoplastic human endometrial tissue contain a specific cytoplasmic component that reversibly binds progesterone. The specific binding component contains many of the traits by which steroid hormone receptors are characterized. These traits include a specific affinity to progesterone or R5020, a progestational agent of high affinity for the progesterone receptor, and a limited number of binding sites which are saturated at physiological levels of the steroid hormone. The characteristics of the receptor are similar to those found in tissues of the reproductive tract of several species including man (43,51-56). The possibility that progesterone binding to receptor protein has been mistaken for binding to cortisol binding globulin (CBG) was ruled out by several experiments:

1. Cortisol was added to eliminate binding of ^{3}H-progesterone to CBG.

2. The competition studies indicated that cortisol was not an effective inhibitor of progesterone binding, whereas unlabeled progesterone reduced ^{3}H-progesterone binding to almost basal levels. These findings are in good agreement with the results of other studies in which the high specificity of the binding site of the receptor protein for progesterone is reported (18, 19,24-27).

3. The isoelectric point of the labeled progesterone receptor complex (approximately pH 4.8) was considerably more acid than that of CBG. CBG ^{3}H-progesterone complex focused as a distinct peak at about pH 5.8, as seen by Westphal (57).

4. It is interesting that progesterone interacts at low-salt concentration with two specific proteins (as seen, for example, on density gradient centrifugation), namely, a 4S CBG-like protein and an 8S protein, whereas R5020 was almost exclusively bound to the 8S protein. These data reflect the progestin specificity of the 8S binding component of endometrial cytosol.

The binding to the 4S region is partially specific for progesterone as demonstrated by the relative displacement potency of various steroids, whereas the binding of R5020 is probably unspecific as reflected by its very low binding in this region. Similar observations have been made by Kontula et al. (58), Young and Cleary (24), and Philibert and Raynaud (27).

When the characteristics of the binding proteins from normal and neoplastic endometrium were compared, no differences could be found between these two tissues with regard to their affinity for progesterone and R5020, their sedimentation properties, elution profiles from agarose gel filtration and ion exchange chromatography, as well as electrophoretic patterns from isoelectric focusing and gel electrophoresis. These findings revealed a striking similarity between progesterone receptors of normal endometrium and endometrial carcinoma. Some differences were, however, found when the specific progesterone binding capacity was studied. In normal human endometrium the concentration of cytoplasmic binding sites for progesterone varies according to the phase of the menstrual cycle.

Variations in the level of receptor proteins are of particular interest since the hormone receptors have a direct effect on nuclear control. Progesterone receptor concentration was highest at midcycle. These findings are in good agreement with those of Crocker et al. (17), Haukkamaa and Luukkainen (18), Bayard et al. (28), and MacLaughlin and Richardson (26). However, Rao et al. (23) did not find changes in the progesterone receptor level during the menstrual cycle. Whether or not the binding site concentration in the target cells during the menstrual cycle is influenced by serum levels of steroid hormones remains to be clarified. In these experiments the charcoal adsorption technique was used to remove endogenous progesterone from the cytosol prior to measuring the level of the binding sites. It is important to note that the charcoal adsorption technique, as shown by Kontula et al. (58), usually results in an underestimation of the receptor concentration due to a slight degrading effect of the above procedure.

The simultaneous measurement of the cytoplasmic progesterone receptor concentration and the progesterone and estradiol concentrations in serum and endometrial tissues clearly demonstrated that the increase of the estradiol concentration correlates with the increase of the progesterone binding capacity in the target cell. Contrary to this, no correlation between the serum or tissue progesterone levels and the progesterone receptor level was observed.

The first findings clearly agree with the results of earlier experiments using animal models. These studies showed that estradiol induces the synthesis of the progesterone receptors in the mammalian uterus (59–61). The same may be true for humans since the midcycle estrogen peak coincides with the highest level of the cytoplasmic progesterone receptor. In addition, it was observed that progesterone increases the inactivation rate of its own receptor (62). We could confirm these results for the second half of the cycle but not for the first.

The results of the progesterone binding studies done on the 30 endometrial carcinomas demonstrated that there is a strong correlation between the amount of progesterone receptors contained in a tumor specimen and the degree of differentiation of this tumor. The concentration of progesterone binding sites in undifferentiated tumors was low, whereas well-differentiated tumors contained a relatively high concentration of progesterone binding sites. For estradiol receptors the contrary was the case. Similar results for progesterone binding in human endometrial carcinoma were reported recently by MacLaughlin and Richardson (26). These findings may explain why in undifferentiated carcinomas treatment with progestins (e.g., chlormadinone acetate) achieves little, whereas in well-differentiated tumors long remissions (2 years and more), even in patients with metastases, have been achieved (32–34).

For the wide range of steroid receptor concentrations in the different types of endometrial carcinomas, the following might be considered: Firstly, the total progesterone receptor concentration is an integrated value for the whole tumor. Since the tumor is a mixture of various cell types, the receptor concentration will vary inversely with the proportion of connective tissue and specific neoplastic cells. Secondly, it is possible that the receptor concentration varies from cell to cell and a decrease in receptor concentration in a given cell might reflect its degree of dedifferentiation. Finally, it is possible that a given endometrial carcinoma contains a mixture of cells which contain "normal" levels of progesterone receptors and those that do not have receptors.

The results of the investigations of the 17β-HSD in normal and neoplastic endometrium should be discussed in connection with the findings for the progesterone receptor. Although in normal endometrium specific enzyme activity in subcellular fractions depended on the phase of the cycle, in endometrial carcinoma it depended on the degree of differentiation of the tumors. The highest values of 17β-HSD activity were found in mitochondria and microsomes of early secretory endometrium (factor 10 as compared to proliferative endometrium) and in particulate fractions of well-differentiated carcinomas (factor 10 to $\leqslant 10$ as compared to undifferentiated carcinomas). These findings are consistent with observations of Tseng and Gurpide (63) for normal endometrium.

Furthermore, there is a close correlation between the progesterone receptor level and the specific 17β-HSD activity in the normal endometrial cell.

One must not overlook the fact that in well-differentiated endometrial carcinoma, which possesses a relatively high concentration of progesterone receptors, there is a steep increase of the 17β-HSD activity after gestagen treatment *in vivo*, similar to the normal endometrial cell after ovulation (64). It may be speculated that the 17β-HSD activity is regulated by the progesterone receptor concentration, whereas the latter in turn is dependent on estradiol concentration. Since a high intracellular activity of the 17β-HSD (second phase of the cycle) leads to a decrease in the intracellular concentration of estradiol, the progesterone receptor concentration also decreases.

On the basis of the above hypothesis, the measurement of 17β-HSD activity could be a simple and rapid method to determine the intactness of the above-named chain of reactions.

ACKNOWLEDGMENTS

We thank the surgeons and pathologists at the institute for their cooperation in preparing the tissue specimens. The skillful assistance of Mrs. B.

Pollow is gratefully acknowledged. This investigation was supported in part by a grant from the Deutsche Forschungsgemeinschaft.

REFERENCES

1. Jensen, E. V., DeSombre, E. R., and Jungblut, P. W. (1967): Estrogen receptors in hormone-responsive tissues and tumors. In: *Endogenous Factors Influencing Host-Tumor Balance,* edited by R. W. Wissler, T. L. Dao, and S. Wood, Jr., pp. 15–30, 68. University of Chicago Press, Chicago.
2. Jensen, E. V., Block, G. E., Smith, S., and DeSombre, E. R. (1973): Hormonal dependency of breast cancer. Recent results. *Cancer Res.,* 42:55–62.
3. Maas, H., Engel, B., Hohmeister, H., Lehmann, F., and Trams, G. (1972): Estrogen receptors in human breast cancer tissue. *Am. J. Obstet. Gynecol.,* 113:377–382.
4. McGuire, W. L. (1973): Estrogen receptors in human breast cancer. *J. Clin. Invest.,* 52:73–77.
5. Savlov, E. D., Wittliff, J. L., Hilf, R., and Hall, T. C. (1974): Correlations between certain biochemical properties of breast cancer and response to therapy: A preliminary report. *Cancer,* 33:303–309.
6. Terenius, L., Johansson, H., Rimsten, A., and Thoren, L. (1974): Malignant and benign mammary disease: Estrogen binding in relation to clinical data. *Cancer,* 33:1364–1368.
7. Hähnel, R., and Twaddle, E. (1973): Estimation of the association constant of the estrogen receptor complex in human breast cancer. *Cancer Res.,* 33:559–566.
8. Hähnel, R., Twaddle, E., and Vivian, A. B. (1971): Estrogen receptors in human breast cancer. 2. In vitro binding of estradiol by benign and malignant tumors. *Steroids,* 18:681–707.
9. Feherty, P., Farrer-Brown, G., and Kellie, A. E. (1971): Estradiol receptors in carcinoma and benign disease of the breast. *Br. J. Cancer,* 25:697–710.
10. Leung, B. S., Manaugh, L. C., and Wood, D. C. (1973): Estradiol receptors in benign and malignant disease of the breast. *Clin. Chim. Acta,* 46:69–76.
11. Wittliff, J. L., Hilf, R., and Brooks, W. F., Jr. (1971): Specific bindings of estradiol-17β by normal and neoplastic breast tissues from humans. *Proc. Am. Assoc. Cancer Res.,* 12:47.
12. Wittliff, J. L., Hilf, R., Brooks, W. F., Jr., Savlov, E. D., Hall, T. C., and Orlando, R. A. (1972): Specific estrogen-binding capacity of the cytoplasmic receptor in normal and neoplastic tissues of humans. *Cancer Res.,* 32:1983–1992.
13. McGuire, W. L., and Dela Garza, M. (1973): Improved sensitivity in the measurement of estrogen receptor in human breast cancer. *J. Clin. Endocrinol. Metab.,* 37:986–989.
14. McGuire, W. L., Chamness, G. C., Costlow, M. E., and Shepherd, R. E. (1974): Hormone dependence in breast cancer. *Metabolism,* 23:75–100.
15. McGuire, W. L., and Chamness, G. C. (1973): Studies on the estrogen receptor in breast cancer. *Adv. Exp. Med. Biol.,* 36:113–136.
16. Gabb, R. G., and Stone, G. M. (1974): Uptake and metabolism of tritiated estradiol and estrone by human endometrial and myometrial tissue in vitro. *J. Endocrinol.,* 62:109.
17. Crocker, S. G., Milton, P. J. D., and King, R. J. B. (1974): Uptake of (6,7-^{3}H)oestradiol-17β by normal and abnormal human endometrium. *J. Endocrinol.,* 62:145.
18. Haukkamaa, M., and Luukkainen, T. (1974): The cytoplasmic progesterone receptor of human endometrium during the menstrual cycle. *J. Steroid Biochem.,* 5:447.
19. Laumas, K. R., and Uniyal, J. P. (1974): Binding of (14,15-^{3}H)D, 1-norgestrel to receptor proteins in human endometrium and myometrium. *J. Steroid Biochem.,* 5:330 (abstract 143).
20. Milton, P., Crocker, S. G., and King, R. J. B. (1974): (15-^{3}H)D,1-norgestrel to receptor proteins in human normal, hyperplastic and carcinomatous human endometrium. *J. Steroid Biochem.,* 5:329 (abstract 142).
21. Richardson, G. S., MacLaughlin, D. T., and Scully, R. E. (1974): Progesterone-specific binding by cytosols of human endometria. *J. Steroid Biochem.,* 5:329 (abstract 141).
22. Tseng, L., and Gurpide, E. (1974): Nuclear concentration of estriol in superfused human endometrium; competition with estradiol. *J. Steroid Biochem.,* 5:273.

23. Rao, B. R., Wiest, W. G., and Allen, W. M. (1974): Progesterone "receptor" in human endometrium. *Endocrinology,* 95:1275-1281.
24. Young, P. C. M., and Cleary, R. E. (1974): Characterization and properties of progesterone binding components in human endometrium. *J. Clin. Endocrinol. Metab.,* 39:425-439.
25. Pollow, K., Lübbert, H., Boquoi, E., Kreuzer, G., and Pollow, B. (1975): Characterization and comparison of receptors for 17β-estradiol and progesterone in human proliferative endometrium and endometrial carcinoma. *Endocrinology,* 96:319-328.
26. MacLaughlin, D. T., and Richardson, G. S. (1976): Progesterone binding by normal and abnormal human endometrium. *J. Clin. Endocrinol. Metab.,* 42:667-678.
27. Philibert, D., and Raynaud, J. P. (1974): Binding of progesterone and R5020, a highly potent progestin, to human endometrium and myometrium. *Contraception,* 10:457-466.
28. Bayard, F., Damilano, S., Robel, P., and Baulieu, E.-E. (1975): Recepteurs de l'oestradiol et de la progesterone dans l'endometre humain au cours du cycle menstruel. *Endocrinologie,* 281:1341-1344.
29. Woll, E., Hertig, A. T., Smith, G. V. S., and Johnson, L. C. (1948): The ovary in endometrial carcinoma with notes on the morphological history of the aging ovary. *Am. J. Obstet. Gynecol.,* 56:617.
30. Sherman, A. J., and Woolf, R. B. (1959): An endocrine basis for endometrial carcinoma. *Am. J. Obstet. Gynecol.,* 77:233-242.
31. Dallenbach-Hellweg, G. (1964): Das Karzinom des Endometriums und seine Vorstufen. *Verh. Dtsch. Ges. Pathol.,* 48:81.
32. Nordqvist, S. (1971): Effect of progesterone on human endometrial carcinoma in different experimental systems. *Acta Obstet. Gynecol. Scand.* [*Suppl.*], 19:25.
33. Kistner, R. W., and Griffiths, C. T. (1968): Use of progestational agents in the management of metastatic carcinoma of the endometrium. *Clin. Obstet. Gynecol.,* 11:439.
34. Boquoi, E., and Kreuzer, G. (1973): Histomorphologische Untersuchungen des Endometriumkarzinoms unter Gestagen-Therapie (Chlormadinonazetat). *Geburtshilfe Frauenheilkd.,* 33:697-708.
35. Noyes, R. W., Hertig, A. T., and Rock, J. (1950): Dating the endometrial biopsy. *Fertil. Steril.,* 1:3-25.
36. Pollow, K., Lübbert, H., Boquoi, E., Kreuzer, G., Jeske, R., and Pollow, B. (1975): Studies on 17β-hydroxysteroid dehydrogenase in human endometrium and endometrial carcinoma. 1. Subcellular distribution and variations of specific enzyme activity. *Acta Endocrinol. (Kbh.),* 79:134-145.
37. Pollow, K., Lübbert, H., Jeske, R., and Pollow, B. (1975): Studies on 17β-hydroxysteroid dehydrogenase in human endometrium and endometrial carcinoma. 2. Characterization of the soluble enzyme from secretory endometrium. *Acta Endocrinol. (Kbh.),* 79:146-156.
38. Pollow, K., Lübbert, H., Boquoi, E., and Pollow, B. (1975): Progesterone metabolism in normal human endometrium during the menstrual cycle and in endometrial carcinoma. *J. Clin. Endocrinol. Metab.,* 41:729-737.
39. Pollow, K., Lübbert, H., and Pollow, B. (1976): On the mitochondrial 17β-hydroxysteroid dehydrogenase from human endometrium and endometrial carcinoma: Characterization and intramitochondrial distribution. *J. Steroid Biochem.,* 7:45-50.
40. Pollow, K., Lübbert, H., and Pollow, B. (1975): Studies on 17β-hydroxysteroid dehydrogenase in human endometrium and endometrial carcinoma. 3. Partial purification and characterization of the microsomal enzyme. *Acta Endocrinol. (Kbh.),* 80:355-364.
41. Martin, R. G., and Ames, B. N. (1961): A method for determining the sedimentation behavior of enzymes: application to protein mixtures. *J. Biol. Chem.,* 236:1372-1379.
42. Vesterberg, D., and Svensson, H. (1966): Isoelectric fractionation analysis and characterization of ampholytes in natural pH gradients. IV. Further studies on the resolving power in connection with separation of myoglobins. *Acta Chem. Scand.,* 20:820-829.
43. Milgrom, E., and Baulieu, E.-E. (1970): Progesterone in uterus and plasma. I. Binding in rat uterus 105,000 g supernatant. *Endocrinology,* 87:276-287.
44. Davies, I. J. (1973): The measurement of cytoplasmic steroid "receptor" proteins. In: *Molecular Techniques and Approaches in Developmental Biology,* edited M. J. Chrispeels, pp. 39-54. J. Wiley and Sons, New York.
45. Scatchard, G. (1949): The attraction of proteins for small molecules and ions. *Ann. N.Y. Acad. Sci.,* 51:660-672.

46. Lowry, O. H., Rosebrough, N. J., Farr, A. L., and Randall, R. J. (1951): Protein measurement with the Folin phenol reagent. *J. Biol. Chem.*, 193:265–275.
47. England, B. G., Niswender, G. D., and Midgley, A. M., Jr. (1974): Radioimmunoassay of estradiol-17β without chromatography. *J. Clin. Endocrinol. Metab.*, 38:42–50.
48. Hoffman, B., Kyrein, H. J., and Ender, M. L. (1973): An efficient procedure for the determination of progesterone by radioimmunoassay applied to bovine peripheral plasma. *Horm. Res.*, 4:302–310.
49. Schmidt-Gollwitzer, M., and Pachaly, J. (1974): Radioimmunoassay Paket. Computer program. Freie Universität, Berlin.
50. Korenman, S. G. (1970): Relation between estrogen inhibitory activity and binding to cytosol of rabbit and human uterus. *Endocrinology*, 87:1119–1123.
51. O'Malley, B. W., McGuire, W. L., Kohler, P. O., and Korenman, S. G. (1969): Studies on the mechanism of steroid hormone regulation of synthesis of specific proteins. *Recent Prog. Horm. Res.*, 25:105–160.
52. Stone, G. M., and Baggett, B. (1965): The uptake of some tritiated estrogenic and nonestrogenic steroids by the mouse uterus and vagina in vivo and in vitro. *Steroids*, 6:277–299.
53. Smith, J. A., Martin, L., King, R. J. B., and Vertes, M. (1970): Effects of oestradiol-17β and progesterone on total and nuclear protein synthesis in epithelial and stromal tissues of the mouse uterus and of progesterone on the ability of these tissues to bind oestradiol-17β. *Biochem. J.*, 119:773–784.
54. Milgrom, E., and Baulieu, E.-E. (1970): Progesterone in the uterus and the plasma. II. The role of hormone availability and metabolism on selective binding to uterus protein. *Biochem. Biophys. Res. Commun.*, 40:723–730.
55. McGuire, J. L., and DeDella, C. (1971): In vitro evidence for a progesterone receptor in the rat and rabbit uterus. *Endocrinology*, 88:1099–1103.
56. Faber, L. E., Sandmann, M. L., and Stavely, H. E. (1972): Progesterone binding in uterine cytosols of the guinea pig. *J. Biol. Chem.*, 247:8000–8004.
57. Westphal, U. (Ed.) (1971): *Steroid-Protein Interactions.* Springer Verlag, Berlin.
58. Kontula, K., Jänne, O., Vihko, R., de Jager, E., de Visser, J., and Zeelen, F. (1975): Progesterone-binding proteins: in vitro binding and biological activity of different steroidal ligands. *Acta Endocr. (Kbh.)*, 78:574–592.
59. Feil, P. D., Glasser, S. R., Toft, D. O., and O'Malley, B. W. (1972): Progesterone binding in the mouse and rat uterus. *Endocrinology*, 91:738–746.
60. Milgrom, E., Atger, M., Perrot, M., and Baulieu, E.-E. (1972): Progesterone in uterus and plasma: VI. Uterine progesterone receptors during the estrus cycle and implantation in the guinea pig. *Endocrinology*, 90:1071–1078.
61. O'Malley, B. W. (1974): A specific progesterone receptor in the hamster uterus: physiologic properties and regulation during the estrous cycle. *Endocrinology*, 94:1041–1053.
62. Kontula, K. (1975): Progesterone-binding protein in human myometrium. Binding site concentration in relation to endogenous progesterone and estradiol-17β levels. *J. Steroid Biochem.*, 6:1555–1561.
63. Tseng, L., and Gurpide, E. (1974): Estradiol and 20α-dihydroprogesterone dehydrogenase activities in human endometrium during the menstrual cycle. *Endocrinology*, 94:419–423.
64. Pollow, K., Boquoi, E., Lübbert, H., and Pollow, B. (1975): Effect of gestagen therapy upon 17β-hydroxysteroid dehydrogenase in human endometrial adenocarcinoma. *J. Endocrinol.*, 67:131–132.

Subject Index

Agarose gel filtration, of progesterone receptors in human uterus, 315, 320
Aldosterone, and progestin receptor levels in mouse uterus, 238
Ammonium sulfate precipitation
 of progesterone receptors, 106, 159
 of R5020 binding in breast tumors, 162, 167
Androgen receptors, in MCF-7 cells, 109
Association rate, for hormone-receptor complexes, 288-290

Breast tissue
 benign lesions of, steroid receptors in, 145, 178
 glucose oxidation in, 211-223
 progestin receptors in goat lactating tissue, 23-37
Breast tumors
 DMBA-induced, see DMBA-induced mammary tumors
 estrogen receptors in, see Estrogen receptors
 hormone-dependent
  estrogen, progesterone, and prolactin affecting, 96-97
  receptors in, 67-68, 76-78
 human, see Human breast tumors and MCF-7 cell studies, see MCF-7 cell lines
 progesterone receptors in, see Progesterone receptors
 prolactin receptors in, 86
  ovariectomy affecting, 97
  RU16117 affecting, 92
 R5020 binding in, see R5020

Calusterone, in breast cancer, responses to, 151
CBG
 and chromatography of progesterone binding components, 55
 progesterone binding to, 110, 114, 115
Chick oviducts, progesterone receptors in, 23, 52, 105, 156
Chlormadinone acetate, competition for R5020 binding, in human breast tumors, 180
Chromatography of progesterone binding
 in mammary tissue, 54-55
 in uterine cytosol, 315, 320-322
Cortexolone
 competition for binding sites in MCF-7 cells, 198
 and thymidine incorporation in MCF-7 cells, 200
Corticosteroid-binding globulin, see CBG
Cortisol
 binding to lymphocytes, 207
 competition for binding sites in MCF-7 cells, 198, 202-203
 and dydrogesterone binding to serum, 161
 and thymidine incorporation in MCF-7 cells, 200
Cyproterone acetate, as inhibitor of R5020 binding, 109

Dexamethasone
 binding to lymphocytes, 205-206
 binding to MCF-7 cell extracts, 35, 44, 196-200
 competition for progesterone receptors
  in endometrial cytosol, 305
  in serum from endometrial cancer patients, 306
 and progesterone binding
  to hepatoma tissue culture cells, 110
  to human mammary tumor cytosol, 114
 and R5020 binding in breast tumors, 35, 44, 64
 and thymidine incorporation in MCF-7 cells, 200
Dextran-coated charcoal assay of receptors
 in human breast cancer, 117, 172
 in uterus, 25-26, 61-62, 73-74, 86, 228-229, 263, 281-282, 334
Diethylstilbestrol
 and estradiol binding in human breast tumors, 180
 and progesterone binding to uterine cytosol, 161-162, 167
Dihydrotestosterone
 binding inhibition by progestin, 109
 competition for binding sites
  in lactating mammary gland, 45
  in MCF-7 cells, 198, 201
 and progestin receptor levels in mouse uterus, 237
 receptors in prostate cytosol, 309
Dimethylbenzanthracene, see DMBA-induced mammary tumors
Dissociation from progesterone receptor for progesterone

Dissociation from progesterone receptor
for progesterone *(contd.)*
    in breast tumors, 182-183
    in uterine cytosol, 188, 233, 241, 290-292
for R5020, 222, 240, 241
    in breast tumor, 182-183
    in uterine cytosol, 188, 233, 290-292, 305
DMBA-induced mammary tumors, 74-76, 85-99
    control of hormone receptors in, 96-99
    dual effect of estrogens in, 91
    estrogen receptors in, see Estrogen receptors
    ovariectomy affecting, 88, 93
    pregnancy affecting, 103-104
    progesterone receptors in, 110-113
    prolactin receptors in, 86
    R5020 binding in, 35, 44-45, 89
        progesterone affecting, 96
        RU16117 affecting, 93
    RU16117 affecting, 88-94
        dosage in, 91
        inhibition from, 88-92
        regression from, 92-94
    stimulation by progesterone, 94-96, 104
        after ovariectomy, 96-97
DU-41164
    binding to human serum, 160
    binding to uterine progesterone receptors, 166
Dydrogesterone, binding to human serum, 160-161, 166

Electrofocusing, of R5020-receptor complex, 301, 306, 309, 315-316, 322
Electrophoresis, of progesterone receptors in human uterus, 316, 323-324
Endocrine status, and progesterone receptors in uterus, 235-240, 282
    in humans, 325-328
Endocrine therapy in breast tumors
    responses to, 86, 118-119, 136-138
    and role of steroid hormone receptors, 59-68, 118-119, 136-138, 151-152, 155-156
Endometrial carcinoma, and progestin receptors in uterus, 299-311, 329, 333, 334-335
Estradiol
    affinity for receptors in endometrial carcinoma, 329
    binding to DMBA-induced mammary tumors, 89
    progesterone affecting, 96
    RU16117 affecting, 93
    binding to rat uterus in pregnancy, 251
    binding sites correlated with R2858 binding sites, in human breast

tumor cytosol, 183-184
combined with progesterone, and tumor remission, 104-105
competition for receptors
    in endometrial cytosol, 305
    in serum from endometrial cancer patients, 306
dissociation from binding site
    in breast tumor, 182
    in uterine cytosol, 188
dual effect on DMBA-induced breast tumors, 91
and gestagen receptors in breast tumors, 79-82, 83
and glucose oxidation in mammary tissue
    in human breast cancer, 219
    in pregnant rats, 217
and inhibition of R5020 binding in mammary tumors, 45
and progesterone receptors
    in DMBA-induced mammary tumors, 96, 99, 112-133
    in uterus, 105, 156, 276-277, 282
    in serum, and progesterone receptors in human uterus, 326-328
and tumor growth after ovariectomy, 96-97
and tumor regression, 104
Estrogen receptors
    assays with 8S and 4S forms, 135, 305, 319, 333
    in breast lesions, benign, 145, 178
    dextran-coated charcoal assay for, 117, 172
    in DMBA-induced tumors, 74-78
        distribution of levels of, 77
        estradiol and ovariectomy affecting, 79-82, 83
        ovariectomy affecting, 78-79, 82-83, 97
        RU16117 affecting, 92
    exchange assay for, in mouse breast tumors, 59-68
    and glucose oxidation in mammary tissue, 211-224
    in human breast tumors, 49, 135, 141-152, 163-165, 214-215
        concentrations of, 145-146
        dissociation rates of complexes, 182
        in male cancer, 145, 175, 177
        menopausal status affecting, 146-147, 148, 150-151, 174-175
        and response to therapy, 118-119, 136-138, 151-152, 155-156
    and responsiveness of glucose oxidation to steroids, 220-221
    specificity of binding in, 49, 180-181
    sucrose gradient analysis of, 132

SUBJECT INDEX

and thymidine labeling index, 165, 167
in lactating goat tissue, 34
in lymph nodes, 163
in MCF-7 cells, 108-109
nafoxidine binding to, 109
prolactin affecting, 96, 99
sucrose gradient assay of, 117, 132, 158
tamoxifen binding to, 109
in uterine cytosol, 165
cancer affecting, 165
Estrous cycle, and receptors in uterus, 239, 241, 245-257, 266-267
Ethynylestradiol, competition for estradiol binding in human breast tumors, 180
Exchange assay for gestagen receptors
in human breast tumors, 173, 187-189
in mouse mammary tumors, 59-68
for progesterone receptors in uterus, 241

Glucocorticoid(s)
activity of R5020, 205, 208
binding properties compared to progesterone, 53-55
competition for R5020 binding sites, 44-46
Glucocorticoid receptors
in lactating goat tissue, 34
in lymphocytes, 194-195, 205-208
in MCF-7 cells, 110, 194
Glucose
metabolism in uterus, steroids affecting, 223
oxidation in mammary tissue, 211-224
in human breast cancer, 218-220
in pregnant rats, 216-217
Glycerol
and association rate of hormone receptor complexes, 288
and dissociation rate of hormone receptor complexes, 291, 294
Growth hormone, binding to DMBA-induced tumors, 90
Gynecomastia, steroid receptors in, 177, 178

Hepatoma tissue culture cells
aminotransferase production in, progestins and glucocorticoids affecting, 110
progesterone interaction with, 201
Human breast tumors
endocrine therapy in, responses to, 86, 118-119, 136-138
estrogen receptors in, 49, 135, 141-152, 163-165, 214-215, *see also* Estrogen receptors
glucocorticoid receptors in MCF-7 cells, 110, 194

and glucose oxidation in mammary tissue, 218-220
and plasma contamination of cytosol, 184, 188
and plasma sex steroid binding protein levels, 178
progesterone receptors in, 106, 107, 113-119, 141-152, 163-165, 214-215, *see also* Progesterone receptors
prolactin receptors in, 86
R5020 binding in, 43, 45, 49-52, 214-215
ammonium sulfate precipitation affecting, 162, 167
in MCF-7 cells, 35, 44, 198, 202-205
thymidine labeling index of, 156, 159
and steroid receptors in tissue, 165, 167
Human serum, *see* Serum
Human uterus
estrogen receptors in, 165
17β-hydroxysteroid dehydrogenase activity in, 330-331, 335
progesterone receptors in, 165, 287-296, 313-335, *see also* Progesterone receptors
Hydrocortisone, and R5020 binding in breast tumors, 44, 64, 110, 114
17β-Hydroxysteroid dehydrogenase, endometrial, 330-331, 335
in carcinoma, 331
menstrual cycle affecting, 330
Hypophysectomy, and tumor response to estrogen-progesterone combination, 105
Hypothalamus, R5020 binding to progestin receptor in, 13

ICI 46474, competition for estradiol binding in human breast tumors, 180
Isoelectric focusing, of R5020-receptor complex, 301, 306, 309, 315-316, 322

Lactation
and glucose oxidation in mammary tissue, 222
and progesterone levels in serum, 256, 257
and progestin receptors in goat mammary tissue, 23-37
Luteinizing hormone levels
R5020 affecting, 17-19
RU16117 affecting, 90, 94
Lymph nodes, estrogen and progesterone receptors in, 163
Lymphocytes, glucocorticoid receptors in, 194-195, 205-208

MCF-7 cell lines, 108-110
dexamethasone binding to, 35, 44, 196-200

MCF-7 cell lines *(contd.)*
  glucocorticoid receptors in, 194, 195-205
  progesterone receptors in, 203-205
  R5020 binding in, 35, 43, 44
Menopausal status, and hormone receptors in human breast cancers, 146-147, 148, 150-151, 174-175
Menstrual cycle
  and 17β-hydroxysteroid dehydrogenase in endometrium, 330
  and progesterone receptors in human uterus, 296, 325, 328, 334
  and thymidine labeling index of human breast tumors, 156
Methyltrienolone, 9, 10, 11
Moxestrol, *see* R2858

Nafoxidine
  binding to estrogen receptor, 109
  progesterone-like effects of, 108
Norgestrel, competition for R5020 binding in human breast tumors, 180

Ovariectomy
  with adrenalectomy, and breast tumor remission in humans, 136-138
  and growth of DMBA-induced mammary tumors, 88, 93, 97
  and hormone receptor levels in mammary tumors, 78-79, 82-83, 97
  and plasma prolactin levels, 90
  and prolactin effects on tumor growth, 96-97, 104
Oviducts, chick, progesterone receptors in, 23, 52, 105, 156

Phosphoprotein phosphorylation, steroid hormones affecting, 223
Pituitary
  R5020 binding to progestin receptor in, 13
  and tumor response to estrogen-progesterone after hypophysectomy, 105
Plasma contamination, of breast tumor cytosol, 184, 188
Pregnancy
  and glucose oxidation in rat mammary tissue, 216-217
  and growth of DMBA-induced breast tumors, 103-104
  and progesterone levels in tissues and serum, 241-242
  and progestin receptor levels in uterus, 239-240, 241, 245-257, 267-268
Progesterone
  binding to CBG, 110, 114, 115
  binding to human serum, 115
  binding to mammary tissue, 42

binding sites correlated with R5020 binding sites in tumor cytosol, 184-185
binding to uterine cytosols, 106-107
  diethylstilbestrol affecting, 161-162, 167
clinical effects in breast cancer, 103-105
competition for binding sites in MCF-7 cells, 198, 201
as competitive inhibitor of specific R5020 binding, 33-34, 37, 45
dissociation from binding site
  in breast tumors, 182-183
  in uterine cytosol, 188, 233, 241, 290-292
and dydrogesterone binding to serum, 161
with estrogen, and tumor remission, 104-105
and glucose oxidation in mammary tissue
  in human breast cancer, 219
  in pregnant rats, 217
interaction with androgen receptors, 109
interaction with estrogen receptors, 108-109
interaction with glucocorticoid receptors, 109-110
interaction with hepatoma tissue cells, 201
sedimentation properties in cytosol, 48-49
in serum, 115
  and endometrial progesterone in humans, 296, 328
  in puerperium in rats, 256, 257
  radioimmunoassay of, 174, 229
  and tissue levels in pregnancy, 241-242
  and stimulation of DMBA-induced tumors, 94-96, 104
  after ovariectomy, 96-97
Progesterone receptors
  ammonium sulfate precipitation of, 106, 159
  assays with 8S and 4S peaks, 114, 115-116, 117, 136
  in breast lesions, benign, 145, 178
  in breast tumors in mice, 59-68
  in chick oviducts, 23, 52, 105, 156
  dextran-coated charcoal assay of, 25-26, 61-62, 73-74, 86, 228-229, 263, 281-282, 334
  in DMBA-induced tumors, 74-78, 110-113
    distribution of levels of, 77
    estradiol and ovariectomy affecting, 79-82, 83
    estrogens affecting, 99

SUBJECT INDEX

ovariectomy affecting, 78-79, 82-83, 97
RU16117 affecting, 92
and glucose oxidation in mammary tissue, 211-224
in human breast tumors, 106, 107, 113-119, 135, 141-152, 163-165, 214-215
   concentrations of, 145-146
   dissociation rates of complexes, 182-183
   distribution of, 165, 167
   in male cancer, 145, 175, 177
   menopausal status affecting, 146-147, 148, 150-151, 174-175
   and response to therapy, 118-119, 136-138, 151-152, 155-156
   and responsiveness of glucose oxidation to steroids, 220-221
   specificity of binding to tumor cytosol, 180-181
   sucrose gradient analysis of, 132
   and thymidine labeling index, 165, 167
in human uterus, 165, 287-296, 313-335
   agarose gel filtration of, 315, 320
   and association rate constant for progesterone or R5020, 288-290
   chromatography of, 315, 320-322
   electrophoresis of, 316, 323-324
   endocrine factors affecting, 325-328
   in endometrial carcinoma, 165, 299-311, 329, 333, 334-335
   menstrual cycle affecting, 296, 325, 328
   and serum progesterone levels, 296
   sucrose gradient analysis of, 106, 161, 229, 249, 301, 315, 318-320, 333
in lactating goat tissue, 23-37
in lymph nodes, 163
in MCF-7 cell lines, 203-205
overview of research in, 1-8
R5020 as tag for, 1, 9-20, *see also* R5020
in uterus of animals, 50, 52, 86, 105-106, 227-241
   cytosol receptor assay of, 261-263, 273-277
   DU-41164 binding to, 166
   endocrine status affecting, 235-240, 282
   estradiol affecting, 276-277, 282
   in estradiol-primed castrated mice, 229-235, 236-238
   estrogen affecting, 96, 99, 103, 105
   in estrus and diestrus, 239, 241, 245-257, 266-267
   exchange assay of, 241
   nuclear receptor assay of, 263-266, 277-279

   in pregnant mice, 239-240, 241
   in pregnant rats, 245-257, 267-268
   species differences in, 241
Prolactin
   affecting estrogen receptors, 96, 99
   binding to DMBA-induced mammary tumors, 89
   ovariectomy affecting, 90
   progesterone affecting, 96
   RU16117 affecting, 90, 93
   and mammary tumor growth, 67
   after ovariectomy, 96-97, 104
   and progesterone receptors in DMBA tumors, 112-113
   receptors in breast tumors, 86
   ovariectomy affecting, 97
   RU16117 affecting, 92
   serum levels in pregnancy and puerperium, in rats, 255, 256
Promestone, *see* R5020
Prostate
   cytosol receptors for dihydrotestosterone, 309
   hyperplasia studies with R1881, 11
Puerperium, and progesterone levels in serum, in rats, 256, 257

R1881, 9, 10, 11
R2858, 9, 10, 11
   binding sites correlated with estradiol binding sites in breast tumor cytosol, 183-184
   competition for estradiol binding in human breast tumors, 180
   dissociation from binding site
      in breast tumor, 182
      in uterine cytosol, 188
   and progestin receptor levels in mouse uterus, 238
   in tissue receptor assays, 173
R5020, 1, 9-20
   antiestrogenic activity of, 16, 108, 109
   antiovulatory activity of, 14
   in assay of cytosol receptor in rat uterus, 261-263, 273-277
   in assay of nuclear receptor in rat uterus, 263-266, 277-279
   binding in breast tumors
      ammonium sulfate precipitation affecting, 162, 167
      and dexamethasone activity, 35, 44, 64
      in humans, 43, 45, 49-52, 86, 114-115, 143-144, 162, 167, 214-215
      in MCF-7 cells, 35, 43, 44, 198, 202-205
      in mice, 63-68
   binding in DMBA-induced mammary tumors, 35, 44-45, 89

R5020
  binding in DMBA-induced mammary, tumors *(contd.)*
    progesterone affecting, 96
    RU16117 affecting, 93
  binding sites correlated with progesterone binding sites in tumor cytosol, 184-185
  binding in uterus, 13, 39, 86, 106-107, 161-162, 287-296
    diethylstilbestrol affecting, 162
    in endometrial carcinoma, 299-311
    in estradiol-primed castrated mice, 229-235
    in pregnant rats, 251-256
  biochemical studies of, 12-14
  biological activity of, 14-20
  compared to triamcinolone binding, 53-55
  dextran-coated charcoal assay of, 25-26
  dissociation from receptor, 222, 240, 241
    in breast tumor, 182-183
    in uterine cytosol, 188, 233, 290-292, 305
  glucocorticoid activity of, 205, 208
  and glucose oxidation in mammary tissue
    in human breast cancer, 219
    in pregnant rats, 217
  in goat lactating mammary glands, 23-37
  interaction with binding sites in mammary tissue, 39-56
  interrelationships with steroid hormones, 108-110
  isoelectric focusing of complex with receptor, 301, 306, 309, 315-316, 322
  and luteinizing hormone secretion, 17-19
  progestational activity of, 14
  purification of receptor by ammonium precipitation, 27
  serum binding of
    in goats, 31-32
    in humans, 46-48, 115, 306
  specificity of binding to receptors, 13-14, 32-34, 43, 44-46, 281
  competitive inhibitors of, 33-34, 35, 37, 45
  and cytosol protein concentrations, 27-28
  temperature stability of, 30
  time dependence of, 28, 42
  sucrose density gradient centrifugation of, 26-27, 30, 41, 48, 86
  sedimentation properties in cytosol, 30, 48-52
  therapeutic uses of, 19
  and tyrosine aminotransferase activity in hepatoma tissue culture cells, 19-20

Radioimmunoassay
  for estradiol in serum, 318
  for progesterone in serum, 174, 229, 318
RU16117
  competition for estradiol binding in human breast tumors, 180
  and inhibition of DMBA-induced tumors, 88-92
  dose affecting, 91
  and luteinizing hormone levels, 90, 94
  and plasma prolactin levels, 90
  and progestin receptor level in mouse uterus, 237
  and regression of DMBA-induced tumors, 92-94

Serum
  goat, R5020 binding to, 31-32
  human
    DU-41164 binding to, 160
    dydrogesterone binding to, 160-161, 166
    estradiol levels in, and progesterone receptors in human uterus, 326-328
    progesterone binding to, 115, *see also* Progesterone, in serum
    R5020 binding to, 46-48, 115, 306
Sex steroid binding protein, plasma levels in human breast cancer, 178
Specificity of R5020 binding, *see* R5020
Sucrose gradient analysis of hormone-receptor complexes, 172
  assays with 8S and 4S forms
    in endometrial cancer, 305
    in goat mammary gland, 49
    in human breast cancer, 114, 115-116, 117, 132, 136, 163
  estrogen receptors in human breast cancer, 135
  progesterone receptors in breast tissue, 48-49
    in human cancer, 135
  progesterone receptors in uterus, 106, 161, 229, 249, 301, 315, 318-320, 333
  for R5020, 26-27, 30, 41, 48, 86

Tamoxifen
  binding to estrogen receptor, 109
  in breast cancer, responses to, 151
  competition for estradiol binding, in human breast tumors, 180
  progesterone-like effects of, 108
Testosterone
  competition for progesterone receptors in endometrial cytosol, 305
  in serum from endometrial cancer patients, 306

and progestin receptor levels in mouse uterus, 237
Thymidine
  incorporation in lymphocytes, agents affecting, 207-208
  incorporation in MCF-7 cells, agents affecting, 200
  labeling index of human breast tumors, 156, 159
  and steroid receptors in tissue, 165, 167
Transcortin, *see* CBG
Triamcinolone acetonide
  binding compared to R5020 binding, 53-55
  binding to lactating mammary gland cytosol, 42
  as competitive inhibitor of specific R5020 binding, 33-34, 35, 37, 45
  and progesterone or R5020 binding, 110
Tyrosine aminotransferase in hepatoma tissue culture cells
  progestins and glucocorticoids affecting, 110
  R5020 affecting, 19

Uterus
  estrogen binding in, 51, 165
  glucose metabolism in, steroids affecting, 223
  progesterone receptors in, *see* Progesterone receptors
  R5020 binding in, *see* R5020